A Bibliography
for Finite Elements

A Bibliography
For Finite Elements

Compiled by

J. R. WHITEMAN

School of Mathematical Studies, Brunel University,
Uxbridge, Middlesex, England

1975

ACADEMIC PRESS
London · New York · San Francisco

A Subsidiary of Harcourt Brace Jovanovich, Publishers

ACADEMIC PRESS INC. (LONDON) LTD.
24/28 Oval Road
London NW1

United States Edition published by
ACADEMIC PRESS INC.
111 Fifth Avenue
New York, New York 10003

Library of Congress Catalog Card Number: 75–587
ISBN: 0–12–747260–6

Printed in Great Britain
by Unwin Brothers Limited
The Gresham Press, Old Woking, Surrey, England

Preface

In April 1975 I organised at Brunel University an international conference on the Mathematics of Finite Elements and Applications. For that conference I prepared a bibliography for finite element methods, of some four hundred references, based on the collection that I had accumulated through studying finite element techniques. That bibliography took the form of a Mathematics Department Technical Report which was given to all delegates attending the conference. In the intervening period since April 1972 the demand for the report has been so phenomenal that it has been quite impossible for us to satisfy all the requests we have received. In addition my collection of references has grown so that there are now approximately two thousand two hundred. I therefore decided that for the second conference on the Mathematics of Finite Elements and Applications, which will be held at Brunel University in April 1975, if I were again to produce a bibliography, it would have to be done through a publisher. Consequently I persuaded Academic Press to publish this bibliography and the aim is simply to provide workers in the field with a concise list of papers, together with a comprehensive key-word listing, as cheaply as possible.

This bibliography is thus based on my own collection of references and has been compiled without any outside assistance. I am aware that it does contain inaccuracies, which I have obviously tried to reduce as much as possible, accepting that this is a solo effort. There is the possibility of a second edition, and I would of course be pleased to receive notice of existing errors and of additional references for inclusion. The listings have been produced on the computer at Brunel University and I am pleased to express my gratitude to the members of the Brunel University Computer Unit, particularly Margaret Clifton, Janet Gray and Norman Bonney, for the cheerful and charming way in which they have assisted me. I also acknowledge the backing that has been provided by Brunel University through their allowing me to use the computer.

With any bibliography it is very difficult to define the boundaries of the fields which should be included. Thus I have included such topics as classical analysis, functional analysis, approximation theory, fluids, diffusion and aeronautical, civil, mechanical, electrical and nuclear engineering applications. I have not included the vast literature on the solution of systems of linear and nonlinear equations, and I have also avoided references involving splines when they are not used for solving differential equations. One of the interesting aspects of finite element applications is the number of packages which have been produced, and as a consequence I have listed all the packages which I have encountered during my studies. There will be some omissions from my list, as there are of important papers in the bibliography; to the authors of these I tender my apologies and ask for their understanding.

The bibliography is made up of two parts. The first of these consists of a numbered listing of all complete references in alphabetical order by first author. Throughout this part the sequence is primarily determined alphabetically, with chronological order being a secondary factor. The second part is an alphabetical listing of a key word search done on the titles of the publications. Here the occurrence of a key word causes a part of the relevant title to be printed in a format centred on the key word, together with the number of the entry in the author listing.

I realise that no list of references can be complete and that it becomes quickly less so with the passage of time. However, I hope that this bibliography will be of use to those interested in finite elements.

J. R. WHITEMAN
BRUNEL UNIVERSITY   JANUARY 1975

Finite Element Packages and Programs

APACHE
ASAS
ASKA
BERDYNE
BERGEN
BERSAFE
CHILES
CONCOR
DYNAPLAS
DYSAFE
EASE
ELFED
EURCYL
EXLASTIC
FACE
FAMSOR
FINEL
FINESSE
HEATRAN
HERMES
INGA
MARK
MATUS
MESHGEN
MESH – 3D
NASTRAN
NEWPACK
NONSAP
NORSAM
OSCAR
PAFEC
SAFE
SAMMSOR
SESAM
SESAM – 69
SLADE
STARDYNE
STRAP
STRUDL
TAFEST
TITUS
TRIM

These are included in the Key Word Listing.

Author Listing

1   AALTO S.K.,   THE REPRESENTATION OF FUNCTIONALS ON SUBSPACES BY
        FUNCTIONALS HAVING CERTAIN SUPPORT SETS. SIAM.J.NUMER.
        ANAL.9, 35-39, 1972.
2   ABADIE J.,   NONLINEAR PROGRAMMING. NORTH HOLLAND, AMSTERDAM,1967.
3   ABE T.,   ELASTIC DEFORMATION OF POLYCRYSTALLINE METALS.
        III INFLUENCE OF NEIGHBOUR GRAINS. BULL. JAPANESE S.M.E.
        16, 1540-1549, 1973.
4   ABEL J.F. AND POPOV E.P.,   STATIC AND DYNAMIC FINITE ELEMENT ANALYSIS OF
        SANDWICH STRUCTURES. PROC.2ND.CONF.MATRIX METHODS IN STRUCTURAL
        MECHANICS, WRIGHT-PATTERSON AFB, OHIO, AFFDL-TR68-150,1968.
5   ABRAMOWITZ M., AND STEGUN I.A.,   HANDBOOK OF MATHEMATICAL FUNCTIONS.
        DOVER, NEW YORK, 1964.
6   ABSI E.,   FINITE ELEMENT METHOD. ANNALES DE L'INSTITUT TECHNIQUE DU
        BATIMENT ET DES TRAVAUX PUBLICS N262, 1593-1620,1969.
7   ABSI E.,   LA THEORIE DES EQUIVALENCES ET SON APPLICATION A L'ETUDE
        DES OUVRAGES D'ART. ANALES DE L'INSTITUT DU BATIMENT ET
        DES TRAVAUX PUBLICS; THEORIE ET METHODES DE CALCUL,
        NO.153, 1972.
8   ABU-SHUMAYS I.K., AND BAREIS E.H.,   SINGULAR ELEMENTS IN
        VARIATIONAL AND FINITE ELEMENT TRANSPORT COMPUTATIONS.
        TRANS.AMER.NUCL.SOC.17, 236-237, 1973.
9   ACHIESER N.I.,   THEORY OF APPROXIMATION. UNGAR, NEW YORK, 1956.
10 ACHUTARAMAYYA G. AND SCOTT W.D.,   FRACTURE SURFACE ENERGIES BY
        FINITE ELEMENT STRESS ANALYSIS. AMERICAN CERAMIC. SOC.
        BULLETIN 52, 709, 1973.
11 ADELMAN H.M., AND CATHERINES D.S.,   CALCULATION OF TEMPERATURE
        DISTRIBUTIONS IN THE THIN SHELLS OF REVOLUTION BY THE
        FINITE ELEMENT METHOD. N.A.S.A. TECHNICAL NOTE, TN
        D-6100, 1971.
12 ADINI A., AND CLOUGH R.W.,   ANALYSIS OF PLATE BENDING BY THE FINITE
        ELEMENT METHOD. NAT.SCI.FOUND.REPT. G 7337, UNIV. OF
        CALIFORNIA, BERKELEY, 1961.
13 ADLER A., AND GALLAGHER R.H.,   THE USE OF FINITE METHODS IN HEAT
        FLOW ANALYSIS. BELL AEROSYSTEMS CORPORATION, REPORT
        9500-920134, 1968.
14 AGARWAL B.D.,   MICROMECHANICS ANALYSIS OF COMPOSITE MATERIALS USING
        FINITE ELEMENT METHODS. REPORT COO-1794-16, ILLINOIS INSTITUTE
        OF TECHNOLOGY, CHICAGO, 1972.
15 AGARWAL R.K., AND BOSHKOV S.H.,   STRESSES AND DISPLACEMENTS
        AROUND A CIRCULAR TUNNEL IN A THREE LAYER MEDIUM, 1 AND
        2. INT.J. ROCK MECH. AND MINING SCIENCE 6, 519-528 AND
        529-540, 1969.
16 AGMON S.,   LECTURES ON ELLIPTIC BOUNDARY VALUE PROBLEMS. VAN NOSTRAND,
        PRINCETON, NEW JERSEY, 1965.
17 AGUIRRE-RAMIREZ G., AND ODEN J.T.,   FINITE ELEMENT TECHNIQUE
        APPLIED TO HEAT CONDUCTION IN SOLIDS WITH TEMPERATURE
        DEPENDENT THERMAL CONDUCTIVITY. A.S.M.E. PAPER 69-
        WA/HT-34, N.P.13, 1969.
18 AHLBERG J.H.,   SPLINE APPROXIMATION AND COMPUTER-AIDED DESIGN.
        PP.275-289 OF F.L. ALT AND M. RUBINOFF (EDS.), ADVANCES
        IN COMPUTERS VOL.10, ACADEMIC PRESS, NEW YORK, 1970.
19 AHLBERG J.H.,   THE SPLINE APPROXIMATION AS AN ENGINEERING TOOL.
        PP. 1-18 OF G.M.L. GLADWELL (ED.), COMPUTER AIDED ENGINEERING,
        UNIVERSITY OF WATERLOO PRESS, 1971.
20 AHLBERG J.H., AND NILSON E.N.,   ORTHOGONALITY PROPERTIES OF SPLINE
        FUNCTIONS. J.MATH.ANAL.APPL. 11., 321-337, 1965.
21 AHLBERG J.H., AND NILSON E.N.,   THE APPROXIMATION OF LINEAR FUNCTIONALS.
        SIAM J. NUMER.ANAL. 3, 173-182, 1966.
22 AHLBERG J.H., NILSON E.N. AND WALSH J.L.,   THE THEORY OF SPLINES AND
        THEIR APPLICATIONS. ACADEMIC PRESS, NEW YORK, 1967.
23 AHLFORS L.V.,   COMPLEX ANALYSIS. MCGRAW HILL, NEW YORK, 1953.
24 AHLFORS L.V., KRA I., MASKIT B., AND NIRENBERG L. (EDS.),
        CONTRIBUTIONS TO ANALYSIS. ACADEMIC PRESS, NEW YORK, 1974.

25 AHLIN A.C., A BIVARIATE GENERALIZATION OF HERMITE'S INTERPOLATION
        FORMULA. MATH.COMP.18, 264-273, 1964.
26 AHMAD S., FINITE ELEMENT METHOD FOR WAVEGUIDE PROBLEMS.
        ELECTRONIC LETTERS 4, 387-389, 1968.
27 AHMAD S., PSEUDO ISOPARAMETRIC FINITE ELEMENTS FOR SHELL AND
        PLATE ANALYSIS. RECENT ADVANCES IN STRESS ANALYSIS,
        ROYAL AERO SOC., 6-21, 1969.
28 AHMAD S., ANDERSON R.G., AND ZIENKIEWICZ O.C., VIBRATION OF THICK,
        CURVED, SHELLS WITH PARTICULAR REFERENCE TO TURBINE BLADES.
        J. STRAIN ANALYSIS 5, 200-206, 1970.
29 AHMAD S., IRONS B.M. AND ZIENKIEWICZ O.C., CURVED THICK SHELL AND
        MEMBRANE ELEMENTS WITH PARTICULAR REFERENCE TO AXISYMMETRIC
        PROBLEMS. PROC.2ND CONF.MATRIX METHODS IN STRUCTURAL MECHANICS,
        WRIGHT-PATTERSON AFB,OHIO,AFFDL-TR-68-150,1968.
30 AHMAD S., IRONS B.M. AND ZIENKIEWICZ O.C., ANALYSIS OF THICK AND THIN
        SHELL STRUCTURES BY GENERAL CURVED ELEMENTS. INT.J.NUMER.METH.
        ENG. 2,419-452, 1970.
31 AHMED S. AND DALY P., FINITE ELEMENT METHODS FOR INHOMOGENEOUS
        WAVEGUIDES. PROC. I.E.E. 116, 1661-1664, 1969.
32 AHMED S., AND DALY P., WAVEGUIDE SOLUTIONS BY THE FINITE ELEMENT
        METHOD. RADIO AND ELECTRON.ENG.38, 217-223, 1969.
33 AHN C.S., MATHEMATICAL MODEL OF HEATED EFFUENTS IN COASTAL WATER.
        TRANS. AMERICAN NUCL.SOC.16, 16-17,1973.
34 AKIN J.E., A FINITE ELEMENT COLLOCATION SOLUTION OF DIFFERENTIAL
        EQUATIONS. INTERN.CONF. ON VARIATIONAL METHODS IN ENG.,
        SOUTHAMPTON UNIVERSITY, 1972.
35 AKIN J.E., A LEAST SQUARES-FINITE ELEMENT SOLUTION OF NONLINEAR
        OPERATORS. PP. 153-162 OF J.R. WHITEMAN (ED.), THE MATHEMATICS
        OF FINITE ELEMENTS AND APPLICATIONS. ACADEMIC PRESS,
        LONDON, 1973.
36 AKIN J.E., FENTON D.L. AND STODDART W.C.T., THE FINITE ELEMENT
        METHOD - A BIBLIOGRAPHY OF ITS THEORY AND APPLICATIONS.
        REPORT 72-1, DEPARTMENT OF ENGINEERING MECHANICS,
        UNIVERSITY OF TENNESSEE, KNOXVILLE,1972.
37 AKIN J.E., AND SEN GUPTA S.R., FINITE ELEMENT APPLICATIONS OF
        THE LEAST SQUARES METHOD. INTERN.CONF. ON VARIATIONAL
        METHODS IN ENG., SOUTHAMPTON UNIVERSITY, 1972.
38 AKYUZ A., NATURAL CO-ORDINATE SYSTEMS. AN AUTOMATIC INPUT DATA
        GENERATION SCHEME FOR A FINITE ELEMENT METHOD. NUCLEAR ENG. AND
        DESIGN 11, 195-207, 1970.
39 AKYUZ F.A., AND MERWIN J.E., SOLUTION OF NONLINEAR PROBLEMS
        OF ELASTOPLASTICITY BY THE FINITE ELEMENT METHOD. AIAA.
        J.6, 1825-1831, 1968.
40 AKYUZ F.A., AND UTKU S., AN AUTOMATIC NODE-RELABELLING SCHEME FOR
        BANDWIDTH MINIMIZATION OF STIFFNESS MATRICES. AIAA J.6 , 728-
        730, 1968.
41 ALBASINY E.L., AND HOSKINS W.D., CUBIC SPLINE SOLUTIONS TO TWO-
        POINT BOUNDARY VALUE PROBLEMS. COMP.J.12, 151-153, 1969.
42 ALBASINY E.L., AND HOSKINS W.D., THE NUMERICAL CALCULATION OF
        ODD-DEGREE POLYNOMIAL SPLINES WITH EQUISPACED KNOTS. J.
        INST.MATH.APPLICS.7, 384-397, 1971.
43 ALBASINY E.L. AND HOSKINS W.D., EXPLICIT ERROR BOUNDS FOR PERIODIC
        SPLINES OF ODD ORDER ON A UNIFORM MESH. J. INST.MATH.APPLICS.12,
        305-318,1973.
44 ALBERT I.G.F., BERECHNUNG VON PLATTEN UND RIPPENPLATTEN NACH DER
        METHODE DER ENDLICHEN ELEMENTE. DISSERTATION NO.4/81, E.T.H.
        ZURICH, 1971.
45 ALBRECHT J., AND COLLATZ L. (EDS.), NUMERISCHE METHODEN BEI
        DIFFERENTIALGLEICHUNGEN UND MIT FUNKTIONALANALYTISCHEN
        HILFSMITTELN. I.S.N.M. 19, BIRKHAUSER VERLAG, BASEL,
        1974.
46 ALDABBAGH A., GOODMAN J.R., AND BODIG J., FINITE ELEMENT METHOD
        FOR WOOD MECHANICS. J.STRUCT.DIV.ASCE.98, ST3, 351-366,1971.
47 ALDSTEDT E., SHELL ANALYSIS USING PLANAR TRIANGULAR ELEMENTS. FINITE
        ELEMENT METHODS, TAPIR, TRONDHEIM, NORWAY,1969.
48 ALLEN M., NASTRAN EXPERIENCES OF FORT WORTH OPERATION. CONVAIR

AEROSPACE DIVISION OF GENERAL DYNAMICS- NASTRAN USERS
EXPERIENCES, NASA TECHNICAL NOTE, TN X-2378, 1971.

49 ALLIK H., AND HUGHES T.J.R., FINITE ELEMENT METHOD FOR PIEZOELECTRIC
VIBRATION. INT.J.NUMER.METH.ENG.2, 151-158, 1970.

50 ALLMAN D.J., TRIANGULAR FINITE ELEMENTS FOR PLATE BENDING WITH
CONSTANT AND LINEARLY VARYING BENDING MOMENTS. IUTAM
COLLOQUIUM ON HIGH SPEED COMPUTING OF ELASTIC STRUCTURES,
UNIVERSITY OF LIEGE, 1970.

51 ALLMAN D.J., FINITE ELEMENT ANALYSIS OF PLATE BUCKLING USING A
MIXED VARIATIONAL PRINCIPLE. PROC. 3RD CONF. MATRIX METHODS
IN STRUCTURAL MECHANICS, WRIGHT-PATTERSON AFB, OHIO, 1971.

52 ALLWOOD R.J., MATRIX METHODS OF STRUCTURAL ANALYSIS. IN REID
(ED.), LARGE SPARSE SETS OF LINEAR EQUATIONS, ACADEMIC
PRESS, LONDON, 1971.

53 ALLWOOD R.J. AND CORNES G.M.M., A POLYGONAL FINITE ELEMENT FOR
PLATE BENDING PROBLEMS USING THE ASSUMED STRESS APPROACH.
INT.J.NUMER.METH.ENG.1, 135-150, 1969.

54 ALTMAN W., AND VENANCIO-FILHO F., STABILITY OF PLATES USING A
MIXED FINITE ELEMENT FORMULATION. COMPUTERS AND STRUCTURES
4, 437-444, 1974.

55 ALUJEVIC A., AND HEAD J.L., APPLICATION OF THREE-DIMENSIONAL FINITE
ELEMENTS FOR ANALYSIS OF IRRADIATION INDUCED STRESSES IN
NUCLEAR REACTOR FUEL ELEMENTS. PP.415-426 OF J.R. WHITEMAN
(ED.), THE MATHEMATICS OF FINITE ELEMENTS AND APPLICATIONS.
ACADEMIC PRESS, LONDON, 1973.

56 ALUJEVIC A., AND HEAD J.L., FUEL ELEMENT STRESSES AT
DISCONTINUITIES AND INTERACTIONS BY FINITE ELEMENTS IN
TWO-DIMENSIONS. ATOMKERNENERGIE 21, 75-80, 1973.

57 ALUJEVIC A., HEAD J.L., RODGERS I.H., SOME MODELS FOR THE ANALYSIS OF
STRESSES IN A TUBULAR FUEL PIN FOR HIGH TEMPERATURE REACTORS BY
FINITE ELEMENTS. PAPER D1/1 PROCEEDINGS 2ND STRUCTURAL MECH IN
REACTOR TECH.CONF., BERLIN, 1973.

58 AMANN H., ZUM GALERKIN VERFAHREN FUR DIE HAMMERSTEINSCHE
GLEICHUNG. ARCH.RAT.MECH.ANAL.35, 114-121, 1969.

59 AMANN H., UBER DIE KONVERGENZGESCHWINDIGKEIT DES GALERKIN-VERFAHRENS
FUR DIE HAMMERSTEINSCHE GLEICHUNG. ARCH.RAT. MECH. ANAL. 37,
33-47, 1970.

60 AMES W.F., NONLINEAR PROBLEMS OF ENGINEERING. ACADEMIC PRESS, NEW YORK,
1964.

61 AMES W.F., NONLINEAR PARTIAL DIFFERENTIAL EQUATIONS IN ENGINEERING.
ACADEMIC PRESS, NEW YORK, 1965.

62 ANAND S.C., AND SHAW H., NONLINEAR ANALYSIS OF A ROLLING CONTACT
PROBLEM BY FINITE ELEMENTS. PROC.CONF.COMP.METHODS IN NON-
LINEAR MECH., AUSTIN, TEXAS, 1974.

63 ANDERHEGGEN E., FINITE ELEMENT PLATE BENDING EQUILIBRIUM ANALYSIS.
J.ENG.MECH.DIVISION, A.S.C.E. 95, 1969.

64 ANDERHEGGEN E., PROGRAMME ZUR METHODE DER FINITEN ELEMENTE.
INSTITUT FUR BANSTATIK, EIDGENOSSISCHE TECHNISCHE
HOCHSCHULE, ZURICH, 1969.

65 ANDERHEGGEN E., A CONFORMING TRIANGULAR FINITE ELEMENT PLATE BENDING
SOLUTION. INT.J.NUMER.METH.ENG.2, 259-264,1970.

66 ANDERHEGGEN E., STARR-PLASTISCHE TRAGLASTBERECHNUNGEN MITTELS DER
METHODE DER FINITEN ELEMENTE. REPORT 0032, INSTUT FUR BAUSTATIK,
EIDGENOESSISCHE TECHNISCHE HOCHSCHULE, ZURICH, 1971.

67 ANDERSON C.A., AND ZIENKIEWICZ O.C., SPONTANEOUS IGNITION, FINITE ELEMENT
SOLUTIONS FOR STEADY AND TRANSIENT CONDITIONS. REPORT LA-5037,
LOS ALAMOS SCIENTIFIC LABORATORY, NEW MEXICO, 1973.

68 ANDERSON G.P., RUGGLES V.P. AND STIBOR G.S., USE OF FINITE ELEMENT
COMPUTER PROGRAMS IN FRACTURE MECHANICS. INT.J.FRACT.MECH.7,
63-76, 1971.

69 ANDERSON N., AND ARTHURS A.M., COMPLEMENTARY VARIATIONAL PRINCIPLES
FOR (PHI)IV = F(PHI). PROC. CAMB. PHIL. SOC. 68, 173-177, 1970.

70 ANDERSON N., AND ARTHURS A.M., COMPLEMENTARY VARIATIONAL PRINCIPLES
FOR BIHARMONIC BOUNDARY VALUE PROBLEMS. NUOVO CIMENTO LETT 2,
503-507, 1971.

71 ANDERSON N., AND ARTHURS A.M., COMPLEMENTARY VARIATIONAL PRINCIPLES

FOR STEADY HEAT CONDUCTION WITH MIXED BOUNDARY
CONDITIONS. J.ENG.MATH.6, 23-30, 1972.

72 ANDERSON N., AND ARTHURS A.M., EXTREMUM PRINCIPLES FOR SOME NONLINEAR
HEAT TRANSFER PROBLEMS. J. ENG. MATH. 6, 331-339, 1972.

73 ANDERSON R.G., SOME APPLICATIONS OF BEAM, COLUMN AND PLATE THEORY
TO PROBLEMS OF REACTOR FUEL ELEMENT DEFORMATIONS. PAPER
D2/1, PROCEEDINGS 2ND STRUCTURAL MECH IN REACTOR TECH.
CONF., BERLIN, 1973.

74 ANDERSON R.G., IRONS B.M., AND ZIENKIEWICZ O.C., VIBRATION AND
STABILITY OF PLATES USING FINITE ELEMENTS. INT.J.SOLIDS
STRUCT. 4, 1031-1035, 1968.

75 ANDERTON G.L., CONNACHER N.E., DOUGHERTY C.S. AND HANSEN S.D.,
ANALYSIS OF THE 747 AIRCRAFT WING BODY INTERSECTION. PROC.2ND
CONF. MATRIX METHODS IN STRUCTURAL MECHANICS, WRIGHT-PATTERSON
AFB.,OHIO, AFFDL-TR-68-150, 1968.

76 ANDO Y., YAGAWA G., AND OKABAYASHI K., APPLICATION OF THE FINITE
ELEMENT METHOD TO THE ANALYSIS OF FRACTURE OF CYLINDRICAL
SHELLS. PAPER G5/4, PROC.2ND STRUCTURAL MECH. IN REACTOR
TECH. CONF., BERLIN, 1973.

77 ANG A.H., AND LOPEZ L.A., DISCRETE MODEL ANALYSIS OF ELASTIC
PLASTIC PLATES. J.ENG.MECH.DIVISION, A.S.C.E. 94, EM1,
271-293,1973.

78 ANGEL E., DISCRETE INVARIANT IMBEDDING AND ELLIPTIC BOUNDARY
VALUE PROBLEMS OVER IRREGULAR REGIONS. J.MATH.ANAL.APPL.
23, 471-484, 1968.

79 ANQUES L., BERGER H., OHAYON R., AND VALID R., VIBRATIONS OF TANKS
PARTIALLY FILLED WITH LIQUIDS. PROCEEDINGS OF SYMPOSIUM
ON FINITE ELEMENT METHODS IN FLOW PROBLEMS, SWANSEA, 1974.

80 ANSELONE P.M., COLLECTIVELY COMPACT OPERATOR APPROXIMATION THEORY.
PRENTICE HALL, NEW JERSEY, 1971.

81 ANSORGE R. AND TORING W. (EDS.), NUMERISCHE LOSUNG NICHTLINEARER
PARTIALLER DIFFERENTIAL UND INTEGRODIFFERENTIALGLEICHUNGEN.
LECTURE NOTES IN MATHEMATICS, NO. 267. SPRINGER-VERLAG,
BERLIN, 1972.

82 ANTES H., BICUBIC FUNDAMENTAL SPHERES IN PLATE BENDING. INT.J.NUMER.
METH.ENG.8, 503-511, 1974.

83 APPA RAO T.S.V.R., ASSUMED STRESS HYBRID FINITE ELEMENT MODEL
FOR THE ANALYSIS OF AN AXISYMMETRIC THICK-WALLED PRESSURE VESSEL,
PP.315-335, VOL.6 OF PROCEEDINGS 1ST STRUCTURAL MECH. IN REACTOR
TECH. CONF., BERLIN, 1971.

84 APPERT K., BERGER D., AND GRUBER R., NUMERICALLY DETERMINED
INSTABILITY GROWTH RATES FOR A BELT PINCH. PHYS.LETT.
(NETHERLANDS) 46A, 339-340, 1974.

85 ARAL M.M., APPLICATION OF FINITE ELEMENT ANALYSIS IN FLUID
MECHANICS. PH.D.THESIS. GEORGIA INSTITUTE OF TECHNOLOGY,
ATLANTA, 1971.

86 ARAL M.M., MAYER P.G. AND SMITH C.V., FINITE ELEMENT GALERKIN METHOD
SOLUTIONS TO SELECTED ELLIPTIC AND PARABOLIC DIFFERENTIAL
EQUATIONS. PROC.3RD CONF.MATRIX METHODS IN STRUCTURAL MECHANICS
WRIGHT-PATTERSON AFB,OHIO,1971.

87 APALDSEN P.O., THE APPLICATION OF THE SUPERELEMENT METHOD IN
ANALYSIS AND DESIGN OF SHIP STRUCTURES AND MACHINERY
COMPONENTS. NATIONAL SYMPOSIUM ON COMPUTERIZED STRUCTURAL
ANALYSIS AND DESIGN. WASHINGTON D.C., 1972.

88 ARALDSEN PER O., AN EXAMPLE OF LARGE-SCALE STRUCTURAL ANALYSIS.
COMPARISON BETWEEN FINITE ELEMENT CALCULATION AND FULL
SCALE MEASUREMENTS ON THE OIL TANKER ESSO NORWAY. COMPUTERS
AND STRUCTURES 4, 69-94, 1974.

89 ARCHER J.S., CONSISTENT MASS MATRIX FOR DISTRIBUTED MASS SYSTEMS.
J.STRUCT. DIV.A.S.C.E. 89, 161-178, 1963.

90 ARCHER J.S., CONSISTENT MATRIX FORMULATION FOR STRUCTURAL ANALYSIS
USING FINITE ELEMENT TECHNIQUES. AIAA J.3, 1910-1918, 1965.

91 ARCHER J.S., AND RUBIN C.P., IMPROVED LINEAR AXISYMMETRIC SHELL FLUID
MODEL FOR LAUNCH VEHICLE LONGITUDAL RESPONSE ANALYSIS.
PROC. 1ST CONF.MATRIX METHODS IN STRUCTURAL MECHANICS,
WRIGHT-PATTERSON AFB, OHIO, AFFDL TR66-80, 1965.

92 ARGYRIS J.H., ENERGY THEOREMS AND STRUCTURAL ANALYSIS.
    BUTTERWORTH, LONDON 1960.

93 ARGYRIS J.H., MATRIX METHODS OF STRUCTURAL ANALYSIS. PROC.14TH MEETING
    OF AGARD, AGARDOGRAPHY 72, 1962.

94 ARGYRIS J.H., RECENT ADVANCES IN MATRIX METHODS OF STRUCTURAL ANALYSIS.
    PROGRESS IN AERONAUTICAL SCIENCES,PERGAMON PRESS,OXFORD,1964.

95 ARGYRIS J.H., CONTINUA AND DISCONTINUA. PROC.1ST CONF. MATRIX METHODS
    IN STRUCTURAL MECHANICS, WRIGHT-PATTERSON AFB, OHIO, AFFDL
    TR66-80, 1965.

96 ARGYRIS J.H., ELASTO-PLASTIC MATRIX DISPLACEMENT ANALYSIS OF
    THREE-DIMENSIONAL CONTINUA. J.ROY.AERO.SOC,69, 633-635,
    1965.

97 ARGYRIS J.H., MATRIX ANALYSIS OF THREE-DIMENSIONAL SOLIDS - SMALL AND
    LARGE DISPLACEMENTS. AIAA.JL.3, 45-51, 1965.

98 ARGYRIS J.H., MATRIX DISPLACEMENT ANALYSIS OF ANISOTROPIC SHELLS BY
    TRIANGULAR ELEMENTS. J.ROYAL AERO. SOC. 69, 801-805,1965.

99 ARGYRIS J.H., REINFORCED FIELDS OF TRIANGULAR ELEMENTS WITH LINEARLY
    VARYING STRAIN; EFFECTS OF INITIAL STRAINS. J. ROYAL AERO.SOC.69,
    799-801,1965.

100 ARGYRIS J.H., THREE DIMENSIONAL ANISOTROPIC AND INHOMOGENEOUS
    ELASTIC MEDIA MATRIX ANALYSIS FOR SMALL AND LARGE
    DISPLACEMENTS. INGENIEUR ARCHIV 34, 33-35, 1965.

101 ARGYRIS J.H., TETRAHEDRON ELEMENTS WITH LINEARLY VARYING STRAIN FOR THE
    MATRIX DISPLACEMENT METHOD. J.ROYAL AERO.SOC.69, 877-880,1965.

102 ARGYRIS J.H. TRIANGULAR ELEMENTS WITH LINEARLY VARYING STRAIN FOR THE
    MATRIX DISPLACEMENT METHOD. J.ROYAL AERO.S.C.69,711-713,1965.

103 ARGYRIS J.H., ARBITRARY QUADRILATERAL SPAR WEBS FOR THE MATRIX
    DISPLACEMENT METHOD. J.ROYAL AERO.SOC.70,359-362,1966.

104 ARGYRIS J.H., MEMBRANE PARALLELOGRAM ELEMENT WITH LINEARLY VARYING EDGE
    STRAIN FOR THE MATRIX DISPLACEMENT METHOD. J.ROYAL AERO,SOC,70,
    599-604,1966.

105 ARGYRIS J.H., A TAPERED THICKNESS TRIM 6 ELEMENT FOR THE MATRIX
    DISPLACEMENT METHOD. J.ROYAL AERO.SOC.70,1040-1043,1966,

106 ARGYRIS J.H., MATRIX DISPLACEMENT ANALYSIS OF PLATES AND SHELLS,
    INGENIEUR-ARCHIV.35,102-142,1966.

107 ARGYRIS J.H., THE HERMES 8 ELEMENT FOR THE MATRIX DISPLACEMENT METHOD.
    J.ROYAL AERO.SOC.72, 613-617,1968.

108 ARGYRIS J.H., THE TET 20 AND TEA8 ELEMENTS FOR THE MATRIX DISPLACEMENT
    METHOD. J.ROYAL AERO.SOC.72, 618-623,1968.

109 ARGYRIS J.H., THE TUBA FAMILY OF PLATE ELEMENTS FOR THE MATRIX
    DISPLACEMENT METHOD. AERO.J.ROYAL AERO. SOC. 72,
    701-709, 1968.

110 ARGYRIS J.H., FINITE ELEMENTS IN TIME AND SPACE. AERONAUTICAL J.73,
    1041-1044,1969.

111 ARGYRIS J.H., THE IMPACT OF THE DIGITAL COMPUTER ON ENGINEERING
    SCIENCES. AERO.J.ROYAL AERO.SOC.74, 14-41, 111-127, 959-
    964 AND 1041-1046, 1970.

112 ARGYRIS J.H., ANGELOPOULOS T. AND BICHAT B., A GENERAL METHOD FOR
    THE SHAPE FINDING OF LIGHTWEIGHT TENSION STRUCTURES. COMP.
    METH.APPL. MECH. ENG. 3, 135-149, 1974.

113 ARGYRIS J.H., BORNLAND O.E., GRIEGER I., SORENSEN M., A SURVEY OF
    THE APPLICATION OF FINITE ELEMENT METHODS TO STRESS
    ANALYSIS PROBLEMS WITH PARTICULAR EMPHASIS ON THEIR
    APPLICATION TO NUCLEAR ENGINEERING PROBLEMS. PROC.CONF.
    APPLICATION OF FINITE ELEMENT METHODS TO STRESS ANALYSIS
    PROBLEMS IN NUCLEAR ENGINEERING. ISPRA, ITALY, 1971.

114 ARGYRIS J.H., BUCK K.E., GRIEGER I., AND WILLAM K.J., DISCUSSION OF FINITE
    ELEMENT ANALYSIS OF PRESTRESSED CONCRETE REACTOR VESSELS, PP.
    181-201 OF C.E. KESLER (ED.), CONCRETE FOR NUCLEAR REACRORS VOL.1,
    AMERICAN CONCRETE INSTITUTE, DETROIT, 1972.

115 ARGYRIS J.H., BUCK.K.E., SCHARPF D.W., KILBER H.M. AND MARECZEK G.,
    SOME NEW ELEMENTS FOR THE MATRIX DISPLACEMENT METHOD.
    PROC.2ND CONF.MATRIX METHODS IN STRUCTURAL MECHANICS,
    WRIGHT-PATTERSON AFB, OHIO, AFFDL-TR-68-150, 1968.

116 ARGYRIS J.H., BUCK K.E., SCHARPF D.W. AND WILLIAM K., NONLINEAR
    METHODS OF STRUCTURAL ANALYSIS. PAPER M2/2, PROCEEDINGS

1ST STRUCTURAL MECH. IN REACTOR TECH.CONF., BERLIN, 1971.

117 ARGYRIS J.H., BUCK K.E., SCHARPF D.W., AND WILLIAM K.J., LINEAR METHODS OF STRUCTURAL ANALYSIS. NUCL.ENG.DESIGN 19, 139-167, 1972.

118 ARGYRIS J.H., BUCK K.E., SCHARPF D.W., AND WILLIAM K.J., NONLINEAR METHODS OF STRUCTURAL ANALYSIS. NUCL.ENG.DESIGN 19, 169-179, 1972.

119 ARGYRIS J.H., AND CHAN A.S.L., APPLICATION OF FINITE ELEMENTS IN SPACE AND TIME. INGENIEUR ARCHIV.41,235-257,1972.

120 ARGYRIS J.H., AND CHAN A.S.L., STATIC AND DYNAMIC ELASTO-PLASTIC ANALYSIS BY THE METHOD OF FINITE ELEMENTS IN SPACE AND TIME. PP. 147-175 OF A. SAWCZUK (ED.), FOUNDATIONS OF PLASTICITY. NOORDHOFF, LEYDEN, 1972.

121 ARGYRIS J.H., AND CHAN A.S.L., APPLICATION OF FINITE ELEMENTS IN SPACE AND TIME. INGENIEUR ARCHIV 41, 235-257, 1972.

122 ARGYRIS J.H., DUNNE P.C., AND ANGELOPOULOS T., DYNAMIC RESPONSE BY LARGE STEP INTEGRATION. EARTHQUAKE ENG. AND STRUCTURAL DYNAMICS 2, 185-203, 1973.

123 ARGYRIS J.H., DUNNE P.C., AND ANGELOPOULOS T., NONLINEAR OSCILLATIONS USING THE FINITE ELEMENT TECHNIQUE. COMP. METH. APPL. MECH.ENG.2, 203-250, 1973.

124 ARGYRIS J., DUNNE P.C., AND ANGELOPOULOS T., FINITE ELEMENT DIFFICULTIES IN NONLINEAR AND INCOMPRESSIBLE MATERIAL. PROC.CONF.COMP.METHODS IN NONLINEAR MECHANICS, AUSTIN, TEXAS, 1974.

125 ARGYRIS J.H., DUNNE P.C., ANGELOPOULOS T., AND BICHAT B., LARGE NATURAL STRAINS AND SOME SPECIAL DIFFICULTIES DUE TO NON-LINEARITY AND INCOMPRESSIBILITY IN FINITE ELEMENTS. COMP.METH.APPL.MECH.ENG.4, 219-278, 1974.

126 ARGYRIS J.H., FAUST G., SZIMMAT J., WARNKE P., AND WILLAM K.J., RECENT DEVELOPMENTS IN THE FINITE ELEMENT ANALYSIS OF PRESTRESSED CONCRETE REACTOR VESSELS. PAPER H1/1, PROCEEDINGS 2ND STRUCTURAL MECH. IN REACTOR TECH. CONF., BERLIN, 1973.

127 ARGYRIS J.H., AND FRIED I., THE LUMINA ELEMENT FOR THE MATRIX DISPLACEMENT METHOD. AERO.J.ROYAL AERO.SOC.72, 514-517, 1968.

128 ARGYRIS J.H., AND FRIED I., THE PUBA FAMILY OF PLATE ELEMENTS FOR THE MATRIX DISPLACEMENT METHOD. ISD REPORT NO.56, 1968.

129 ARGYRIS J.H., FRIED I., AND SCHARPF D.W., THE HERMES EIGHT ELEMENT FOR THE MATRIX DISPLACEMENT METHOD. AERO.J.ROY.AERO.SOC.72, 613-617,1968.

130 ARGYRIS J.H., FRIED I., AND SCHARPF D.W., THE TUBA FAMILY OF PLATE ELEMENTS FOR THE MATRIX DISPLACEMENT METHOD. AERO. J.ROY.AERO. SOC. 72, 618-623, 1968.

131 ARGYRIS J.H., FRIED I., AND SCHARPF D.W., THE TUBA FAMILY OF PLATE ELEMENTS FOR THE MATRIX DISPLACEMENT METHOD. AERO.J. ROYAL AERO.SOC.72, 701-709, 1968.

132 ARGYRIS J.H., HAASE M., AND MALEJANNAKIS G.A., NATURAL GEOMETRY OF SURFACES WITH SPECIFIC REFERENCE TO THE MATRIX DISPLACEMENT ANALYSIS OF SHELLS. I, II AND III. KONINKLIJKE NEDERLANDSE AKAD. VAN WETENSCHAPPEN B 76, 361-410, 1973.

133 ARGYRIS J.H., AND KELSEY S., ENERGY THEOREMS AND STRUCTURAL ANALYSES. BUTTERWORTH, LONDON, 1960.

134 ARGYRIS J.H., KELSEY S. AND KAMEL W.H., MATRIX METHODS IN STRUCTURAL ANALYSIS; A PRECIS OF RECENT DEVELOPMENTS. PROC.14TH MEETING OF STRUCTURES AND MATERIALS PANEL, AGARD, 1963.

135 ARGYRIS J.H., AND LOCHNER N., ON THE APPLICATION OF THE SHEBA SHELL ELEMENT. COMP.METH.APPL.MECH.ENG.1, 317-347,1972.

136 ARGYRIS J.H., AND MARECZEK G., FINITE ELEMENT ANALYSIS OF SLOW INCOMPRESSIBLE VISCOUS FLUID MOTION.

137 ARGYRIS J.H., AND MARECZEK G., POTENTIAL FLOW ANALYSIS BY FINITE ELEMENTS. INGENIEUR ARCHIV 41, 1-25, 1972.

138 ARGYRIS J.H., MARECZEK G. AND SHARPF D.W., TWO AND THREE DIMENSIONAL FLOW USING FINITE ELEMENTS. AERONAUTICAL J.73, 961-964,1969.

139 ARGYRIS J.H., AND SCHARPF D.W., THE SHEBA FAMILY OF SHELL ELEMENTS FOR THE MATRIX DISPLACEMENT METHOD. AERO.J.ROYAL AERO. SOC. 72, 873-883, 1968.

140 ARGYRIS J.H., AND SCHARPF D.W., FINITE ELEMENTS IN TIME AND SPACE.

AERO.J.ROYAL.AERO.SOC.73, 1041-1044,1969.

141   ARGYRIS J.H., AND SCHARPF D.W.,   SOME GENERAL CONSIDERATIONS ON
       THE NATURAL MODE TECHNIQUE; I: SMALL DISPLACEMENTS, II:
       LARGE DISPLACEMENTS.   AERO.J.ROYAL AERO.SOC.73, 218-226 AND
       361-368, 1969.

142   ARGYRIS J.H., AND SCHARPF D.W.,   THE SHEBA FAMILY OF SHELL
       ELEMENTS FOR THE MATRIX DISPLACEMENT METHOD.   III, LARGE
       DISPLACEMENTS.   AERO.J.ROYAL.AERO.SOC.73, 423-426, 1969.

143   ARGYRIS J.H., AND SHARPF D.W.,   FINITE ELEMENT FORMULATION OF THE
       INCOMPRESSIBLE LUBRICATION PROBLEM.   NUCL.ENG.DES.11, 225-229,
       AND 230-236, 1970.

144   ARGYRIS J.H., AND SCHARPF D.W.,   FINITE ELEMENT THEORY OF PLATES
       AND SHELLS INCLUDING SHEAR STRAIN EFFECTS.   PP. 253-292
       OF PROC. IUTAM SYMPOSIUM, UNIVERSITY OF LIEGE, 1971.

145   ARGYRIS J.H., AND SCHARPF D.W.,   FINITE ELEMENTS IN TIME AND SPACE. AERO.
       J.ROY.AERO.SOC.73,377-384,1972.

146   ARGYRIS J.H., AND SCHARPF D.W.,   LARGE DEFLECTIONS ANALYSIS OF
       PRESTRESSED NETWORKS.   J.STRUCTURAL DIV. A.S.C.E.
       633-654, 1972.

147   ARGYRIS J.H., AND SCHARPF D.W.,   MATRIX DISPLACEMENT ANALYSIS OF SHELLS
       AND PLATES INCLUDING TRANSVERSE SHEAR STRAIN EFFECTS.   COMP.METH.
       APPL.MECH.ENG.1, 81-139,1972.

148   ARGYRIS J.H., AND SCHARPF D.W.,   METHODS OF ELASTOPLASTIC ANALYSIS.
       Z.A.M.P. 23, 517-552, 1972.

149   ARGYRIS J.H., AND WILLIAM K.,   CALCULATIONS FOR PRESTRESSED
       CONCRETE PRESSURE VESSELS.   REPORT ISD-104, INST. FUR
       STATIK UND DYNAMIK DER LUFT UND RAUMFAHRTKONSTRUKTIONEN,
       UNIVERSITAT STUTTGART, 1971.

150   ARGYRIS J.H., AND WILLIAM K.J.,   SOME CONSIDERATIONS FOR THE
       EVALUATION OF FINITE ELEMENT MODELS.   PAPER M2/1, PROC.
       2ND STRUCTURAL MECH. IN REACTOR TECH.CONF., BERLIN, 1973.

151   ARLETT P.L., BAHRANI A.K. AND ZIENKIEWICZ O.C.,   APPLICATION OF FINITE
       ELEMENTS TO THE SOLUTION OF HELMHOLTZ'S EQUATION.   PROC.INST.
       ELEC.ENG.115, 1762-1766,1968.

152   ARMEN H.,   PLASTICITY THEORY AND FINITE ELEMENT APPLICATIONS. PP.393-438
       OF J.T. ODEN, R.W. CLOUGH, AND Y. YAMOMOTO (EDS.), ADVANCES IN
       COMPUTATIONAL METHODS IN STRUCTURAL MECHANICS AND DESIGN,
       UNIVERSITY OF ALABAMA PRESS, HUNTSVILLE, 1972.

153   ARMEN H., ISAKSON G. AND PIFKO A.,   DISCRETE ELEMENT METHODS FOR THE
       PLASTIC ANALYSIS OF STRUCTURES SUBJECTED TO CYCLIC LOADING.
       INT.J.NUMER.METH.ENG.2, 189-206, 1970.

154   ARMEN H., PIFKO A. AND LEVINE H.S.,   A FINITE ELEMENT METHOD FOR THE
       PLASTIC BENDING ANALYSIS OF STRUCTURES.   PROC.2ND.CONF.MATRIX
       METHODS IN STRUCTURAL MECHANICS, WRIGHT-PATTERSON AFB,OHIO,
       AFFDL-TR-68-150,1968.

155   ARMSTRONG W.H.,   ANALYSIS OF NONLINEAR THERMOVISCOELASTICITY PROBLEMS
       BY THE FINITE ELEMENT METHOD.   M.SC.THESIS, UNIVERSITY OF
       ALABAMA IN HUNTSVILLE, 1971.

156   ARSENIN V.Y.,   BASIC EQUATIONS AND SPECIAL FUNCTIONS OF
       MATHEMATICAL PHYSICS.   ILIFFE, LONDON, 1968.

157   ARTHURS A.H.,   EXTREMUM PRINCIPLES FOR A CLASS OF BOUNDARY VALUE
       PROBLEMS.   PROC.CAMB.PHIL SOC.65, 803-806, 1969.

158   ARTHURS A.H.,   COMPLEMENTARY VARIATIONAL PRINCIPLES.   OXFORD UNIVERSITY
       PRESS, OXFORD, 1970.

159   ARTHURS A.H.,   CANONICAL APPROACH TO BIHARMONIC VARIATIONAL PROBLEMS.
       QUART.APPL.MATH.28, 135-138, 1970.

160   ARTHURS A.H.,   A NOTE ON KOMKOV'S CLASS OF BOUNDARY VALUE PROBLEMS
       AND ASSOCIATED VARIATIONAL PRINCIPLES. J.MATH.ANAL.APPL.33,
       402-407, 1971.

161   ARTHURS A.H.,   DUAL EXTREMUM PRINCIPLES AND ERROR BOUNDS FOR A CLASS
       OF BOUNDARY VALUE PROBLEMS.   J.MATH.ANAL.APPL.41, 781-795, 1973.

162   ARTHURS A.H.,   ON OPERATOR EQUATIONS AND VARIATIONAL PRINCIPLES.   PROC.
       CAMB.PHIL.SOC. 74, 277-279, 1973.

163   ARTHURS A.H., AND COLES C.W.,   ERROR BOUNDS AND VARIATIONAL METHODS FOR
       NON-LINEAR DIFFERENTIAL AND INTEGRAL EQUATIONS.   J.INST.MATH.
       APPLICS.7, 324-330,1971.

164   ARTHURS A.M., AND REEVES R.I.,   MAXIMUM AND MINIMUM PRINCIPLES FOR A CLASS
           OF PARTIAL DIFFERENTIAL EQUATIONS.   J.INST.MATH.APPLICS.14,
           1-7, 1974.
165   ASHWELL D.G., AND SABIR A.B.,   LIMITATIONS OF CERTAIN CURVED FINITE
           ELEMENTS WHEN APPLIED TO ARCHES.   INT. J. MECH. SCI.13,
           133-139, 1971.
166   ASHWELL D.G., AND SABIR A.B.,   A NEW CYLINDRICAL SHELL FINITE
           ELEMENT BASED ON SIMPLE INDEPENDENT STRAIN FUNCTIONS.
           INT. J.MECH.SCI. 14, 171-183, 1972.
167   ASHWELL D.G., SABIR A.B. AND ROBERTS T.M.,
           FURTHER STUDIES IN THE APPLICATION OF CURVED
           FINITE ELEMENTS TO CIRCULAR ARCHES.
           INT.J.MECH.SCI.13, 507-517,1971.
168   ASHWELL D.G., AND SABIR A.B.,   A CORRECTED ASSESSMENT OF THE
           CYLINDRICAL SHELL FINITE ELEMENT OF BOGNER, FOX AND
           SCHMIT WHEN APPLIED TO ARCHES.   INT.J.MECH.SCI.15,
           325-327, 1973.
169   ASHWELL D.G., AND SABIR A.B.,   ON THE FINITE ELEMENT CALCULATION
           OF STRESS DISTRIBUTIONS IN ARCHES.   INT.J.MECH.SCI.16,
           21-29, 1974.
170   ASPLUND E.,   AVERAGED NORMS.   ISRAEL J.MATH.5, 227-233, 1967.
171   ATKINSON B., CARD C.C.H. AND IRONS B.M.,   APPLICATION OF THE
           FINITE ELEMENT METHOD TO CREEPING FLOW PROBLEMS.   TRANS.
           INST.CHEM.ENGRS.48, 276-284,1970.
172   ATLURI S.,   ON THE HYBRID STRESS FINITE ELEMENT MODEL FOR
           INCREMENTAL ANALYSIS OF LARGE DEFLECTION PROBLEMS.   INT.
           J. SOLIDS STRUCTURES 9, 1177-1191, 1973.
173   ATLURI S., AND PIAN T.H.H.,   THEORETICAL FORMULATION OF FINITE
           ELEMENT METHODS IN LINEAR ELASTIC ANALYSIS OF GENERAL
           SHELLS.   J.STRUCT.MECH.1, 1-41, 1972.
174   AUBIN J.P.,   BEHAVIOUR OF THE ERROR OF THE APPROXIMATE SOLUTIONS OF
           BOUNDARY VALUE PROBLEMS FOR LINEAR ELLIPTIC OPERATORS BY
           GALLERKIN'S AND FINITE DIFFERENCE METHODS.   ANN.SCUO.NORM.
           PISA 21, 599-637, 1967.
175   AUBIN J.P.,   APPROXIMATION DES ESPACES DE DISTRIBUTIONS ET DES
           OPERATEURS DIFFERENTIELS.   BULL.SOC.MATH.FR.MEMOIRE 12, 1-139,
           1967.
176   AUBIN J.P.,   BEST APPROXIMATION OF LINEAR OPERATORS IN HILBERT
           SPACES.   SIAM J. NUMER. ANAL. 5, 518-521, 1968.
177   AUBIN J.P.,   EVALUATION DES ERREURS DE TRONCATURE DES
           APPROXIMATIONS DES ESPACES DE SOBOLEV.   J.MATH.ANAL.APPL.21,
           356-368, 1968.
178   AUBIN J.P.,   INTERPOLATION ET APPROXIMATION OPTIMALES ET "SPLINE
           FUNCTIONS".   J.MATH.ANAL.APPL.24,1-24, 1968.
179   AUBIN J.P.,   REMARKS ABOUT THE CONSTRUCTION OF OPTIMAL SUBSPACES
           OF APPROXIMANTS OF A HILBERT SPACE.   J. APPROX. THEORY
           4, 21-36, 1971.
180   AUBIN J.P.,   APPROXIMATION OF ELLIPTIC BOUNDARY-VALUE PROBLEMS.
           WILEY-INTERSCIENCE,NEW YORK   1972.
181   AUBIN J.P., AND BURCHARD H.G.,   SOME ASPECTS OF THE METHOD OF THE
           HYPERCIRCLE APPLIED TO ELLIPTIC VARIATIONAL PROBLEMS.
           MATHEMATICS RESEARCH CENTER, TECHNICAL REPORT NO.1005,
           UNIVERSITY OF WISCONSIN, MADISON, 1969, ALSO IN NUMERICAL
           SOLUTION OF PARTIAL DIFFERENTIAL EQUATIONS II, SYNSPADE 1970,
           HUBBARD (ED.), ACADEMIC PRESS, NEW YORK, 1971.
182   AUSLAENDER G.J.,   A NOVEL, HERMITE TYPE, INTERPOLATION PROCEDURE.
           SIAM J.NUMER.ANAL.8, 465-472, 1971.
183   AZIZ A.K., (ED.),   LECTURE SERIES IN DIFFERENTIAL EQUATIONS, VOL.2,
           VAN NOSTRAND MATHEMATICAL STUDY NO.19, PRINCETON, NEW JERSEY,1969.
184   BABUSKA I.,   THE STABILITY OF THE DOMAIN OF DEFINITION WITH RESPECT
           TO BASIC PROBLEMS OF THE THEORY OF PARTIAL DIFFERENTIAL EQUATIONS,
           ESPECIALLY WITH RESPECT TO THE THEORY OF ELASTICITY, I AND II.
           CZECHOSLOVAK MATH.J., 75-105 AND 165-203, 1961.
185   BABUSKA I. (ED.),   DIFFERENTIAL EQUATIONS AND THEIR APPLICATIONS,
           ACADEMIC PRESS, NEW YORK, 1963.
186   BABUSKA I.,   NUMERICAL SOLUTION OF BOUNDARY VALUE PROBLEMS BY PERTURBED

VARIATIONAL PRINCIPLE. TECH.NOTE BN-624, UNIVERSITY OF
MARYLAND, INSTITUTE FOR FLUID DYNAMICS AND APPLIED MATHEMATICS,
1969.

187 BABUSKA I., APPROXIMATION BY HILL FUNCTIONS. COMMENTATIONES
MATHEMATICAE UNIVERSITATIS CAROLINAE, PRAGUE, 11, 4, 1970.

188 BABUSKA I., COMPUTATION OF DERIVATIVES IN THE FINITE ELEMENT METHOD.
TECH.NOTE BN-650, UNIVERSITY OF MARYLAND, INSTITUTE FOR FLUID
DYNAMICS AND APPLIED MATHEMATICS, 1970.

189 BABUSKA I., FINITE ELEMENT METHOD FOR DOMAINS WITH CORNERS. COMPUTING 6,
264-273, 1970.

190 BABUSKA I., THE FINITE ELEMENT METHOD FOR ELLIPTIC DIFFERENTIAL
EQUATIONS. TECH.NOTE BN-653, INSTITUTE FOR FLUID DYNAMICS
AND APPLIED MATHEMATICS, UNIVERSITY OF MARYLAND, 1970.

191 BABUSKA I., THE FINITE ELEMENT METHOD FOR ELLIPTIC EQUATIONS WITH
DISCONTINUOUS COEFFICIENTS. COMPUTING 5, 207-213, 1970.

192 BABUSKA I., THE FINITE ELEMENT METHOD FOR INFINITE DOMAINS, I. TECH.
NOTE BN-670, UNIVERSITY OF MARYLAND, INSTITUTE FOR FLUID
DYNAMICS AND APPLIED MATHEMATICS, 1970. (ALSO TO APPEAR IN
MATH.COMP.)

193 BABUSKA I., A REMARK TO THE FINITE ELEMENT METHOD. COMMENTATIONES
MATHEMATICAE UNIVERSITATIS CAROLINAE,PRAGUE, 12,2,1971.

194 BABUSKA I., APPROXIMATION BY HILL FUNCTIONS II. TECH.NOTE BN-708,
UNIVERSITY OF MARYLAND, INSTITUTE FOR FLUID DYNAMICS AND
APPLIED MATHEMATICS, 1971.

195 BABUSKA I., ERROR BOUNDS FOR THE FINITE ELEMENT METHOD. NUMERISCHE
MATH.16, 322-333, 1971.

196 BABUSKA I., THE FINITE ELEMENT METHOD FOR ELLIPTIC DIFFERENTIAL
EQUATIONS. IN B.E.HUBBARD (ED.), NUMERICAL SOLUTION OF PARTIAL
DIFFERENTIAL EQUATIONS-II, SYNSPADE 1970, ACADEMIC PRESS,
NEW YORK, 1971.

197 BABUSKA I., THE RATE OF CONVERGENCE FOR THE FINITE ELEMENT METHOD.
SIAM J. NUMER.ANAL.8, 304-315, 1971.

198 BABUSKA I., THE FINITE ELEMENT METHOD WITH LAGRANGIAN MULTIPLIERS.
TECH.NOTE BN-724, UNIVERSITY OF MARYLAND, INSTITUTE OF FLUID
DYNAMICS AND APPLIED MATHEMATICS, 1972.

199 BABUSKA I., THE FINITE ELEMENT METHOD WITH LAGRANGIAN MULTIPLIERS.
NUMER.MATH.20, 179-192,1973

200 BABUSKA I., THE FINITE ELEMENT METHOD WITH PENALTY. MATH.COMP.
27, 221-229, 1973.

201 BABUSKA I., NUMERICAL SOLUTION OF PARTIAL DIFFERENTIAL EQUATIONS.
Z.A.M.M. 54, 3-10, 1974.

202 BABUSKA I., SOLUTION OF PROBLEMS WITH INTERFACES AND SINGULARITIES.
TECH. NOTE BN-789, INSTITUTE OF FLUID DYNAMICS AND APPLIED
MATHEMATICS, UNIVERSITY OF MARYLAND, 1974.

203 BABUSKA I., SOLUTION OF THE INTERFACE PROBLEMS BY HOMOGENIZATION II.
TECHNICAL NOTE BN-787, INSTITUTE FOR FLUID DYNAMICS AND APPLIED
MATHEMATICS, UNIVERSITY OF MARYLAND, 1974.

204 BABUSKA I., THE CONNECTION BETWEEN FINITE-DIFFERENCE LIKE
METHODS AND THE METHODS BASED ON INITIAL VALUE PROBLEMS
FOR ODE. TECHNICAL NOTE BN-706, INSTITUTE FOR FLUID
DYNAMICS AND APPLIED MATHEMATICS, UNIVERSITY OF MARYLAND,
1974.

205 BABUSKA I., AND AZIZ A.K., SURVEY LECTURES ON THE MATHEMATICAL
FOUNDATIONS OF THE FINITE ELEMENT METHOD. PP. 1-345 OF
A.K. AZIZ (ED.), THE MATHEMATICAL FOUNDATIONS OF THE
FINITE ELEMENT METHOD WITH APPLICATIONS TO PARTIAL
DIFFERENTIAL EQUATIONS. ACADEMIC PRESS, NEW YORK, 1972.

206 BABUSKA I., AND KELLOGG R.B., NUMERICAL SOLUTION OF THE NEUTRON
DIFFUSION EQUATION IN THE PRESENCE OF CORNERS AND INTERFACES.
TECH NOTE BN-720, UNIVERSITY OF MARYLAND, INSTITUTE OF FLUID
DYNAMICS AND APPLIED MATHEMATICS, 1971.

207 BABUSKA I., AND KELLOGG R.B., MATHEMATICAL AND COMPUTATIONAL
PROBLEMS IN REACTOR CALCULATIONS. PP. VII 67-94 OF
PROCEEDINGS CONFERENCE MATHEMATICAL MODELS AND
COMPUTATIONAL TECHNIQUES FOR ANALYSIS OF NUCLEAR SYSTEMS,
ANN ARBOR, MICHIGAN, 1973.

208   BABUSKA I., AND KELLOGG R.B.,   NON-UNIFORM ERROR ESTIMATES FOR THE
               FINITE ELEMENT METHOD.   TECHNICAL NOTE BN-790, INSTITUTE FOR
               FLUID DYNAMCIS AND APPLIED MATHEMATICS, UNIVERSITY OF MARYLAND
               1974.
209   BABUSKA I., PRAGER M., AND VITASEK E.,   NUMERICAL PROCESSES IN
               DIFFERENTIAL EQUATIONS.   INTERSCIENCE, LONDON, 1966.
210   BABUSKA I., AND ROSENZWEIG M.B.,   A FINITE ELEMENT SCHEME FOR
               DOMAINS WITH CORNERS. NUMER.MATH.20, 1-21, 1972.   ALSO TECH.
               NOTE BN-720, INSTITUTE OF FLUID DYNAMICS AND APPLIED
               MATHEMATICS, UNIVERSITY OF MARYLAND, 1971.
211   BABUSKA I., SEGETH K. AND SEGETHOVA J.,   NUMERICAL EXPERIMENTS AND
               PROBLEMS CONNECTED WITH THE FINITE ELEMENT METHOD.   TO
               APPEAR.
212   BABUSKA I., AND ZLAMAL M.,   NONCONFORMING ELEMENTS IN THE FINITE
               ELEMENT METHOD WITH PENALTY.   SIAM J. NUMER. ANAL. 10,
               863-875, 1973.   ALSO TECH.REPORT. BN-729, INSTITUTE OF FLUID
               DYNAMCIS AND APPLIED MATHEMATICS, UNIVERSITY OF MARYLAND,
               1972.
213   BACKLUND J.,   FINITE ELEMENT ANALYSIS OF NONLINEAR STRUCTURES.
               PH.D. THESIS, DEPT. OF STRUCTURAL MECHANICS, CHALMERS
               INSTITUTE OF TECHNOLOGY,GOTEBORG, 1973.
214   BACKLUND J., AND WENNERSTROM H.,   FINITE ELEMENT ANALYSIS OF
               ELASTO-PLASTIC SHELLS.   INT.J.NUMER.METH.ENG.8, 415-424,
               1974.
215   BACKUS W.E., AND HELLO R.M.,   SOME APPLICATIONS OF FINITE ELEMENT
               ANALYSIS TO SHELL BUCKLING PREDICTION. PROC.2ND CONF. MATRIX
               METHODS IN STRUCTURAL MECHANICS, WRIGHT-PATTERSON AFB,
               OHIO,AFFDL-TR-68-150, 1968.
216   BAKER A.J.,   A NUMERICAL SOLUTION TECHNIQUE FOR A CLASS OF TWO-DIMENSIONAL
               PROBLEMS IN FLUID DYNAMICS FORMULATED WITH THE USE OF FINITE
               ELEMENTS.   BELL AEROSYSTEMS CO., TECH.NOTE TCTN-1005,1969.
217   BAKER G.A.,   FINITE ELEMENT - GALERKIN METHODS FOR HYPERBOLIC
               EQUATIONS WITH DISCONTINUOUS COEFFICIENTS.   TECHNICAL
               REPORT, DEPARTMENT DE MATHEMATIQUES, ECOLE POLYTECHNIQUE
               FEDERALE DE LAUSANNE, 1973.
218   BAKER A.J.,   FINITE ELEMENT SOLUTION ALGORITHM FOR VISCOUS
               INCOMPRESSIBLE FLUID DYNAMICS INT.J.NUMER.METH.ENG.6,
               89-101, 1973.
219   BAKER A.J.,   FINITE ELEMENT SOLUTION THEORY FOR THREE-DIMENSIONAL
               BOUNDARY FLOWS. COMP.METH.APPL.MECH.ENG.4, 367-386, 1974.
220   BAKER A.J.,   NONLINEAR INITIAL-BOUNDARY VALUE SOLUTIONS BY THE
               FINITE ELEMENT METHOD.   PROC.CONF.COMP.METHODS IN NONLINEAR
               MECH.,   AUSTIN, TEXAS, 1974.
221   BAKER G.A.,   SIMPLIFIED PROOFS OF ERROR ESTIMATES FOR THE LEAST
               SQUARES METHOD FOR DIRICHLET'S PROBLEM.   MATH.COMP.27,
               229-236, 1973.
222   BAKHREBAH S.A., PAUL S.L., AND SCHNOBRICH W.C.,   FINITE ELEMENT
               SOLUTION OF NORMALLY INTERSECTING CYLINDERS.   PAPER G2/5A,
               PROC. 2ND STRUCTURAL MECH. IN REACTOR TECH. CONF., BERLIN,
               1973.
223   BALDWIN J.T., RAZZAQUE A. AND IRONS B.M.,   SHAPE FUNCTION SUBROUTINE FOR
               AN ISOPARAMETRIC THIN PLATE ELEMENT.   INT.J.NUMER.METH.ENG.7,
               431-440,1973.
224   BALMER H., DOLTSINIS J. ST. AND KONIG M.,   ELASTOPLASTIC AND CREEP
               ANALYSIS WITH THE ASKA PROGRAM SYSTEM.   COMP. METH. APPL.
               MECH. ENG.3, 87-104, 1974
225   BALUCH M.H., KORMAN T. AND MURGHEM F.T.,   FINITE ELEMENT ANALYSIS
               OF A THICK WALLED MICROPOLAR CYLINDER LOADED AXISYMMETRICALLY.
               PAPER M3/8,   PROCEEDINGS 2ND STRUCTURAL MECH. IN REACTOR
               TECH. CONF.,   BERLIN, 1973.
226   BANGASH Y.,   THE RIDDLE OF BONDED AND UNBONDED TENDONS IN
               PRESTRESSED CONCRETE REACTOR VESSELS.   CONCRETE 8, 46-48,
               1974.
227   BANITCHOUK N.V., PETROV V.M., AND TCHERNOUSSKO F.L.,   RESOLUTION
               NUMERIQUE DE PROBLEMS AUX LIMITES VARIATIONNEIS PAR LA METHODE DES
               VARIATIONS LOCALES.   J. CALCUL. NUMERIQUE PHYS. MATH, 6, 946-961,

1966.

228 BARKER R.M., LIN F.T., AND DANA J.R., THREE DIMENSIONAL FINITE ELEMENT
     ANALYSIS OF LAMINATED COMPOSITES. NATIONAL SYMP. ON COMPUTERISED
     STRUCTURAL ANALYSIS. GEORGE WASHINGTON UNIVERSITY, 1972.

229 BARKER R.M., AND MCLAUGHLIN T.F., STRESS CONCENTRATIONS NEAR A
     DISCONTINUITY IN FIBROUS COMPOSITES. J. COMPOSITE MATERIALS 5,
     492-503, 1971.

230 BARLA G., AND INNAURATO N., INDIRECT TENSILE TESTING OF
     ANISOTROPIC ROCKS. ROCK MECH. (AUSTRIA) 5, 215-230, 1973.

231 BARNARD A.J., A SANDWICH PLATE FINITE ELEMENT. PP. 401-414 OF
     J.R. WHITEMAN (ED.), THE MATHEMATICS OF FINITE ELEMENTS AND
     APPLICATIONS, ACADEMIC PRESS, LONDON, 1973.

232 BARNHILL R.E., SMOOTH INTERPOLATION OVER TRIANGLES. PP.46-71 OF
     R.E.BARNHILL AND R.F.REISENFELD (EDS.), COMPUTER AIDED
     GEOMETRIC DESIGN. ACADEMIC PRESS, NEW YORK, 1975.

233 BARNHILL R.E., BIRKHOFF G., AND GORDON W.J., SMOOTH INTERPOLATION
     IN TRIANGLES. J. APPROX. THEORY 18, 114-128, 1973.

234 BARNHILL R.E., AND GREGORY J.A., SARD KERNEL THEOREMS ON
     TRIANGULAR AND RECTANGULAR DOMAINS WITH EXTENSIONS AND
     APPLICATIONS TO FINITE ELEMENT ERROR BOUNDS. TECHNICAL
     REPORT TR/11, MATHEMATICS DEPARTMENT, BRUNEL UNIVERSITY,
     1972.

235 BARNHILL R.E., AND GREGORY J.A., INTERPOLATION TO BOUNDARY DATA
     ON N-SIMPLICES. (IN PREPARATION).

236 BARNHILL R.E., GREGORY J.A., AND WHITEMAN J.R., THE EXTENSION AND
     APPLICATION OF SARD KERNEL THEOREMS TO COMPUTE FINITE
     ELEMENT ERROR BOUNDS. PP.749-755 OF K. AZIZ (ED.), THE
     MATHEMATICAL FOUNDATIONS OF THE FINITE ELEMENT METHOD WITH
     APPLICATIONS TO PARTIAL DIFFERENTIAL EQUATIONS. ACADEMIC
     PRESS, NEW YORK, 1972.

237 BARNHILL R.E., AND MANSFIELD LOIS, ERROR BOUNDS FOR SMOOTH
     INTERPOLATION IN TRIANGLES. J.APPROX.THEORY 11,306-318,1974.

238 BARNHILL R.E., AND RIESENFELD R.F., (EDS.), COMPUTER AIDED
     GEOMETRIC DESIGN. ACADEMIC PRESS, NEW YORK, 1975.

239 BARNHILL R.E., AND WHITEMAN J.R., ERROR ANALYSIS OF FINITE
     ELEMENT METHODS WITH TRIANGLES FOR ELLIPTIC BOUNDARY VALUE PROBLEMS.
     PP. 83-112 OF J.R.WHITEMAN (ED.), THE MATHEMATICS OF FINITE
     ELEMENTS AND APPLICATIONS. ACADEMIC PRESS, LONDON, 1973.

240 BARNHILL R.E., AND WHITEMAN J.R., COMPUTABLE ERROR BOUNDS FOR
     THE FINITE ELEMENT METHOD FOR ELLIPTIC BOUNDARY VALUE
     PROBLEMS. PP. 9-28 OF J. ALBRECHT AND L. COLLATZ (EDS.),
     NUMERISCHE METHODEN BEI DIFFERENTIALGLEICHUNGEN UND MIT
     FUNKTIONALANALYTISCHEN HILFSMITTELN. I.S.N.M. 19,
     BIRKHAUSER VERLAG, BASEL, 1974.

241 BARNHILL R.E., AND WHITEMAN J.R., ERROR ANALYSIS OF GALERKIN
     METHODS FOR DIRICHLET PROBLEMS CONTAINING BOUNDARY
     SINGULARITIES. J.INST.MATH.APPLICS.14, XXX-XXX, 1974.

242 BARNHILL R.E., AND WHITEMAN J.R., SINGULARITIES DUE TO RE-
     ENTRANT BOUNDARIES IN ELLIPTIC PROBLEMS. PP. 29-45 OF J.
     ALBRECHT AND L. COLLATZ (EDS.), NUMERISCHE METHODEN BEI
     DIFFERENTIALGLEICHUNGEN UND MIT FUNKTIONALANALYTISCHEN
     HILFSMITTELN. I.S.N.M. 19, BIRKHAUSER VERLAG, BASEL,
     1974.

243 BARNHILL R.E., AND WIXOM J.A., AN ERROR ANALYSIS FOR
     INTERPOLATION OF ANALYTIC FUNCTIONS. SIAM J. NUMER.
     ANAL. 5, 522-529, 1968.

244 BARNHILL R.E., AND WIXOM J.A., QUADRATURES WITH REMAINDERS OF
     MINIMUM NORM. MATH.COMP. 21, 66-75, 1967.

245 BARNSLEY M.F., AND ROBINSON P.D., BIVARIATIONAL BOUNDS. PROC. ROY.
     SOC.LONDON A.338, 527-533, 1974.

246 BARON F., AND VENKATESAN M.S., NONLINEAR ANALYSIS OF CABLE AND TRUSS
     STRUCTURES. J.STRUCT.DIV. A.S.C.E. 97, ST2, 679-710, 1971.

247 BARSOUM R.S., FINITE ELEMENT METHOD APPLIED TO THE PROBLEM OF STABILITY
     OF A NON-CONSERVATIVE SYSTEM. INT.J.NUMER.METH.ENG.3,66-68,1971.

248 BARSOUM R.S., AND GALLAGHER R.H., FINITE ELEMENT ANALYSIS OF TORSIONAL AND
     TORSIONAL-FLEXURAL STABILITY PROBLEMS. INT.J.NUMER.METH.ENG.2,

335-352,1970.

249   BARTHOLOME G., MIKSCH M., NEUBRECH G., AND VASOUKIS G., FRACTURE AND
SAFETY ANALYSIS OF NUCLEAR PRESSURE VESSELS. ENG. FRACTURE MECH.
5, 431-446, 1973.

250   BATDORF W.J., KAPUR S.S. AND SAYER R.B., THE ROLE OF COMPUTER GRAPHICS IN
THE STRUCTURAL DESIGN PROCESS. PROC.2ND CONF.MATRIX METHODS IN
STRUCTURAL MECHANICS, WRIGHT-PATTERSON AFB,OHIO,AFFDL-TR68150,1968.

251   BATHE K.J., AND WILSON E.L., NONSAP - A GENERAL FINITE ELEMENT
PROGRAM FOR NONLINEAR DYNAMIC ANALYSIS OF COMPLEX
STRUCTURES. PAPER M3/1, PROCEEDINGS 2ND STRUCTURAL MECH.
IN REACTOR TECH. CONF., BERLIN, 1973.

252   BATHE K.J., AND WILSON E.L., STABILITY AND ACCURACY OF DIRECT
INTEGRATION METHODS. EARTHQUAKE ENG. AND STRUCT.
DYNAMICS 1, 283-291, 1973.

253   BATOZ J.L., AND DHATT G., DEVELOPMENT OF TWO SIMPLE SHELL ELEMENTS.
AIAA J.10, 237-238, 1972.

254   BATOZ J.L., AND DHATT G., BUCKLING OF DEEP SHELLS. PAPER M5/7,
PROC. 2ND STRUCTURAL MECH. IN REACTOR TECH. CONF., BERLIN,
1973.

255   BAUER F.L., OPTIMALLY SCALED MATRICES. NUMER.MATH.5, 73-87, 1963.

256   BAZANT E., APPLICATION OF THE FINITE ELEMENT METHOD TO THE CALCULATION
OF REACTOR PRESSURE VESSELS. PP.49-63 OF VOL.4 OF PROCEEDINGS
1ST STRUCTURAL MECH. IN REACTOR TECH. CONF., BERLIN, 1971.

257   BAZANT Z.P., MATRIX DIFFERENTIAL EQUATION AND HIGHER-ORDER NUMERICAL
METHODS FOR PROBLEMS OF NON-LINEAR CREEP, VISCOELACTICITY AND
ELASTO-PLASTICITY. INT.J.NUMER.METH.ENG.4,11-15,1972.

258   BAZANT Z.P., THREE DIMENSIONAL HARMONIC FUNCTIONS NEAR TERMINATION
OR INTERSECTION OF GRADIENT SINGULARITY LINES: A GENERAL
NUMERICAL METHOD. INT.J.ENG.SCIENCE 12, 221-243, 1974.

259   BAZELEY G.P., CHEUNG Y.K., IRONS B.M. AND ZIENKIEWICZ O.C., TRIANGULAR
ELEMENTS IN BENDING-CONFORMING AND NON-CONFORMING SOLUTIONS.
PROC.1ST CONF.MATRIX METHODS IN STRUCTURAL MECHANICS,
WRIGHT-PATTERSON AFB, OHIO, AFFDL TR66-80, 1965.

260   BEGIS D., AND GLOWINSKI R., APPLICATION DE LA METHODE DES ELEMENTS
FINIS A LA RESOLUTION D'UN PROBLEM DE DOMAINE OPTIMAL.
MANUSCRIPT.

261   BELL K., A REFINED TRIANGULAR PLATE BENDING FINITE ELEMENT. INT.J.
NUMER. METH. ENG. 1, 101-122, 1969.

262   BELL K., TRIANGULAR PLATE BENDING ELEMENTS. PP.213-254 OF FINITE
ELEMENT METHODS, TAPIR, TRONDHEIM, NORWAY, 1969.

263   BELL K., ON THE QUINTIC TRIANGULAR PLATE BENDING ELEMENT. REPORT
72-2, DIVISION OF STRUCTURAL MECHANICS, UNIVERSITY OF
TRONDHEIM, 1972.

264   BELYTSCHKO T., FINITE ELEMENT APPROACH TO HYDRODYNAMICS AND
MESH STABILIZATION. PROC.CONF.COMP.METHODS IN NONLINEAR
MECH., AUSTIN,TEXAS, 1974.

265   BELYTSCHKO T., AND HODGE P.G., PLANE STRESS LIMIT ANALYSIS BY FINITE
ELEMENTS. J.ENG.MECH.DIV.A.S.C.E. 96, 931-944, 1970.

266   BELYTSCHKO T., AND HSIEH B.J., NON-LINEAR TRANSIENT FINITE ELEMENT
ANALYSIS WITH CONVECTED CO-ORDINATES. INT.J.NUMER.METH.ENG.7,
255-271, 1973.

267   BELYTSCHKO T., HSIEH B.J., AND KENNEDY J.M., CONVECTED CO-ORDINATES
FOR TRANSIENT NONLINEAR FINITE ELEMENT ANALYSIS. PAPER
M3/9, PROCEEDINGS 2ND STRUCTURAL MECH. IN REACTOR TECH.
CONF., BERLIN, 1973.

268   BELYTSCHKO T., AND VELEBIT M., FINITE ELEMENT METHOD FOR ELASTIC PLASTIC
PLATES. J.ENG.MECH.DIV,A.S.C.E. 98, 227-242, 1972.

269   BENNETT J.M., AND COOLEY P.C., LINEAR ANALYSIS OF STRUCTURES BY THE
METHOD OF CONJUGATE GRADIENTS. TECH.REPT.51, BASSER
COMPUTING CENTRE, UNIVERSITY OF SYDNEY, 1964.

270   BENTON S.H., GLOBAL VARIATIONAL SOLUTIONS OF HAMILTON-JACOBI BOUNDARY
VALUE PROBLEMS. J.DIFFERENTIAL EQUATIONS 13, 468-480,1973.

271   BENZLEY S.E., REPRESENTATION OF SINGULARITIES WITH ISOPARAMETRIC FINITE
ELEMENTS. INT.J.NUMER.METH.ENG.8, 537-545, 1974.

272   BENZLEY S.E., AND BEISINGER Z.E., CHILES A FINITE ELEMENT
COMPUTER PROGRAM THAT CALCULATES THE INTENSITIES OF

LINEAR ELASTIC SINGULARITIES. REPORT SLA-73-894, SANDIA
LABORATORIES, ALBUQUERQUE, 1973.

273 BENZLEY S.E., KEY S.W., AND HUTCHINSON J.R., A DYNAMIC SHELL
THEORY COUPLING THICKNESS STRESS WAVE EFFECTS WITH GROSS
STRUCTURAL RESPONSE. TRANS. ASME SER.E.40, 731-735, 1973.

274 BEREZIN I.S., AND ZHIDKOV N.P., COMPUTING METHODS. VOLS. 1 AND 2.
PERGAMON PRESS, OXFORD, 1965.

275 BERG S., NONLINEAR FINITE ELEMENT ANALYSIS OF REINFORCED
CONCRETE PLATES. REPORT 73-1. DIVISION OF STRUCTURAL
MECHANICS, NORWEIGEN INSTITUTE OF TECHNOLOGY, TRONDHEIM,
1973.

276 BERG S., BERGAN P.G. AND HOLAND I., NONLINEAR FINITE ELEMENT
ANALYSIS OF REINFORCED CONCRETE PLATES. PAPER M3/5,
PROCEEDINGS 2ND STRUCTURAL MECH IN REACTOR TECH CONF.,
BERLIN, 1973.

277 BERGAN P.G., AND AAMODT B., FINITE ELEMENT ANALYSIS OF CRACK PROPAGATION
IN THREE-DIMENSIONAL SOLIDS UNDER CYCLIC LOADING. PAPER G5/5,
PROCEEDINGS 2ND STRUCTURAL MECH. IN REACTOR TECH. CONF., BERLIN,
1973.

278 BERGAN P.G., AND ALDSTEDT E., SESAM A PROGRAMMING SYSTEM FOR
FINITE ELEMENT PROBLEMS. TECHNICAL UNIVERSITY OF NORWAY,
TRONDHEIM, 1968.

279 BERGAN P.G., AND CLOUGH R.W., ELASTIC-PLASTIC ANALYSIS OF PLATES
USING THE FINITE ELEMENT METHOD. PROC. 3RD CONF. MATRIX
METHODS IN STRUCTURAL MECHANICS, WRIGHT-PATTERSON AFB,
OHIO, 1971.

280 BERGAN P.G., AND CLOUGH R.W., CONVERGENCE CRITERIA FOR ITERATIVE
PROCESSES. AIAA J.10, 1107-1108, 1972.

281 BERGAN P.G., AND CLOUGH R.W., LARGE DEFLECTION ANALYSIS OF
PLATES AND SHALLOW SHELLS USING THE FINITE ELEMENT
METHOD. INT. J.NUMER.METH.ENG.5, 543-556, 1973.

282 BERGAN P.G., AND SOREIDE T., A COMPARATIVE STUDY OF DIFFERENT
NUMERICAL SOLUTION TECHNIQUES APPLIED TO A NONLINEAR
STRUCTURE PROBLEM. COMP.METH.APPL.MECH.ENG.2, 185-202,
1973.

283 BERGER A.E., ERROR ESTIMATES FOR THE FINITE ELEMENT METHOD.
PH.D. THESIS, MASSACHUSETTS INSTITUTE OF TECHNOLOGY,1972.

284 BERGER A.E., TWO TYPES OF PIECEWISE QUADRATIC SPACES AND THEIR ORDER
OF ACCURACY FOR POISSON'S EQUATION.
PP.757-761 OF A.K. AZIZ (ED.), THE MATHEMATICAL FOUNDATIONS OF
THE FINITE ELEMENT METHOD WITH APPLICATIONS TO PARTIAL
DIFFERENTIAL EQUATIONS. ACADEMIC PRESS, NEW YORK,1972.

285 BERGER A.E., L2 ERROR ESTIMATES FOR FINITE ELEMENTS WITH INTERPOLATED
BOUNDARY CONDITIONS. NUMER.MATH.21, 345-349, 1973.

286 BERGER A., SCOTT R., AND STRANG G., APPROXIMATE BOUNDARY CONDITIONS IN
THE FINITE ELEMENT METHOD. PROCEEDINGS SYMPOSIUM ON NUMERICAL
ANALYSIS, ROME, SYMPOSIA MATHEMATICA 10, ACADEMIC PRESS, NEW
YORK, 1972.

287 BERGER D., RAPPAZ J., AND TROYON F., NUMERICAL STUDY OF THE
MAGNETOHYDRODYNAMIC STABILITY OF THE INFINITE BELT-PINCH.
HELVETIA PHYS. ACTA 46, 447, 1973.

288 BERGMAN S., AND SCHIFFER M., KERNEL FUNCTIONS AND ELLIPTIC
DIFFERENTIAL EQUATIONS IN MATHEMATICAL PHYSICS.
ACADEMIC PRESS, NEW YORK, 1953.

289 BERK A.D., VARIATIONAL PRINCIPLES FOR ELECTROMAGNETIC RESONATORS
AND WAVEGUIDES. IRE TRANS. ANTENNAS AND PROPAGATION,
AP-4, 104-111, 1956.

290 BERKOVITZ L.D., AND POLLARD H., A NON-CLASSICAL VARIATIONAL
PROBLEM ARISING FROM AN OPTIMAL FILTER PROBLEM, I AND II,
ARCH.RAT.MECH.ANAL.26, 281-304, AND 38, 161-172, 1970.

291 BERKOWITZ S., AND GARNER F.J., THE CALCULATION OF MULTIDIMENSIONAL
AND GRAM-CHARIER COEFFICIENTS. MATH. COMP. 24, 537-545,
1970.

292 BERNAL M.J.M., AND WHITEMAN J.R., NUMERICAL TREATMENT OF BIHARMONIC
BOUNDARY VALUE PROBLEMS WITH RE-ENTRANT BOUNDARIES.
COMP.J.13, 87-91, 1970.

293   BERNAL M.J.M., AND WHITEMAN J.R., INTRODUCTION TO VARIATIONAL
          AND FINITE ELEMENT METHODS. VOLUME IN SERIES ON
          COMPUTATIONAL MATHEMATICS AND APPLICATIONS. ACADEMIC
          PRESS, LONDON, (TO APPEAR).

294   BERNSTEIN D.L., EXISTENCE THEOREMS IN PARTIAL DIFFERENTIAL EQUATIONS.
          ANN.OF MATH.STUDIES NO.23, PRINCETON UNIVERSITY PRESS,
          PRINCETON, NEW JERSEY, 1950.

295   BERS L., JOHN F. AND SHECHTER M., PARTIAL DIFFERENTIAL EQUATIONS,
          WILEY-INTERSCIENCE, NEW YORK, 1964.

296   BESOV O.V., ON SOME FAMILIES OF FUNCTIONAL SPACES; IMBEDDING AND
          EXTENSION THEOREMS. (RUSSIAN). DOKL.AKAD.NAUK.SSSR 126,
          1163-1165, 1969.

297   BESOV O.V., THE BEHAVIOUR OF DIFFERENTIABLE FUNCTIONS ON NONSMOOTH
          MANIFOLDS. TRUDY MAT.INST.STEKLOV. 117, 3-10, 1972.

298   BEST G.C., A FORMULA FOR CERTAIN TYPES OF STIFFNESS MATRICES
          OF STRUCTURAL ELEMENTS. AIAA J, 1, 212-213, 1963.

299   BEST G.C., A GENERAL FORMULA FOR STIFFNESS MATRICES OF STRUCTURAL
          ELEMENTS. AIAA J.1, 1920-1921, 1963.

300   BETTESS P., A NOTE ON SOME VARIATIONAL STATEMENTS FOR THE SLOW FLOW OF A
          NAVIER-POISSON FLUID. INT.J.NUMER.METH.ENG.8, 17-25,1974.

301   BHANDARI D.R., AND ODEN J.T., A LARGE DEFORMATION ANALYSIS OF CRYSTALLINE
          ELASTIC-VISCOPLASTIC MATERIALS. PAPER L3/10, PROCEEDINGS 2ND
          STRUCTURAL MECH. IN REACTOR TECH. CONF., BERLIN, 1973.

302   BHATTACHARYA R.K., BICUBIC SPLINE INTERPOLATION AS A METHOD FOR
          TREATMENT OF FIELD DATA. GEOPHYSICS 34, 402-423, 1969.

303   BICKLEY W.G., FINITE DIFFERENCE FORMULAE FOR THE SQUARE LATTICE.
          Q.J. MECH.APPL.MATH.1, 35-42, 1948.

304   BICKLEY W.G., PIECEWISE CUBIC INTERPOLATION AND TWO POINT
          BOUNDARY PROBLEMS. COMP.J.11, 206-208, 1968.

305   BIFFLE J.H., FINITE ELEMENT ANALYSIS FOR WAVE PROPAGATION IN ELASTIC-
          PLASTIC SOLIDS. REPORT 73-4, TEXAS INSTITUTE FOR COMPUTATIONAL
          MECHANICS, UNIVERSITY OF TEXAS, AUSTIN, 1973.

306   BIRKHOFF G., GENERAL MEAN VALUE AND REMAINDER THEOREMS WITH
          APPLICATION TO MECHANICAL DIFFERENTIATION AND QUADRATURE.
          TRANS.AMERICAN MATH.SOC.7, 107-136, 1960.

307   BIRKHOFF G., PIECEWISE BICUBIC INTERPOLATION AND APPROXIMATION IN
          POLYGONS. IN SCHOENBERG (ED.), APPROXIMATIONS WITH SPECIAL
          EMPHASIS ON SPLINE FUNCTIONS. ACADEMIC PRESS, NEW YORK, 1969.

308   BIRKHOFF G., THE NUMERICAL SOLUTION OF ELLIPTIC EQUATIONS. REGIONAL
          CONFERENCE SERIES IN APPLIED MATHEMATICS, NO.1, SIAM,
          PHILADELPHIA, 1971.

309   BIRKHOFF G., TRICUBIC POLYNOMIAL INTERPOLATION, PROC.NAT.ACAD.SCI.
          U.S.A.68, 1162-64, 1971.

310   BIRKHOFF G., ANGULAR SINGULARITIES OF ELLIPTIC PROBLEMS. J.APPROX.
          THEORY 6, 215-230, 1972.

311   BIRKHOFF G., PIECEWISE ANALYTIC INTERPOLATION AND APPROXIMATION
          IN TRIANGULATED POLYGONS. PP.363-385 OF A.K.AZIZ (ED.),
          THE MATHEMATICAL FOUNDATIONS OF THE FINITE ELEMENT METHOD
          WITH APPLICATIONS TO PARTIAL DIFFERENTIAL EQUATIONS.
          ACADEMIC PRESS, NEW YORK, 1972.

312   BIRKHOFF G., INTERPOLATION TO BOUNDARY DATA IN TRIANGLES. J.
          MATH.ANAL.APPL.42, 474-484, 1973.

313   BIRKHOFF G., CAVENDISH J.C., AND GORDON W.J., MULTIVARIATE
          APPROXIMATION BY LOCALLY BLENDED UNIVARIATE INTERPOLANTS.
          PROC.NAT.ACAD.SCI.U.S.A.71, 3423-3425, 1974.

314   BIRKHOFF G., AND DE BOOR C., ERROR BOUNDS FOR SPLINE INTERPOLATION.
          J.MATH.MECH 13, 827-836,1964.

315   BIRKHOFF G., AND DE BOOR C., PIECEWISE POLYNOMIAL INTERPOLATION AND
          APPROXIMATION. IN H.L.GARABEDIAN (ED.), APPROXIMATION OF
          FUNCTIONS, ELSEVIER, NEW YORK, 1965.

316   BIRKHOFF G., DE BOOR C., SWARTZ B., AND WENDROFF B., RAYLEIGH-RITZ
          APPROXIMATION BY PIECEWISE POLYNOMIALS. SIAM J.NUMER.ANAL.,
          3,188-203,1966.

317   BIRKHOFF G., AND FIX G., ACCURATE EIGENVALUES COMPUTATIONS FOR ELLIPTIC
          PROBLEMS. IN BIRKHOFF AND VARGA (EDS.), NUMERICAL SOLUTION OF
          FIELD PROBLEMS IN CONTINUUM MECHANICS. AMS, PROVIDENCE,

RHODE ISLAND, 1969.

318 BIRKHOFF G., AND FIX G., RAYLEIGH-RITZ APPROXIMATION BY
     TRIGONOMETRIC POLYNOMIALS. INDIAN MATH.J.9, 269-277,
     1967.

319 BIRKHOFF G., AND FIX G., FINITE ELEMENT METHODS IN REACTOR DESIGN
     CALCULATIONS. TRANS. AMERICAN NUCL.SOC.14, 199-XXX,1971.

320 BIRKHOFF G., AND GARABEDIAN H.L., SMOOTH SURFACE INTERPOLATION. J.
     MATH. PHYS 39, 353-368, 1960.

321 BIRKHOFF G. AND MANSFIELD L., COMPATIBLE TRIANGULAR FINITE ELEMENTS.
     J.MATH.ANAL.APPLICS.47, 531-553, 1974.

322 BIRKHOFF G., MANSFIELD LOIS, AND WIXOM J., SOME FEW-PARAMETER
     FAMILIES OF COMPATIBLE TRIANGULAR FINITE ELEMENTS.

323 BIRKHOFF G., AND PRIVER A., HERMITE INTERPOLATION ERRORS FOR DERIVATIVES.
     J.MATH AND PHYS. 46, 440-447, 1967.

324 BIRKHOFF G., SCHULZ M.H. AND VARGA R.S., PIECEWISE HERMITE
     INTERPOLATION IN ONE AND TWO VARIABLES WITH APPLICATIONS
     TO PARTIAL DIFFERENTIAL EQUATIONS. NUMER.MATH.11, 232-256,
     1968.

325 BIRKHOFF G., AND VARGA R.S., DISCRETISATION ERRORS FOR WELL-SET
     CAUCHY PROBLEMS. J.MATH.PHYS.44, 1-23, 1965.

326 BIRKHOFF G., AND VARGA R.S., (EDS.), NUMERICAL SOLUTION OF FIELD PROBLEMS
     IN CONTINUUM MECHANICS. SIAM-AMS PROCEEDINGS II, AMERICAN
     MATHEMATICAL SOCIETY, PROVIDENCE, 1969.

327 BIRMAN M.S., AND SKVORTSOV, ON THE SUMMABILITY OF HIGHEST ORDER
     DERIVATIVES OF THE SOLUTION OF THE DIRICHLET PROBLEM IN A
     DOMAIN WITH PIECEWISE SMOOTH BOUNDARY. IZV.VYSSH.UCHEBN,
     ZAVED.MATHEMATIKA, 3, 12-21,1962.

328 BLACKBURN W.S., CALCULATION OF STRESS INTENSITY FACTORS AT CRACK
     TIPS USING SPECIAL FINITE ELEMENTS. PP. 327-336 OF J.R.
     WHITEMAN(ED.), THE MATHEMATICS OF FINITE ELEMENTS AND
     APPLICATIONS. ACADEMIC PRESS, LONDON, 1973.

329 BLAIR J., APPROXIMATE SOLUTION OF ELLIPTIC AND PARABOLIC BOUNDARY
     VALUE PROBLEMS.
     PH.D. THESIS, UNIVERSITY OF CALIFORNIA,BERKELEY,1971.

330 BLAIR, J.J. ERROR BOUNDS FOR THE SOLUTION OF NONLINEAR TWO-POINT
     BOUNDARY VALUE PROBLEMS BY GALERKIN'S METHOD.
     NUMER.MATH.19, 99-109,1972.

331 BLOKHOV V.V., INVESTIGATION OF FLEXURAL STIFFNESS OF RECTANGULAR
     PLATES BY THE FINITE ELEMENT METHOD. (RUSSIAN). IZV.
     VUZ. MASHINOSTR 1, 5-9, 1973.

332 BOGNER F.K., FOX R.L., AND SCHMIT L.A., THE GENRATION OF INTERELEMENT
     COMPATIABLE STIFFNESS AND MASS MATRICES BY USE OF INTERPOLATION
     FORMULAE. PROC 1ST CONF.MATRIX METHODS IN STRUCTURAL MECHANICS,
     WRIGHT-PATTERSON AFB, OHIO, AFFDL TR66-80,1965.

333 BOGNER F.K., FOX R.L., AND SCHMIT L.A., A CYLINDRICAL SHELL DISCRETE
     ELEMENT, AIAA JOURNAL 5, 745-750, 1967.

334 BOHL E., MONOTONIE: LOSBARKEIT UND NUMERIK BEI OPERATORGLEICHUNGEN,
     SPRINGER-VERLAG, BERLIN, 1974.

335 BOHN G.J., DYNAMIC ANALYSIS OF REACTOR INTERNAL STRUCTURES WITH IMPACT
     BETWEEN COMPONENTS. NUCL.SCI.ENG.47, 391-408, 1972.

336 BOISSERIE J.M., GENERATION OF TWO - AND THREE -DIMENSIONAL FINITE
     ELEMENTS. INT. J.NUMER.METH.ENG.3,327-348,1971.

337 BOLLAND G.B., FLEXIBILITY AND STIFFNESS MATRICES FOR AN OPEN-TUBE
     WARPING CONSTRAINT FINITE ELEMENT. PP. 387-399 OF J.R.
     WHITEMAN (ED.), THE MATHEMATICS OF FINITE ELEMENTS.
     ACADEMIC PRESS, LONDON, 1973.

338 BONNEROT R., AND JAMET P., A SECOND ORDER FINITE ELEMENT METHOD
     FOR THE ONE-DIMENSIONAL STEFAN PROBLEM. INT.J.NUMER.
     METH.ENG.8, 811-820, 1974.

339 BONNES G., DHATT G., GIROUX Y.M. AND ROBICHAUD L.P.A., CURVED
     TRIANGULAR ELEMENTS FOR THE ANALYSIS OF SHELLS. PROC.2ND
     CONF. MATRIX METHODS IN STRUCTURAL MECHANICS, WRIGHT-PATTERSON
     AFB, OHIO, AFFDL-TR-68-150, 1968.

340 BOSARGE W.E., AND JOHNSON O.G., NUMERICAL PROPERTIES OF THE
     RITZ-TREFFTZ ALGORITHM FOR OPTIMAL CONTROL. COMM. ACM
     14, 402-406, 1971.

341   BOSARGE W.E., JOHNSON O.G., MCKNIGHT R.S., AND TIMLAKE W.P., THE
          RITZ-GALERKIN PROCEDURE FOR NONLINEAR CONTROL PROBLEMS.
          J. NUMER.ANAL. (TO APPEAR).
342   BOSARGE W.E. AND SMITH C.L., NUMERICAL PROPERTIES OF A MULTIVARIATE
          RITZ-TREFFTZ METHOD. I.B.M. J. RESEARCH AND DEVELOPMENT 16,
          303-400, 1972.
343   BOSSHARD W., EIN NEUES VOLLVERTRAGLICHES ENDLICHES ELEMENT FUR
          PLATTENBEIGUNG. ABHANDLUNGEN DER INTERNATIONALEN
          VEREIGIGUNG FUR BRUCKENBAU UND HOCHBAU, ZURICH,1968.
          ALSO INT.ASSOC.BRIDGES STRUCT.ENG.BULL,28, 27-40,1968.
344   BOUSQUET R.D., YATES D.N., SABLE W.W. AND VINSON T.J., THE
          DEVELOPMENT OF COMPUTER GRAPHICS FOR LARGE SCALE FINITE
          ELEMENT CODES. PROC. 3RD CONF. MATRIX METHODS IN
          STRUCTURAL MECHANICS, WRIGHT-PATTERSON AFB., OHIO, 1971.
345   BOYD T.J.M., GARDNER G.A., AND GARDNER L.R.T., NUMERICAL STUDY OF
          HYDROMAGNETIC STABILITY USING THE FINITE ELEMENT METHOD. NUCL.
          FUSION 13, 764-766, 1973.
346   BRAESS D., RATIONALE INTERPOLATION, NORMALITAT UND MONOSPLINES.
          NUMER.MATH.22, 219-232, 1974.
347   BRAMBLE J.H.,(ED.), NUMERICAL SOLUTION OF PARTIAL DIFFERENTIAL
          EQUATIONS. ACADEMIC PRESS, NEW YORK, 1966.
348   BRAMBLE J.H., LECTURE NOTES ON VARIATIONAL METHODS FOR THE NUMERICAL
          SOLUTION OF ELLIPTIC PROBLEMS. DEPARTMENT OF COMPUTER SCIENCES,
          CHALMERS INSTITUTE OF TECHNOLOGY, AND UNIVERSITY OF GOTEBORG,
          SWEDEN, 1970.
349   BRAMBLE J.H., DUPONT T., AND THOMEE V., HIGHER ORDER POLYGONAL DOMAIN
          GALERKIN APPROXIMATIONS IN DIRICHLET'S PROBLEM. TECHNICAL REPORT
          1213, MATHEMATICS RESEARCH CENTER, UNIVERSITY OF WISCONSIN,
          MADISON, 1972.
350   BRAMBLE J.H., AND HILBERT S.R., ESTIMATION OF LINEAR FUNCTIONALS ON
          SOBOLEV SPACES WITH APPLICATION TO FOURIER TRANSFORMS AND SPLINE
          INTERPOLATION. SIAM J.NUMER.ANAL.7, 112-124, 1970.
351   BRAMBLE J.H., AND HILBERT S.R., BOUNDS FOR A CLASS OF LINEAR
          FUNCTIONALS WITH APPLICATIONS TO HERMITE INTERPOLATION.
          NUMER.MATH. 16, 362-369, 1971.
352   BRAMBLE J.H., AND NITSCHE J.A., A GENERALISED RITZ-LEAST-SQUARES
          METHOD FOR DIRICHLET PROBLEMS. SIAM J. NUMER.ANAL.10,
          81-93, 1973.
353   BRAMBLE J.H., AND OSBORN J.E., APPROXIMATION OF STEKLOV EIGENVALUES
          OF NON-SELFADJOINT SECOND ORDER ELLIPTIC OPERATORS. PP.
          387-408 OF A.K. AZIZ (ED.), THE MATHEMATICAL FOUNDATIONS
          OF THE FINITE ELEMENT METHOD WITH APPLICATIONS TO
          PARTIAL DIFFERENTIAL EQUATIONS. ACADEMIC PRESS, NEW
          YORK, 1972.
354   BRAMBLE J.H., AND OSBORN J.E., RATE OF CONVERGENCE ESTIMATES FOR
          NON-SELFADJOINT EIGENVALUE APPROXIMATIONS. TECHNICAL
          REPORT 1232, MATHEMATICS RESEARCH CENTER, UNIVERSITY OF
          WISCONSIN, MADISON, 1972.
355   BRAMBLE J.H., AND SCHATZ A.H., RAYLEIGH-RITZ-GALERKIN METHODS FOR
          DIRICHLET'S PROBLEM USING SUBSPACES WITHOUT BOUNDARY
          CONDITIONS. COMM.PURE.APPL.MATH., 23, 653-676, 1970.
356   BRAMBLE J.H., AND SCHATZ A.H., LEAST SQUARES METHODS FOR 2MTH
          ORDER ELLIPTIC BOUNDARY-VALUE PROBLEMS. MATH.COMP.25,
          113, 1-32, 1971.
357   BRAMBLE J.H., AND SCHATZ A.H., ON THE NUMERICAL SOLUTION OF
          ELLIPTIC BOUNDARY VALUE PROBLEMS BY LEAST SQUARES APPROXIMATION
          OF THE DATA. IN HUBBARD (ED.), NUMERICAL SOLUTION
          OF PARTIAL DIFFERENTIAL EQUATIONS II, SYNSPADE 1970,
          ACADEMIC PRESS, NEW YORK, 1971.
358   BRAMBLE J.H., AND THOMEE V., SEMIDISCRETE-LEAST SQUARES METHODS FOR
          A PARABOLIC BOUNDARY VALUE PROBLEM. MATH.COMP.26, 633-648, 1972.
359   BRAMBLE J.H. AND ZLAMAL M., TRIANGULAR ELEMENTS IN THE FINITE
          ELEMENT METHOD. MATH.COMP.24,809-820,1970.
360   BRAMLETTE T.T. AND LEONARD J.W., FINITE ELEMENT ANALYSIS OF THE
          UNSTEADY LINEARIZED BOLTZMANN EQUATION. BELL AEROSYSTEMS
          CO., REPORT. 9500-920180,1970.

361 BRAMLETTE T.T. AND MALLETT R.H., A FINITE ELEMENT SOLUTION TECHNIQUE
FOR THE BOLTZMANN EQUATION. J. FLUID MECH.42, 177-192,1970.
362 BRAND R.S., AND LAHEY F.J., VARIATIONAL PRINCIPLES FOR STEADY HEAT
CONDUCTION WITH MIXED BOUNDARY CONDITIONS. J.ENG.MATH.3,
119-121, 1969.
363 BRANDES K., NUMERICAL CALCULATION METHODS FOR PRESTRESSED CONCRETE
REACTOR PRESSURE VESSELS. 2ND CONF. PRESTRESSED CONCRETE REACTOR
PRESSURE VESSELS AND THEIR THERMAL ISOLATION, BRUSSELS, 1969.
364 BRANTLY E., ANALYSIS OF FRAMES WITH SHEAR WALLS WITH FINITE
ELEMENTS. PP.567-608 OF PROCEEDINGS SYMP. ON APPLICATION
OF FINITE ELEMENT METHODS IN CIVIL ENGINEERING.
VANDERBILT UNIVERSITY, 1969.
365 BRASHERS M.R., CHAN S.T.K., AND YOUNG V.Y.C., FINITE ELEMENT
ANALYSIS OF TRANSONIC FLOW. PROC.CONF.COMP.METHODS IN
NONLINEAR MECH., AUSTIN, TEXAS, 1974.
366 BRAUCHLI H.J., AND ODEN J.T., CONJUGATE APPROXIMATION FUNCTIONS
IN FINITE ELEMENT ANALYSIS. QUART.OF APPL.MATH.29,
65-90, 1971.
367 BRAUNLEDER B., MODIFIZIERTE RITZ VERFAHREN IN BANACHRAUMEN. REPORT
85, GESELLSCHAFT FUR MATHEMATIK UND DATENVERARBEITUNG,
BONN, 1974.
368 BREBBIA C. AND CONNOR J.J., GEOMETRICALLY NONLINEAR FINITE ELEMENT
ANALYSIS. J.ENG.MECH.DIV.ASCE.98, 463-483, 1969.
369 BREBBIA C.A., AND CONNOR J.J., NUMERICAL METHODS IN FLUID
DYNAMICS. PENTECH PRESS, LONDON 1974.
370 BREZIS H., EQUATIONS ET INEQUATIONS NON LINEAIRES DANS LE ESPACES
VECTORIELS EN DUALITE. ANN. INST.FOURIER, GRENOBLE, 18, 115-175,
1968.
371 BREZIS H., AND SIBONY M., METHOD D'APPROXIMATION ET D'ITERATION POUR
LES OPERATEURS MONOTONES. ARCH.RAT.MECH.ANAL.28, 59-82, 1968.
372 BREZIS H., AND STAMPACCHIA G., SUR LA REGULARITE DE LA SOLUTION
D'INEQUATIONS ELLIPTIQUE. BULL. SOC.MATH.FRANCE 96, 153-180,
1968.
373 BRILLA J., CONVOLUTIONAL VARIATIONAL PRINCIPLES AND METHODS IN
LINEAR VISCOELASTICITY. Z.A.M.M. 54, 46-47, 1974.
374 BRILLA J., FINITE ELEMENT METHOD IN LINEAR VISCOELASTICITY
Z.A.M.M. 54, 47-48, 1974.
375 BROLLIAR R.H., NASTRAN BUCKLING ANALYSIS OF A LARGE STIFFENED
CYLINDRICAL SHELL WITH A CUTOUT. NASTRAN USERS
EXPERIENCES. NASA TM X-2378,1971.
376 BRON J., AND DHATT G., MIXED QUADRILATERAL ELEMENTS FOR BENDING.
AIAA.J.10, 1359-1361,1972.
377 BROWDER F., EXISTENCE AND UNIQUENESS THEOREMS FOR SOLUTIONS OF
NONLINEAR BOUNDARY VALUE PROBLEMS. APPLICATION OF NONLINEAR
PARTIAL DIFFERENTIAL EQUATIONS IN MATHEMATICAL PHYSICS. PROC.
SYMP. AMERICAN MATH.SOC.APPL.MATH 17, 24-49, 1965.
378 BROWDER F.E., EXISTENCE AND APPROXIMATION OF SOLUTIONS OF NONLINEAR
VARIATIONAL INEQUALITIES. PROC. NATIONAL ACAD.SCI. U.S.A. 56,
1080-1086, 1966.
379 BROWDER F.E., ON THE UNIFICATION OF THE CALCULUS OF VARIATIONS AND THE
THEORY OF MONOTONE NONLINEAR OPERATORS IN BANACH SPACES. PROC.
NATIONAL ACAD.SCI.U.S.A. 56, 419-425, 1966.
380 BROWDER F.E., NONLINEAR VARIATIONAL INEQUALITIES AND MAXIMAL MONOTONE
MAPPINGS IN BANACH SPACES. MATH.ANNALEN 175, 89-113, 1968.
381 BROWN E.H., STRUCTURAL ANALYSIS. WILEY, NEW YORK, 1967.
382 BROWN J.E., FINITE ELEMENT SOLUTION TO DYNAMIC STABILITY OF BARS.
AIAA J. 6, 1423-1424, 1968.
383 BRUCH J.C., AND ZYVOLOSKI G., FINITE ELEMENT SOLUTION OF UNSTEADY
AND UNSATURATED FLOW IN POROUS MEDIA. PP.201-211 OF J.R.WHITEMAN
(ED.), THE MATHEMATICS OF FINITE ELEMENTS AND APPLICATIONS.
ACADEMIC PRESS, LONDON, 1973.
384 BRUCH J.C., AND ZYVOLOSKI G., A FINITE ELEMENT WEIGHTED RESIDUAL
SOLUTION TO ONE-DIMENSIONAL FIELD PROBLEMS. INT. J.NUMER.METH.
ENG.6, 577-585, 1973.
385 BRUCH J.C., AND ZYVOLOSKI G., TRANSIENT TWO-DIMENSIONAL HEAT CONDUCTION
PROBLEMS SOLVED BY THE FINITE ELEMENT METHOD. INT.J.NUMER. METH.

ENG.8, 481-494, 1974.

386   BRUSA L., CIACCI R., AND TICOZZI C., TWO-DIMENSIONAL FINITE
ELEMENT COMPUTER CODES FOR STRESS ANALYSIS AND FIELD
PROBLEMS. ENERG.NUCL.(MILAN) 20, 154-170, 1973.

387   BUHLER H., AND BUCHHOLTZ D.W., RELATIONSHIP BETWEEN RESIDUAL
STRESSES AND FLATNESS DEFECTS IN COLD-ROLLED SHEET. ARCH.
EISENHUETTENWES 44, 899-906, 1973.

388   BULLEY R.M., COMPUTATION OF APPROXIMATE POLYNOMIAL SOLUTIONS TO THE
HELMHOLTZ EQUATION USING THE RAYLEIGH-RITZ METHOD.
PH.D.DISSERTATION, UNIVERSITY OF SHEFFIELD, 1968.

389   BURMAN Z.I., ON THE THEORY OF CALCULATION OF THE OVERALL STRENGTH
OF A FUSELAGE BY THE METHOD OF FINITE ELEMENTS. IZV.
VUZ.AVIATS. TEKH.15, 49-55, 1972.

390   BUSHNELL D., ENERGY APPROACHES TO FINITE DIFFERENCE AND FINITE
ELEMENT METHODS. INT.SYMP.NUMERICAL AND COMPUTER
METHODS IN STRUCTURAL MECHANICS. U.S. OFFICE OF NAVAL
RESEARCH, URBANA, ILLINOIS, 1971.

391   BUTLER H.W., A MINIMUM PRINCIPLE FOR HEAT CONDUCTION. PROC. 5TH U.S.
NATIONAL CONGRESS OF APPL.MECH.A.S.M.E., NEW YORK, 1966.

392   BUTLIN G., AND FORD R., A COMPATIBLE TRIANGULAR PLATE BENDING FINITE
ELEMENT. UNIVERSITY OF LEICESTER, ENGINEERING DEPT.,
REPORT 68-15,1968.

393   BUTLIN G.A., AND LECKIE F.A., A STUDY OF FINITE ELEMENTS APPLIED
TO PLATE FLEXURE. NUMERICAL METHODS FOR VIBRATION
PROBLEMS 2, 26-37, I.S.V.R. SOUTHAMPTON, 1966.

394   BUTURLA E.M., AND COTTRELL P.E., TWO-DIMENSIONAL FINITE ELEMENT
ANALYSIS OF SEMI-CONDUCTOR STEADY STATE TRANSPORT EQUATIONS.
PROC.CONF.COMP.METHODS IN NONLINEAR MECH., AUSTIN, TEXAS,
1974.

395   BUTZER P.L., KAHANE J.P., AND SZOKEFALVI-NAGY B., LINEAR
OPERATORS AND APPROXIMATION. I.S.N.M. 20, BIRKHAUSER
VERLAG, 1972.

396   BYERS N.R., AND SCHULTZ R.E., ANALYSIS OF A STUB END BY THE FINITE
ELEMENT METHOD. WELDING J.51, 31-35, 1972.

397   BYKAT A., AUTOMATIC TRIANGULATION OF TWO DIMENSIONAL REGIONS.
REPORT ICSI 420, UNIVERSITY OF LONDON, INSTITUTE OF COMPUTER
SCIENCE, 1972.

398   BYRNE G.D., AND HALL C.A. (EDS.), NUMERICAL SOLUTION OF SYSTEMS OF
NONLINEAR ALGEBRAIC EQUATIONS. ACADEMIC PRESS, NEW YORK, 1973.

399   BYSKOV E., CALCULATION OF STRESS INTENSITY FACTORS USING THE FINITE
ELEMENT METHOD WITH CRACK ELEMENTS. INT. J.FRACTURE MECHANICS
6, 159-168, 1970.

400   CAIRO L., AND KAHAN T., VARIATIONAL TECHNIQUES IN ELECTROMAGNETISM.
GORDON AND BREACH, NEW YORK, 1965.

401   CALLIDINE C.R., A NEW FINITE ELEMENT METHOD FOR ANALYSING
SYMMETRICALLY LOADED THIN SHELLS OF REVOLUTION. INT.J.
NUMER.METH.ENG.6, 475-487, 1973.

402   CAMERON I.G., A DYNAMIC ELASTOPLASTIC ANALYSIS OF THIN SHELLS
OF REVOLUTION UNDER ASYMMETRIC MECHANICAL AND THERMAL
LOADING. PP. 283-297 OF J.R. WHITEMAN (ED.), THE MATHEMATICS
OF FINITE ELEMENTS AND APPLICATIONS. ACADEMIC PRESS, LONDON
1973.

403   CAMPBELL J.S., STRESS ANALYSIS BY THE FINITE ELEMENT DISPLACEMENT
METHOD. REPORT. RD/B/N-834, CENTRAL ELECTRICITY
GENERATING BOARD, BERKELEY LABORATORIES, 1968.

404   CAMPBELL J.S., FINITE ELEMENTS FOR THE REPRESENTATION OF PLANE
AND AXISYMMETRIC STRUCTURAL REINFORCEMENT. REPORT
RD/B/N-1105, CENTRAL ELECTRICITY GENERATING BOARD,
BERKELEY LABORATORIES, 1968.

405   CAMPBELL J.S., FINITE ELEMENT STRESS ANALYSIS OF PLANE AND
AXISYMMETRIC STRUCTURES. REPORT RD/B/N-1209, CENTRAL
ELECTRICITY GENERATING BOARD, BERKELEY LABORATORIES, 1969.

406   CANTIN G., RIGID BODY MOTIONS IN CURVED FINITE ELEMENTS.
AIAA J.8, 1252-1255,1970.

407   CANTIN G., AN EQUATION SOLVER OF VERY LARGE CAPACITY. INT.J.NUMER.
METH.ENG.3, 379-388, 1971.

408   CANTIN G. AND CLOUGH R.W., A CURVED CYLINDRICAL SHELL FINITE
          ELEMENT. AIAA J.6, 1057-1062, 1968.
409   CAPRILI M., CELLA A., AND GHERI G., SPLINE INTERPOLATION TECHNIQUES
          FOR VARIATIONAL METHODS. REPORT B72-1, INSTITUTO DI
          ELABORAZIONE DELA INFORMAZIONE, PISA, 1972.
410   CAPRILI M., CELLA A. AND GHERI G., SPLINE INTERPOLATION TECHNIQUES FOR
          VARIATIONAL METHODS. INT.J.NUMER.METH.ENG.6,565-576,1973.
411   CARLSON R.E., AND HALL C.A., ON PIECEWISE POLYNOMIAL INTERPOLATION IN
          RECTANGULAR POLYGONS. J.APPROX.THEORY 4,37-53,1971.
412   CARLSON R.E., AND HALL C.A., RITZ APPROXIMATIONS TO TWO-DIMENSIONAL
          BOUNDARY VALUE PROBLEMS. NUMER.MATH.18,171-181, 1971.
413   CARLSON R.E., AND HALL C.A., ERROR BOUNDS FOR BICUBIC SPLINE INTERPOLATION.
          (TO APPEAR).
414   CARMICHAEL G.L.T., THE APPLICATION OF THREE DIMENSIONAL FINITE
          ELEMENTS TO THE ANALYSIS OF PODDED BOILER TYPE PRESTRESSED
          CONCRETE PRESSURE VESSELS. NUCLEAR ENG. AND DESIGN 16,
          35-44,1971.
415   CARMIGNANI C., AND CELLA A., ELASTIC ANALYSIS OF CRACKED THIN SHELLS BY
          THE FINITE ELEMENT METHOD. PP.57-76, VOL.5, PROCEEDINGS 1ST
          STRUCTURAL MECH. IN REACTOR TECH.CONF., BERLIN, 1971.
416   CARMIGNANI C., AND CELLA A., STRUCTURAL DYNAMICS BY FINITE
          ELEMENTS, MODAL AND FOURIER ANALYSIS. PAPER K3/2,
          PROCEEDINGS 2ND STRUCTURAL MECH. IN REACTOR TECH. CONF.,
          BERLIN, 1973.
417   CARNAHAN B., LUTHER H.A., AND WILKES J.O., APPLIED NUMERICAL METHODS.
          WILEY, NEW YORK, 1969.
418   CARPENTER W.C., AND GILL P.A.T., AUTOMATED SOLUTIONS OF TIME
          DEPENDENT PROBLEMS. PP. 495-506 OF J.R. WHITEMAN (ED.),
          THE MATHEMATICS OF FINITE ELEMENTS AND APPLICATIONS, ACADEMIC
          PRESS, LONDON, 1973.
419   CARRARA S., AND MCGARRY F., MATRIX AND INTERFACE STRESSES IN A
          DISCONTINUOUS FIBER COMPOSITE MODEL.J.COMPOSITE MATERIALS
          2, 222-243,1968.
420   CARSLAW H.S., AND JAEGER J.C., CONDUCTION OF HEAT IH SOLIDS. OXFORD
          UNIVERSITY PRESS,1967.
421   CARSON W.G., AND NEWTON R.E., PLATE BUCKLING ANALYSIS USING A FULLY
          COMPATIBLE FINITE ELEMENT. AIAA J.8, 527-529, 1969.
422   CARY G.F., AND VARANSI S.R., EXPERIENCE WITH FINITE ELEMENT
          COMPUTATIONS FOR NONLINEAR STRUCTURAL PROBLEMS. PROC.
          CONF.COMP.METHODS IN NONLINEAR MECHANICS, AUSTIN, TEXAS,
          1974.
423   CASCIARO R., AND DI CARLO A., MIXED FINITE ELEMENT MODELS IN
          LIMIT ANALYSIS. PROC.CONF.COMP.METHODS IN NONLINEAR
          MECH., AUSTIN, TEXAS, 1974.
424   CAVENDISH J.C., AUTOMATIC TRIANGULATION OF ARBITRARY PLANAR
          DOMAINS FOR THE FINITE ELEMENT METHOD. INT.J.NUMER.
          METH.ENG.8, 679-696, 1974.
425   CAVENDISH J.C., AND HALL C.A., L-INFINITY CONVERGENCE OF COLLOCATION
          AND GALERKIN APPROXIMATIONS TO LINEAR TWO-POINT PARABOLIC PROBLEMS.
          (TO APPEAR).
426   CEA J., APPROXIMATION VARIATIONELLE DES PROBLEMS AUX LIMITES.
          ANN.INST.FOURIER (GRENOBLE) 14,345-444, 1964.
427   CECCHI M.M., A NUMERICAL STUDY ON NONLINEAR INSTABILITY DUE TO
          A THREE-DIMENSIONAL DISTURBANCE IN A PLANE PARALLEL FLOW.
          PROC.CONF.COMP.METHODS IN NONLINEAR MECH., AUSTIN, TEXAS,
          1974.
428   CECCI M.M. AND CELLA A., A FINITE ELEMENT SOLUTION OF THE STABILITY
          OF SUPERPOSED FLUIDS.
          PROCEEDINGS OF SYMPOSIUM ON FINITE ELEMENT METHODS IN FLOW
          PROBLEMS, SWANSEA 1974.
429   CELLA A., APPROXIMATION TECHNIQUES IN THE FINITE ELEMENT METHOD. REPORT
          N.I. B72-11, CONSIGLIO NAZIONALE DELLE RICERCHE, INSTUIO DI
          ELABORAZIONE DELLA INFORMAZIONE, PISA, 1972.
430   CELLA A., CARMIGNANI C., AND DEPAULIS A., FUNCTIONAL MINIMIZATION
          IN NONLINEAR SOLID MECHANICS. PROC.CONF.COMP. METHODS IN
          NONLINEAR MECHANICS, AUSTIN, TEXAS, 1974.

431 CHAKRABARTI S., TRIGONOMETRIC FUNCTION REPRESENTATIONS FOR RECTANGULAR
         PLATE BENDING ELEMENTS. INT.J.NUMER.METH.ENG.3, 261-273, 1971.
432 CHAN A.S.L. AND FIRMIN A., THE ANALYSIS OF COOLING TOWERS BY THE FINITE
         ELEMENT TECHNIQUE; I: SMALL DISPLACEMENTS. II:LARGE DISPLACEMENTS.
         AERO.J.ROYAL AERO.SOC.74, 826-835 AND 971-982,1970.
433 CHAN H.S.Y., THE COLLAPSE LOAD OF REINFORCED CONCRETE PLATES.
         INT.J.NUMER.METH.ENG.5, 57-64, 1972.
434 CHAN S.K., TUBA I.S., AND WILSON W.K., ON THE FINITE ELEMENT
         METHOD IN LINEAR FRACTURE MECHANICS. ENG.FRACTURE.MECH.
         2, 1-17, 1970.
435 CHANG T-C, AND CRAIG R.R., COMPUTATION OF UPPER AND LOWER BOUNDS
         TO THE FREQUENCIES OF ELASTIC SYSTEMS BY THE METHOD OF
         LEHMANN AND MAEHLY. INT.J.NUMER.METH. ENG.6, 323-332,
         1973.
436 CHARI M.V.K., AND SILVESTER P., ANALYSIS OF TURBOALTERNATOR MAGNETIC FIELDS
         BY FINITE ELEMENTS. IEEE TRANS. PAS.90, 454-464, 1971.
437 CHATTERJEE A. AND SETLUR A.V., A MIXED FINITE ELEMENT FORMULATION FOR
         PLATE PROBLEMS. INT.J.NUMER.METH.ENG.4,67-84, 1972.
438 CHATTOPADHYAY A., AND SETLUR A.V., APPLICATION OF FINITE ELEMENT
         METHOD TO CONTINUUM MECHANICS PROBLEMS. PROC.3RD CONF.MATRIX
         METHODS IN STRUCTURAL MECHANICS, WRIGHT-PATTERSON AFB.,
         OHIO,1971.
439 CHENEY E.W., INTRODUCTION TO APPROXIMATION THEORY. MCGRAW HILL, NEW
         YORK, 1966.
440 CHENG R.T., NUMERICAL INVESTIGATION OF LAKE CIRCULATION AROUND
         ISLANDS BY THE FINITE ELEMENT METHOD. INT.J.NUMER.METH.
         ENG.5, 103-112, 1972.
441 CHENG R.T., NUMERICAL SOLUTION OF THE NAVIER STOKES EQUATIONS
         BY THE FINITE ELEMENT METHOD. PHYSICS OF FLUIDS 15,
         2098-2105, 1972.
442 CHENG R.T.S., ON THE ACCURACY OF CERTAIN CO CONTINUOUS FINITE ELEMENT
         REPRESENTATIONS. INT.J.NUMER.METH.ENG.8, 649-657, 1974.
443 CHENG R.T., AND TUNG C., WIND DRIVEN LAKE CIRCULATION BY THE FINITE
         ELEMENT METHOD. PROC.13TH CONF.GREAT LAKES RESEARCH,
         BUFFALO, NEW YORK, 1970.
444 CHERNUKA M.W., COUPER G.R., LINDBERG G.M. AND OLSON M.D., APPLICATION
         OF THE HIGH PRECISION TRIANGULAR PLATE-BENDING ELEMENT TO
         PROBLEMS WITH CURVED BOUNDARIES. NAT.RES.COUNCIL CANADA,
         AERONAUT.REP.LR-529,1969.
445 CHERNUKA M.W., COUPER G.R., LINDBERG G.M. AND OLSON M.D., FINITE
         ELEMENT ANALYSIS OF PLATES WITH CURVED EDGES. INT.J.NUMER.
         METH.ENG.4, 40-65, 1972.
446 CHEUNG T.-Y. SPLINE APPROXIMATION OF THE CAUCHY PROBLEM D(P+Q) DXP DYQ = F
         (X,Y,U,...),TECHNICAL REPORT 29, COMPUTER SCIENCE DEPARTMENT,
         UNIVERSITY OF WISCONSIN, MADISON, 1968.
447 CHEUNG T.-Y., APPROXIMATE SOLUTIONS AND ERROR BOUNDS FOR QUASILINEAR
         ELLIPTIC BOUNDARY VALUE PROBLEMS. J.COMP.SYSTEM SCIENCES 7,
         306-322, 1973.
448 CHEUNG T.-Y., QUASILINEAR PARABOLIC BOUNDARY VALUE PROBLEMS APPROXIMATE
         SOLUTIONS AND ERROR BOUNDS BY LINEAR PROGRAMMING. SIAM.J.NUMER.
         ANAL.,10, 1061-1079,1973.
449 CHEUNG Y.K., KING I.P., AND ZIENKIEWICZ O.C., GENERAL METHOD OF ANALYSIS
         BASED ON FINITE ELEMENTS. PROC.INST.CIV.ENG.40, 9-36, 1968.
450 CHEUNG Y.K., AND MEDWELL J.C., FINITE ELEMENT METHOD APPLIED TO THE
         SOLUTION OF HEAT CONDUCTION PROBLEMS. REV.ROUMAINE DES SCIENCES
         TECHNIQUES - SERIES DE MECHANIQUE APPLIQUE 14, 361-372,1969.
451 CHEUNG Y.K., SISODIYA R.G., AND GHALI A., FINITE ELEMENT ANALYSIS
         OF SKEW, CURVED BOX-GIRDER BRIDGE. PUBLICATIONS,
         INTERNATIONAL ASSOCIATION OF BRIDGE AND STRUCTURAL
         ENGINEERS 30, 1970.
452 CHOPRA P.S., FINITE ELEMENT FRACTURE MECHANICS ANALYSIS OF CREEP
         RUPTURE OF FUEL ELEMENT CLADDING. PAPER C2/5,
         PROCEEDINGS 2ND STRUCTURAL MECH. IN REACTOR TECH. CONF.,
         BERLIN, 1973.
453 CHRISTIAN J.T. AND BOEHMER J.W., PLANE STRAIN CONSOLIDATION BY
         FINITE ELEMENTS. J.SOIL MECHANICS AND FOUNDATIONS DIV.

A.S.C.E.96, 1435-1457,1970.

454 CHRUSLOV F.J., METHOD OF ORTHOGONAL PROJECTION AND DIRICHLET
BOUNDARY VALUE PROBLEM IN A GRAIN DOMAIN. MAT.SB.88,
38-60, 1972.

455 CHU A.S.L., ANALYSIS AND DESIGN CAPABILITIES OF THE STRUDL PROGRAM.
INT.SYMP.NUMERICAL AND COMPUTER METHODS IN STRUCTURAL MECHANICS,
UNIVERSITY OF ILLINOIS, 1971.

456 CHU S.L., A COMPARISON OF SOME FINITE ELEMENT AND FINITE
DIFFERENCE METHODS FOR A SIMPLE SLOSH PROBLEM. AIAA J.9,
2094-2095,1971.

457 CHU S.L., AGRAWAL P.K., AND SINGH S., FINITE ELEMENT TREATMENT
OF SOIL STRUCTURE INTERACTION PROBLEM FOR NUCLEAR POWER
PLANT UNDER SEISMIC EXCITATION. PAPER K2/4,
PROCEEDINGS 2ND STRUCTURAL MECH. IN REACTOR TECH.CONF.,
BERLIN, 1973.

458 CHU T.C., AND SCHNOBRICH W.C., FINITE ELEMENT ANALYSIS OF
TRANSLATIONAL SHELLS. J.COMPUTERS STRUCTURES 2, 197-222,
1972.

459 CHUGH Y.D., AND HARDY H.R., APPLICATION OF FINITE ELEMENT ANALYSIS TO
UNDERGROUND STORAGE OF NATURAL GAS. TRANS.AM.GEOPHYSICAL UNION
51, 432,1970.

460 CHUNG T.J., AND EIDSON R.L., ANALYSIS OF VISCOELASTOPLASTIC STRUCTURAL
BEHAVIOUR OF ANISOTROPIC SHELLS BY THE FINITE ELEMENT METHOD,
PP.77-89, VOL.5, PROCEEDINGS 1ST STRUCTURAL MECH. IN REACTOR
TECH. CONF., BERLIN, 1971.

461 CHUNG T.J., AND YAGAWA G., INCREMENTAL THERMOMECHANICAL THEORY
OF VISCOELASTOPLASTIC SOLIDS AND SOLUTION BY FINITE
ELEMENTS. PAPER L4/2, PROC.2ND STRUCTURAL MECH IN REACTOR
TECH.CONF., BERLIN, 1973.

462 CIARLET P.G., DISCRETE VARIATIONAL GREEN'S FUNCTION. (TO APPEAR).

463 CIARLET P.G., ORDERS OF CONVERGENCE IN FINITE ELEMENT METHODS,
PP.113-129 OF J.R. WHITEMAN (ED.), THE MATHEMATICS OF FINITE
ELEMENTS AND APPLICATIONS. ACADEMIC PRESS, LONDON, 1973.

464 CIARLET P.G., QUELQUES METHODES D'ELEMENTS FINIS POUR LE PROBLEME D'UNE
PLAQUE ENCASTREE. COLL. INTERNATIONAL SUR LES METHODES DE CALCUL
SCIENTIFIQUE ET TECHNIQUE, I.R.I.A., LE CHESNAY, DECEMBER 1973.

465 CIARLET P.G., CONFORMING AND NONCONFORMING FINITE ELEMENT METHODS
FOR SOLVING THE PLATE PROBLEM. PP.21-32 OF G.A. WATSON
(ED.), PROCEEDINGS OF CONF. ON NUMERICAL SOLUTION OF
DIFFERENTIAL EQUATIONS. LECTURE NOTES IN MATHEMATICS,
NO. 363, SPRINGER-VERLAG, BERLIN, 1974.

466 CIARLET P.G., SUR L'ELEMENT DE CLOUGH ET TOCHER.
(MANUSCRIPT)

467 CIARLET P.G., NATTERER F., AND VARGA R.S., NUMERICAL METHODS OF
HIGHER-ORDER ACCURACY FOR SINGULAR NONLINEAR BOUNDARY VALUE
PROBLEMS. NUMER.MATH. 15, 87-99, 1970.

468 CIARLET P.G., AND RAVIART P.A., GENERAL LAGRANGE AND HERMITE
INTERPOLATION IN RN WITH APPLICATIONS TO FINITE ELEMENT
ARCH.RAT.MECH.ANAL.46,177-199,1972.

469 CIARLET P.G., AND RAVIART P-A., INTERPOLATION THEORY OVER CURVED
ELEMENTS WITH APPLICATIONS TO FINITE ELEMENT METHODS.
COMP.METH. IN APPL.MECH.ENG.1,217-249,1972.

470 CIARLET P.G., AND RAVIART P.A., THE COMBINED EFFECT OF CURVED
BOUNDARIES AND NUMERICAL INTEGRATION IN ISOPARAMETRIC
FINITE ELEMENT METHODS. PP. 409-474 OF A.K. AZIZ (ED.),
THE MATHEMATICAL FOUNDATIONS OF THE FINITE ELEMENT METHOD
WITH APPLICATIONS TO PARTIAL DIFFERENTIAL EQUATIONS,
ACADEMIC PRESS, NEW YORK, 1972.

471 CIARLET P.G., AND RAVIART P.A., MAXIMUM PRINCIPLE AND UNIFORM
CONVERGENCE FOR THE FINITE ELEMENT METHOD. COMP.METH.IN
APPL.MECH.ENG.2, 17-31, 1973.

472 CIARLET P.G., AND RAVIART P.A., A MIXED FINITE ELEMENT METHOD FOR
THE BIHARMONIC EQUATION. SYMPOSIUM ON MATHEMATICAL
ASPECTS OF FINITE ELEMENTS IN PARTIAL DIFFERENTIAL
EQUATIONS. MATHEMATICS RESEARCH CENTER, UNIVERSITY OF
WISCONSIN, MADISON, 1974.

473   CIARLET P.G., AND RAVIART P.A., LA METHODE DES ELEMENTS FINIS
        POUR LES PROBLEMES AUX LIMITES ELLIPTIQUES. REPORT 74006,
        LABORATOIRE ANALYSE NUMERIQUE, UNIVERSITE PARIS, 1974.

474   CIARLET P.G., SCHULTZ M.H., AND VARGA R.S., NUMERICAL METHODS
        OF HIGH-ORDER ACCURACY FOR NONLINEAR BOUNDARY VALUE PROBLEMS I:
        ONE DIMENSIONAL PROBLEM. NUMER.MATH.9, 394-430, 1967.

475   CIARLET P.G., SCHULTZ M.H., AND VARGA R.S., NUMERICAL METHODS OF
        HIGH-ORDER ACCURACY FOR NONLINEAR BOUNDARY VALUE PROBLEMS II:
        NONLINEAR BOUNDARY CONDITIONS. NUMER.MATH.II, 331-345, 1968.

476   CIARLET P.G., SCHULTZ M.H., AND VARGA R.S., NUMERICAL METHODS OF
        HIGH-ORDER ACCURACY FOR NONLINEAR BOUNDARY VALUE PROBLEMS III:
        EIGENVALUE PROBLEMS. NUMER.MATH.12, 120-133, 1968.

477   CIARLET P.G., SCHULTZ M.H., AND VARGA R.S., NUMERICAL METHODS OF
        HIGH-ORDER ACCURACY FOR NONLINEAR BOUNDARY VALUE PROBLEMS IV:
        PERIODIC BOUNDARY CONDITIONS. NUMER.MATH.12, 266-279, 1968.

478   CIARLET P.G., SCHULTZ M.H., AND VARGA R.S., NUMERICAL METHODS OF
        HIGH-ORDER ACCURACY FOR NONLINEAR BOUNDARY VALUE PROBLEMS V:
        MONOTONE OPERATORS. NUMER.MATH.13, 51-77, 1969.

479   CIARLET P.G., AND WAGSCHAL C., MULTIPOINT TAYLOR FORMULAS AND
        APPLICATIONS TO THE FINITE ELEMENT METHOD. NUMER.MATH.17,
        84-100, 1971.

480   CLEMENT PH. AND PINI F., APPROXIMATION BY FINITE ELEMENT FUNCTIONS
        USING REGULARIZATION. TECHNICAL REPORT, DEPARTEMENT DE
        MATHEMATIQUES, ECOLE POLYTECHNIQUE FEDERALE DE LAUSANNE,1974.

481   CLINARD J.A., AND CROWELL J.S., ORNL USER'S MANUAL FOR CREEP-
        PLAST COMPUTER PROGRAM. REPORT ORNL-TM-4062, OAK RIDGE
        NATIONAL LABORATORY, TENNESSEE, 1973.

482   CLOUGH R.W., THE FINITE-ELEMENT IN PLANE STRESS ANALYSIS.
        PROC.2ND A.S.C.E. CONF. ON ELECTRONIC COMPUTATION, PITTSBURG,
        PA., 1960.

483   CLOUGH R.W., THE STRESS DISTRIBUTION OF NORFOLK DAM. RESEARCH REPORT
        100, 19, INSTITUTE OF ENGINEERING, BERKELEY, CALIFORNIA, 1962.

484   CLOUGH R.W., THE FINITE ELEMENT METHOD IN STRUCTURAL MECHANICS.
        IN ZIENKIEWICZ AND HOLISTER (EDS.), STRESS ANALYSIS,
        WILEY, NEW YORK, 1965.

485   CLOUGH R.W., COMPARISON OF THREE DIMENSIONAL FINITE ELEMENTS. SYMP.
        APPLICATIONS OF FINITE-ELEMENT METHODS IN CIVIL ENGINEERING.
        VANDERBILT UNIVERSITY, 1969.

486   CLOUGH R.W., VIBRATION ANALYSIS OF FINITE ELEMENT SYSTEMS. PP.513-525 OF
        J.T. ODEN AND E.R. DE A. OLIVEIRA (EDS.), LECTURES ON FINITE ELEMENT
        METHODS IN CONTINUUM MECHANICS. UNIVERSITY OF ALABAMA PRESS,
        HUNTSVILLE, 1973.

487   CLOUGH R.W., AND BATHE K.J., FINITE ELEMENT ANALYSIS OF DYNAMIC
        RESPONSE. PP.153-180 OF J.T.ODEN, R.W.CLOUGH,AND Y.YAMOMOTO (EDS.),
        ADVANCES IN COMPUTATIONAL METHODS IN STRUCTURAL MECHANICS AND
        DESIGN. UNIVERSITY OF ALABAMA PRESS, HUNTSVILLE, 1972.

488   CLOUGH R.W., AND DUNCAN J.M., FINITE ELEMENT ANALYSES OF RETAINING
        WALL BEHAVIOUR. J. SOIL MECH. AND FOUNDATIONS DIV. A.S.C.E. 97,
        SM12, 1657-1673, 1971.

489   CLOUGH R.W., POWELL G.H., AND GANTAYAT A.N., STRESS ANALYSIS OF B.16.9
        TEES BY THE FINITE ELEMENT METHOD. PP.439-458 OF VOL.3 OF
        PROCEEDINGS 1ST STRUCTURAL MECH. IN REACTOR TECH.CONF., BERLIN,
        1971.

490   CLOUGH R.W., AREAS OF APPLICATION OF THE FINITE ELEMENT METHOD.
        COMPUTERS AND STRUCTURES 4, 17-40, 1974.

491   CLOUGH R.W., AND FELIPPA C.A., A REFINED QUADRILATERAL ELEMENT FOR
        ANALYSIS OF PLATE BENDING. PROC. 2ND CONF. MATRIX METHODS IN
        STRUCTURAL MECHANICS, WRIGHT-PATTERSON AFB, OHIO,
        AFFDL - TR-68-150, 1968.

492   CLOUGH R.W. AND JOHNSON C.P., A FINITE ELEMENT APPROXIMATION FOR THE
        ANALYSIS OF THIN SHELLS. INT.J.SOLIDS STRUCTURES 4, 43-60,
        1968.

493   CLOUGH R.W., AND JOHNSON C.P., FINITE ELEMENT ANALYSIS OF ARBITRARY
        THIN SHELLS. ACI SYMP. ON CONCRETE THIN SHELLS, NEW YORK, 1970.

494   CLOUGH R.W. AND RASHID Y.R., FINITE ELEMENT ANALYSIS OF
        AXISYMMETRIC SOLIDS. J.ENG.MECH.DIV. A.S.C.E. 91, EM1,

      71-85, 1965.

495   CLOUGH R.W., AND TOCHER J.L., FINITE ELEMENT STIFFNESS MATRICES FOR
      THE ANALYSIS OF PLATE BENDING. PROC.1ST CONF.MATRIX MATHODS IN
      STRUCTURAL MECHANICS, WRIGHT-PATTERSON AFB, AFFDL
      TR66-80, 1965.

496   CLOUGH R.W. AND WILSON E.L., DYNAMIC FINITE ELEMENT ANALYSIS
      OF ARBITRARY THIN SHELLS. COMPUTERS AND STRUCTURES 1,
      33-55, 1972.

497   CLOUGH R.W., WILSON E.L., AND KING I.P., LARGE CAPACITY MULTISTORY
      FRAME ANALYSIS PROGRAMS, J.STRUCT.DIV. A.S.C.E. 89, 179-204,1963.

498   CODY W.J., MEINARDUS G., AND VARGA R.S., CHEBYSHEV RATIONAL
      APPROXIMATIONS TO EXP(-X) IN [0, + INFINITY) AND APPLICATIONS
      TO HEAT CONDUCTION PROBLEMS. J.APPROX.THEORY 2, 50-65, 1969.

499   COHEN E., AND MCCALLION H., IMPROVED DEFORMATION FUNCTIONS FOR FINITE
      ELEMENT ANALYSIS OF BEAM SYSTEMS. INT. J. NUMER. METH.ENG.1,
      163-168, 1969.

500   COLLATZ L., NUMERICAL TREATMENT OF DIFFERENTIAL EQUATIONS.
      SPRINGER VERLAG, BERLIN, 1960.

501   COLLATZ L., HERMITEAN METHODS FOR INITIAL VALUE PROBLEMS IN PARTIAL
      DIFFERENTIAL EQUATIONS. PP.41-61 OF J.J.H. MILLER (ED.), TOPICS IN
      NUMERICAL ANALYSIS. ACADEMIC PRESS, LONDON 1973.

502   COLLIER W.D., HEATRAN, A FINITE ELEMENT CODE FOR HEAT TRANSFER
      PROBLEMS. U.K. ATOMIC ENERGY AUTHORITY, REPORT TRG-1807,
      RISLEY,1969.

503   COLLINS R.J., BANDWIDTH REDUCTION BY AUTOMATIC RENUMBERING.
      INT.J.NUMER.METH.ENG.6, 345-356, 1973.

504   COMINI G., GUIDICE S.D., LEWIS R.W., AND ZIENKIEWICZ O.C., FINITE ELEMENT
      SOLUTION OF NONLINEAR HEAT CONDUCTION PROBLEMS WITH SPECIAL
      REFERENCE TO PHASE CHANGE. INT.J.NUMER.METH.ENG.8, 613-624, 1974.

505   CONNOR J.J. AND BREBBIA C., STIFFNESS MATRIX FOR SHALLOW
      RECTANGULAR SHELL ELEMENT. J.ENG.MECH.DIV.A.S.C.E. 93, EM5,
      46-66,1967.

506   CONSTANTINO C.J., FINITE ELEMENT APPROACH TO STRESS WAVE PROBLEMS.
      J.ENG.MECH.DIV.A.S.C.E.93, EM2, 153-176,1967.

507   COOK R.D., EIGENVALUE PROBLEMS WITH A MIXED PLATE ELEMENT. AIAA
      J,7, 982-983, 1969.

508   COOK R.D., STRAIN RESULTANTS IN CERTAIN FINITE ELEMENTS. AIAA
      J.7, 535   , 1969.

509   COOK R.D., MORE ON REDUCED INTEGRATION AND ISOPARAMETRIC ELEMENTS.
      INT.J.NUMER.METH.ENG. 5, 141-142, 1972.

510   COOK D.R., TWO HYBRID ELEMENTS FOR ANALYSIS OF THICK THIN AND
      SANDWICH PLATES. INT.J.NUMER.METH.ENG.5, 277-288, 1972.

511   COOK R.D., A NOTE ON CERTAIN INCOMPATIBLE ELEMENTS. INT.J.NUMER.
      METH.ENG. 6, 146-147, 1973.

512   COOK R.D., AND ABDULLA J.K., SOME PLANE QUADRILATERAL "HYBRID" FINITE
      ELEMENTS. AIAA J.7, 2184-2185, 1969.

513   COOK R.D. AND LADKANY S.G., OBSERVATIONS REGARDING ASSUMED STRESS HYBRID
      PLATE ELEMENTS. INT.J.NUMER.METH.ENG.8, 513-519, 1974.

514   COOK W.A., BODY ORIENTED (NATURAL) CO-ORDINATES FOR GENERATING THREE
      DIMENSIONAL MESHES. INT.J.NUMER.METH.ENG.8, 27-43, 1974.

515   COOK W.L., AUTOMATED INPUT DATA PREPARATION FOR NASTRAN. NASA
      TECHNICAL REPORT TM-X-63607, 1965.

516   COONS S.A., SURFACES FOR COMPUTER AIDED DESIGN OF SPACE FORMS. REPORT
      MAC-TR-41, DESIGN DIVISION DEPARTMENT OF MECHANICAL ENGINEERING,
      M.I.T., 1967.

517   COPUM J.M. AND KRISHNAMURTHY N., A THREE DIMENSIONAL FINITE ELEMENT
      ANALYSIS OF A PRESTRESSED CONCRETE REACTOR VESSEL MODEL.
      PP.63-69 OF APPLICATION OF FINITE ELEMENT METHODS IN CIVIL
      ENGINEERING, A.S.C.E., 1969.

518   COURANT R., VARIATIONAL METHODS FOR THE SOLUTION OF PROBLEMS OF
      EQUILIBRIUM AND VIBRATIONS. BULL. AMER.MATH.SOC.49, 1-23, 1943.

519   COURANT R., DIRICHLET'S PRINCIPLE, INTERSCIENCE, NEW YORK, 1950.

520   COURANT R., REMARKS ABOUT THE RAYLEIGH-RITZ METHOD. IN LANGER(ED),
      BOUNDARY PROBLEMS IN DIFFERENTIAL EQUATIONS, UNIV.OF WISCONSIN
      PRESS, MADISON, 1960.

521   COURANT R., AND HILBERT D., METHODS OF MATHEMATICAL PHYSICS, VOLS.I

AND II, INTERSCIENCE, NEW YORK, 1953-1962.

522 COURTNEY R.L., FINITE ELEMENT MODELING STUDIES IN THE NORMAL MODE
METHOD AND NORMAL MODE SYNTHESIS. N.A.S.A. TECHNICAL REPORT
TN-D-6326,1971.

523 COWPER G.R., FURSHL, A HIGH-PRECISION FINITE ELEMENT FOR SHELLS OF
ARBITRARY SHAPE. REPORT, NATIONAL RESEARCH COUNCIL OF
CANADA, 1972.

524 COWPER G.R., GAUSSIAN QUADRATURE FORMULAS FOR TRIANGLES. INT.J.NUMER.
METH.ENG.7, 405-408,1973.

525 COWPER G.R., KOSKO, E., LINDBERG G.M., AND OLSON M.D., FORMULATION OF
A NEW TRIANGULAR PLATE BENDING ELEMENT. CANADIAN AERO.SPACE
INST. TRANS, 1, 86-90, 1968.

526 COWPER G.R., KOSKO E., LINGBERG G.M., AND OLSON M.D., STATIC AND
DYNAMIC APPLICATIONS OF A HIGH-PRECISION TRIANGULAR PLATE
BENDING ELEMENT. AIAA, J.7, 1957-1965, 1969.

527 COWPER G.R., KOSKO E., LINDBERG G.M. AND OLSON M.D., A SHALLOW
SHELL FINITE ELEMENT OF TRIANGULAR SHAPE. INT.J.SOLIDS
STRUCTURES 6, 1133-1156,1970.

528 COWPER G.R., LINDBERG G.M., AND OLSON M.D., COMPARISON OF TWO
HIGH-PRECISION TRIANGULAR FINITE ELEMENTS FOR ARBITRARY
DEEP SHELLS. PROC 3RD CONF. MATRIX METHODS IN STRUCTURAL
MECHANICS, WRIGHT-PATTERSON AFB., OHIO, 1971.

529 CRAGGS A., COMPUTATION OF THE RESPONSE OF COUPLED PLATE ACOUSTIC
SYSTEMS USING PLATE FINITE ELEMENTS AND ACOUSTIC VOLUME-
DISPLACEMENT THEORY. J.SOUND AND VIBRATION 18, 235-246,
1971.

530 CRANDALL S.H., ENGINEERING ANALYSIS, MCGRAW HILL, NEW YORK, 1956.

531 CRASTAN V., GENERALIZATION OF THE ELEMENT METHOD. NUCL.ENG.DES.15, 113-
120, 1971.

532 CRISFIELD M.A., FINITE ELEMENT METHODS FOR THE ANALYSIS OF MULTICELLULAR
STRUCTURES. PROC. INST.CIV.ENG.48, 413-437, 1971.

533 CROLL J.C.A., THE TREATMENT OF NATURAL BOUNDARY CONDITIONS IN
THE FINITE ELEMENT AND FINITE DIFFERENCE METHODS. INT.
J.NUMER.METH.ENG.5, 443-445, 1973.

534 CROLL J.C.A., AND WALKER A.C., THE FINITE DIFFERENCE AND LOCALIZED
RITZ METHODS. INT.J.NUMER.METH.ENG.3, 155-160, 1971.

535 CRONK M., AN INTERACTIVE COMPUTER GRAPHICS PROGRAM FOR NASTRAN.
NASTRAN USERS EXPERIENCES. N.A.S.A. TECHNICAL REPORT TM-X-
2378, 1971.

536 CROUZEIX M. AND RAVIART P.-A., CONFORMING AND NONCONFORMING FINITE
ELEMENT METHODS FOR SOLVING THE STATIONARY STOKES
EQUATIONS, I. (MANUSCRIPT).

537 CROUZEIX M., AND THOMAS J.M., RESOLUTION NUMERIQUE PAR DES
METHODES D'ELEMENTS FINIS DE PROBLEMES ELLIPTIQUES
DEGENERES. COMPTES RENDUS DE L'ACAD.DES SCIENCES 275,
1115-1118, 1972.

538 CUBITT N.J. AND ROY S.K., APPLICATION OF THE FINITE ELEMENT METHOD TO
CASES REQUIRING THE COMBINATION OF ELEMENTS POSSESSING DIFFERENT
NUMBERS OF DEGREES OF FREEDOM. INT.CONF. VARIATIONAL METHODS
IN ENGINEERING, UNIVERSITY OF SOUTHAMPTON, 1972.

539 CULLEN M.J.P., SOLUTION OF WEATHER PREDICTION PROBLEMS BY
FINITE ELEMENT METHODS. PROC.CONF.COMP.METHODS IN
NONLINEAR MECH., AUSTIN, TEXAS, 1974.

540 CYR N.A., AND TETER R.D., FINITE ELEMENT ELASTIC PLASTIC CREEP ANALYSIS
OF TWO DIMENSIONAL CONTINUUM WITH TEMPERATURE DEPENDANT MATERIAL
PROPERTIES. NATIONAL SYMP. COMPUTERISED MATERIAL PROPERTIES,
COMPUTERISED STRUCTURAL ANALYSIS AND DESIGN, GEORGE
WASHINGTON UNIVERSITY, 1972.

541 CYR N.A., TETER R.D., AND STOCKS B.B., FINITE ELEMENT THERMOELASTIC
ANALYSIS. J.STRUCT DIV.A.S.C.E. 98, ST7, 1585-1603, 1972.

542 CZENDES Z.J., A FINITE ELEMENT METHOD FOR THE GENERAL SOLUTION OF
ORDINARY DIFFERENTIAL EQUATIONS. INT.J.NUMER.METH.ENG. (TO APPEAR),

543 CZENDES Z., GOPINATH A., AND SILVESTER P., GENERALISED MATRIX
INVERSE TECHNIQUES FOR LOCAL APPROXIMATIONS OF OPERATOR
EQUATIONS. PP.189-199 OF J.R. WHITEMAN (ED.), THE
MATHEMATICS OF FINITE ELEMENTS AND APPLICATIONS. ACADEMIC

PRESS, LONDON, 1973.

544   CZENDES Z.J., AND SILVESTER P.,   NUMERICAL SOLUTION OF DIELECTRIC
         LOADED WAVEGUIDES.   I-FINITE ELEMENT ANALYSIS.I.E.E.E.   TRANS.,
         MTT-18, 1124-1131, 1971.

545   DAILEY J.W. AND PIERCE J.G.,   ERROR BOUNDS FOR THE GALERKIN METHOD APPLIED
         TO SINGULAR AND NONSINGULAR BOUNDARY VALUE PROBLEMS,
         NUMER.MATH.19,266-282,1972.

546   DALY P.,   FINITE ELEMENT COUPLING MATRICES. ELECTRON LETT.5,613-615,
         1969.

547   DALY P.,   HYBRID-MODE ANALYSIS OF MICROSTRIP BY FINITE ELEMENT METHODS.
         I.E.E.E. TRANS., MTT-19, 19-25, 1971.

548   DALY P.,   FINITE ELEMENTS FOR FIELD PROBLEMS IN CYLINDRICAL CO-ORDINATES.
         INT.J.NUMER.METH.ENG.6, 169-178,1973.

549   DALY P.,   SINGULARITIES IN TRANSMISSION LINES.   PP. 337-350 OF
         J.R. WHITEMAN (ED.), THE MATHEMATICS OF FINITE ELEMENTS
         AND APPLICATIONS, ACADEMIC PRESS, LONDON, 1973.

550   DALY P.,   FINITE ELEMENT SOLUTIONS FOR AN EQUILATERAL TRIANGLE. INT.J.
         NUMER.METH.ENG.8, 495-501, 1974.

551   DALY P., AND HELPS J.D.,   EXACT FINITE ELEMENT SOLUTIONS TO THE
         HELMHOLTZ EQUATION.   INT.J.NUMER.METH.ENG. 6, 529-542,
         1973.

552   DANHOF R.H.,   SECTOR FINITE ELEMENTS IN THE THEORY OF PLANE ELASTICITY.
         DISSERTATION, UNIVERSITY OF ILLINOIS, URBANA, 1971.

553   DANIEL J.W.,   NEWTON'S METHOD FOR NONLINEAR INEQUALITIES. NUMER.
         MATH.21, 381-388,1973.

554   DANTZIG G.B.,   LINEAR PROGRAMMING AND EXTENSIONS. PRINCETON UNIVERSITY
         PRESS,1963.

555   DARIO N.P., AND BRADLEY W.A.,   A COMPARISON OF FIRST AND SECOND
         ORDER AXIALLY SYMMETRIC FINITE ELEMENTS. INT.J.NUMER.
         METH.ENG.5, 573-583, 1973.

556   DAVIS P.J.,   INTERPOLATION AND APPROXIMATION. BLAISDELL, NEW YORK,1963.

557   DAVID M.,   ANALYSIS OF PRESTRESSED CONCRETE PRESSURE VESSELS· FOR NUCLEAR
         REACTORS IN CZECHOSLOVAKIA. PP.203-219 OF C.E. KESLER (ED.),
         CONCRETE FOR NUCLEAR REACTORS, VOL.I, AMERICAN CONCRETE INSTITUTE,
         DETROIT, 1972.

558   DAVIDSON I.,   RESPONSE OF STRUCTURES TO SEISMIC EXCITATION. J. BRIT.NUCL.
         ENERGY SOC.12, 257-260, 1973.

559   DAWE D.J.,   ON ASSUMED DISPLACEMENTS FOR THE RECTANGULAR PLATE
         BENDING ELEMENT.   J. ROYAL AERO. SOC.71, 722-724, 1967.

560   DAWE D.J.,   A FINITE-DEFLECTION ANALYSIS OF SHALLOW ARCHES BY THE
         DISCRETE ELEMENT METHOD.   INT. J. NUMER.METH ENG.3,
         529-552,1971.

561   DAWE D.J.,   CURVED FINITE ELEMENTS IN THE ANALYSIS OF SHELL STRUCTURES BY
         THE FINITE ELEMENT METHOD. PP.17-36, VOL.5,, PROCEEDINGS 1ST
         STRUCTURAL MECH. IN REACTOR TECH. CONF., BERLIN, 1971.

562   DEAK A.L., AND ATLURI S.,   NONLINEAR HYBRID STRESS FINITE ELEMENT
         ANALYSIS OF LAMINATED SHELLS. PROC.CONF.COMP.METHODS IN
         NONLINEAR MECH., AUSTIN, TEXAS, 1974.

563   DEAK A., AND PIAN T.H.H.,   APPLICATION OF THE SMOOTH-SURFACE
         INTERPOLATION TO THE FINITE ELEMENT METHOD. AIAA J.5,
         187-189, 1967.

564   DEARIEN J.A.,   STRAP A COMPUTER CODE FOR THE STATIC AND DYNAMIC ANALYSIS
         OF REACTOR STRUCTURES. PP.237-253, VOL.3 OF PROCEEDINGS 1ST
         STRUCTURAL MECH IN REACTOR TECH. CONF., BERLIN, 1971.

565   DE BOOR C.,   BICUBIC SPLINE INTERPOLATION. J.MATH.PHYS. 41, 212-218,
         1962.

566   DE BOOR C.,   ON UNIFORM APPROXIMATION BY SPLINES.   J. APPROX.THEORY 1,
         219-262, 1968.

567   DE BOOR C.,   THE METHOD OF PROJECTIONS AS APPLIED TO THE NUMERICAL
         SOLUTION OF TWO-POINT BOUNDARY VALUE PROBLEMS USING CUBIC
         SPLINES. PH.D.THESIS, UNIVERSITY OF MICHIGAN, 1966.

568   DE BOOR C.,   ON CALCULATING WITH B-SPLINES. J.APPROX.THEORY 6, 50-62, 1972.

569   DE BOOR C.,   PACKAGE FOR CALCULATING WITH B-SPLINES. TECHNICAL
         REPORT 1333, MATHEMATICS RESEARCH CENTER, UNIVERSITY OF
         WISCONSIN, MADISON, 1973.

570   DE BOOR C.,   ON BOUNDING SPLINE INTERPOLATION. TECHNICAL REPORT

1378, MATHEMATICS RESEARCH CENTER, UNIVERSITY OF WISCONSIN, MADISON, 1973.

571 DE BOOR C., GOOD APPROXIMATION BY SPLINES WITH VARIABLE KNOTS. II. PP.12-20 OF G.A. WATSON (ED.), PROCEEDINGS OF CONF. ON NUMERICAL SOLUTION OF DIFFERENTIAL EQUATIONS. LECTURE NOTES IN MATHEMATICS, NO. 363, SPRINGER-VERLAG, BERLIN, 1974.

572 DE BOOR C., AND FIX G., SPLINE APPROXIMATION BY QUASI-INTERPOLANTS J.APPROX.THEORY (TO APPEAR)

573 DE BOOR C., AND LYNCH R.E., ON SPLINES AND THEIR MINIMUM PROPERTIES. J. MATH.MECH.15, 953-969, 1966.

574 DE BOOR C., AND SWARTZ B., COLLOCATION AT GAUSSIAN POINTS. SIAM J.NUMER.ANAL. 10, 582-606, 1973.

575 DE DONATO O. AND FRANCHI A., A MODIFIED GRADIENT METHOD FOR FINITE ELEMENT ELASTOPLASTIC ANALYSIS BY QUADRATIC PROGRAMMING. COMP. METHODS IN APPL.MECH.ENG.2,107-132,1973.

576 DE DONATO O. AND MAIER G., FINITE ELEMENT ELASTOPLASTIC ANALYSIS BY QUADRATIC PROGRAMMING: THE MULTISTAGE METHOD. PAPER M2/8, PROC. 2ND STRUCT. MECH. IN REACTOR TECH. CONF., BERLIN, 1973.

577 DE MEERSMAN R., THE METHOD OF GARABEDIAN: SOME EXTENSIONS. NUMER.MATH.11, 257-263, 1968.

578 DELVES L.M., A COMPARISON OF LEAST SQUARES AND VARIATIONAL METHODS FOR THE SOLUTION OF DIFFERENTIAL EQUATIONS. INTERNAL REPORT CSS/73/2/1, DEPARTMENT OF COMPUTATIONAL AND STATISTICAL SCIENCE, UNIVERSITY OF LIVERPOOL,1973.

579 DELVES L.M., AND FREEMAN T.L., ON THE CONVERGENCE RATES OF VARIATIONAL METHODS: HOMOGENEOUS SYSTEMS. REPORT CSS/ 74/9/1, DEPARTMENT OF COMPUTATIONAL AND STATISTICAL SCIENCE, UNIVERSITY OF LIVERPOOL, 1974.

580 DENDY J.E., PENALTY GALERKIN METHODS FOR PARTIAL DIFFERENTIAL EQUATIONS. PH.D.THESIS, RICE UNIVERSITY, 1971.

581 DENDY J., PENALTY GALERKIN METHODS FOR PARTIAL DIFFERENTIAL EQUATIONS. SIAM J.NUMER.ANAL.11, 604-636, 1974.

582 DENDY J., TWO METHODS OF GALERKIN TYPE ACHIEVING OPTIMUM L2 RATES OF CONVERGENCE FOR FIRST ORDER HYPERBOLICS. SIAM J. NUMER. ANAL.11, 637-653, 1974.

583 DENDY J.E. AND FAIRWEATHER G. ALTERNATING-DIRECTION GALERKIN METHODS FOR PARABOLIC AND HYPERBOLIC PROBLEMS ON RECTANGULAR POLYGONS.

584 DENKE P.H., A MATRIX METHOD OF STRUCTURAL ANALYSIS. PP. 445-457, PROC.2ND U.S. NAT.CONGRESS ON APPL.MECH., A.S.M.E., 1954.

585 DEPPE L.O., AND HANSEN K.F., APPLICATION OF THE FINITE ELEMENT METHOD TO TWO-DIMENSIONAL DIFFUSION PROBLEMS. TRANS.AMERICAN NUCL.SOC. 16, 132-133, 1973.

586 DEPPE L.O., AND HANSEN K.F., FINITE ELEMENT METHOD APPLIED TO NEUTRON DIFFUSION PROBLEMS. REPORT COO-2262-1, DEPARTMENT OF NUCLEAR ENGINEERING, MASSACHUSETTS INSTITUTE OF TECHNOLOGY, 1973.

587 DEPREY J., SHELL THEORY IN CARTESIAN COORDINATES: FINITE ELEMENT METHOD. ACADEMIE ROYALE DE BELGIQUE, BULL.DE LA CLASSES DES SCIENCES 56, 1155-1164, 1970.

588 DESAI C.S., NONLINEAR ANALYSIS USING SPLINE FUNCTIONS. J.SOIL MECH. FOUNDATIONS DIV. A.S.C.E. 97, SM10, 1461-1480, 1971.

589 DESAI C.S. (ED.), PROCEEDINGS OF THE SYMPOSIUM ON APPLICATION OF THE FINITE ELEMENT METHOD IN GEOTECHNICAL ENGINEERING. ARMY ENGINEER WATERWAYS EXPERIMENT STATION, VICKSBURG, MISSISSIPPI, 1972.

590 DESAI C.S., A CONSISTENT FINITE ELEMENT TECHNIQUE FOR WORK-SOFTENING BEHAVIOUR. PROC. CONF.COMP.METHODS IN NONLINEAR MECHANICS, AUSTIN, TEXAS, 1974.

591 DESAI C.S., AND ABEL J.F., INTRODUCTION TO THE FINITE ELEMENT METHOD. VAN NOSTRAND, NEW YORK, 1972.

592 DESAI C.S. AND JOHNSON L.D., EVALUATION OF SOME NUMERICAL SCHEMES FOR CONSOLIDATION. INT.J.NUMER.METH.ENG.7, 243-254,1973.

593 DESCLOUX J., ON THE NUMERICAL INTEGRATION OF THE HEAT EQUATION. NUMER.MATH.15, 371-381, 1970.

594 DESCLOUX J., ON FINITE ELEMENT MATRICES. SIAM J. NUMER. ANAL.9, 260-265, 1972.

595 DESCLOUX J., INTERIOR REGULARITY FOR GALERKIN FINITE ELEMENT

APPROXIMATIONS OF ELLIPTIC PARTIAL DIFFERENTIAL EQUATIONS.
(TO APPEAR).

596 DE VRIES G., BERARD G. AND NORRIE D.H., APPLICATION OF THE FINITE ELEMENT
TECHNIQUE TO COMPRESSIBLE FLOW PROBLEMS. MECH.ENG. REPORT,18,
DEPARTMENT OF MECHANICAL ENGINEERING, UNIVERSITY OF CALGARY,1974.

597 DE VRIES G., AND NORRIE D.H., APPLICATION OF THE FINITE
ELEMENT TECHNIQUE TO POTENTIAL FLOW PROBLEMS, 1 AND 2.
MECH. REPORTS 7 AND 8, DEPARTMENT OF MECHANICAL ENGINEERING,
UNIVERSITY OF CALGARY, 1969.

598 DE VRIES G., AND NORRIE D.H., THE APPLICATION OF THE FINITE ELEMENT
TECHNIQUE TO POTENTIAL FLOW PROBLEMS. TRANS.APPLIED MECH.DIV.
ASME 71, APM 22, 798-802, 1971.

599 DE WINDT P., AND REYNEN J., EURCYL, A COMPUTER PROGRAM TO GENERATE
FINITE FINITE ELEMENT MESHES FOR CYLINDER-CYLINDER
INTERSECTIONS. REPORT 5030, JOINT NUCLEAR RESEARCH CENTER,
EUROPEAN ATOMIC ENERGY COMMUNITY, ISPRA, 1973.

600 DHATT G., NUMERICAL ANALYSIS OF THIN SHELLS BY CURVED TRIANGULAR
ELEMENTS BASED ON DISCRETE-KIRCHOFF HYPOTHESIS. PROC. SYMP.
ON APPLICATION OF FINITE ELEMENT METHODS IN CIVIL ENGINEERING,
VANDERBILT UNIVERSITY, 1969.

601 DHATT G.S. AND VENKATASUBBU S., FINITE ELEMENT ANALYSIS OF
CONTAINMENT SHELLS. PAPER J3/6, PROCEEDINGS 1ST
STRUCTURAL MECH. IN REACTOR TECH. CONF., BERLIN, 1971.

602 DIAZ J.B. AND WEINSTEIN A., SCHWARZ' INEQUALITY AND THE METHODS OF
RAYLEIGH-RITZ AND TREFFTZ. J.MATH.PHYS. 26, 133-136, 1947.

603 DI CARLO A., AND PIVA R., FINITE ELEMENT SIMULATION OF THERMALLY
INDUCED FLOW FIELDS. PROC. CONF.COMP. METHODS IN NONLINEAR
MECH., AUSTIN, TEXAS, 1974.

604 DICKEY R.W. (ED.), NONLINEAR ELASTICITY. ACADEMIC PRESS, NEW YORK,
1973.

605 DICKINSON S.M. AND HENSHELL R.D., CLOUGH-TOCHER TRIANGULAR PLATE
BENDING ELEMENT IN VIBRATION. AIAA J. 7, 560-561, 1969.

606 DI GUGLIELMO F., CONSTRUCTION D'APPROXIMATIONS DES ESPACES DE
SOBOLEV SUR DES RESEAUX EN SIMPLEXES. ESTRATTO DA CALCOLE,
13, 51-77, 1969.

607 DI GUGLIELMO F., CONSTRUCTION D'APPROXIMATIONS DES ESPACES DE
SOBOLEV SUR DES RESEAUX EN SIMPLEXES. CALCOLO 6, 279-311, 1969.

608 DILLON E.C., AND O'BRIEN M., SOLUTION BOUNDS IN SOME STRESS PROBLEMS
BY THE HYPERCIRCLE AND FINITE ELEMENT METHODS. INT.CONF.
VARIATIONAL METHODS IN ENG., SOUTHAMPTON UNIVERSITY, 1972.

609 DISTEFANO N., NONLINEAR PROCESSES IN ENGINEERING. ACADEMIC PRESS, NEW
YORK, 1974.

610 DIXON L.C.W., NONLINEAR OPTIMISATION. ENGLISH UNIVERSITIES PRESS,
LONDON, 1972.

611 D'JAKONOV, E.G., ON SOME OPERATOR INEQUALITIES AND THEIR APPLICATIONS.
SOVIET MATH.DOKL.12,921-925,1971.

612 DJUKIC D., AND VUJANOVIC B., ON A NEW VARIATIONAL PRINCIPLE OF
HAMILTONIAN TYPE FOR CLASSICAL FIELD THEORY. ZAMM 51,
611-616, 1971.

613 DLUGACH M.I. AND KOVALCHUK N.V., AN APPLICATION OF THE FINITE
ELEMENT METHOD TO THE CALCULATION OF CYLINDRICAL SHELLS
WITH RECTANGULAR OPENINGS. (RUSSIAN). PRIKLADY MEKH. 9,
35-41, 1974.

614 DOCTORS L.J., AN APPLICATION OF THE FINITE ELEMENT TECHNIQUE TO
BOUNDARY VALUE PROBLEMS OF POTENTIAL FLOW. INT.J.NUMER.
METH.ENG.2, 243-252, 1970.

615 DOKAINISH M.A., AND RAWTANI S., VIBRATION ANALYSIS OF ROTATING
CANTILEVER PLATES. INT.J.NUMER.METH.ENG.3, 233-248, 1971.

616 DONEA J., FINITE ELEMENT ANALYSIS OF RADIATION DAMAGE STRESSES IN
THE GRAPHITE OF MATRIX FUEL ELEMENTS. REPORT EUR-4472E,
JOINT NUCLEAR RESEARCH CENTRE, EUROPEAN ATOMIC ENERGY
COMMISSION, ISPRA, ITALY, 1970.

617 DONEA J., AND GIULIANI S., CODE TAFEST NUMERICAL SOLUTION TO
TRANSIENT HEAT CONDUCTION PROBLEMS USING FINITE ELEMENTS
IN SPACE AND TIME. REPORT EUR-5049, JOINT NUCLEAR
RESEARCH CENTRE, EUROPEAN ATOMIC ENERGY COMMISSION, ISPRA,

ITALY, 1974.

618 DONEA J., GIULIANI S., AND PHILIPPE A., FINITE ELEMENTS IN THE
          SOLUTION OF ELECTROMAGNETIC INDUCTION PROBLEMS. INT.J.
          NUMER.METH.ENG.8, 359-367, 1974.

619 DOUGLAS A., FINITE ELEMENTS FOR GEOLOGICAL MODELLING. NATURE
          226, 1970.

620 DOUGLAS J., A SUPERCONVERGENCE RESULT FOR THE APPROXIMATE
          SOLUTION OF THE HEAT EQUATION BY A COLLOCATION METHOD.
          PP.475-490 OF A.K. AZIZ (ED.), THE MATHEMATICAL FOUNDATIONS
          OF THE FINITE ELEMENT METHOD WITH APPLICATIONS TO PARTIAL
          DIFFERENTIAL EQUATIONS. ACADEMIC PRESS, NEW YORK, 1972.

621 DOUGLAS J., AND DUPONT T., GALERKIN METHODS FOR PARABOLIC EQUATIONS.
          SIAM J. NUMER. ANAL. 7, 575-626, 1970.

622 DOUGLAS J., AND DUPONT T., THE NUMERICAL SOLUTION OF WATERFLOODING
          PROBLEMS IN ENGINEERING. STUDIES IN NUMERICAL
          ANALYSIS 2, 53 - 63, SIAM, PHILADELPHIA, 1970.

623 DOUGLAS J., AND DUPONT T., ALTERNATING-DIRECTION GALERKIN METHODS ON
          RECTANGLES. IN HUBBARD (ED), NUMERICAL SOLUTION OF PARTIAL
          DIFFERENTIAL EQUATIONS - II, SYNSPADE 1970, ACADEMIC PRESS,
          NEW YORK, 1971.

624 DOUGLAS J., AND DUPONT T., GALERKIN METHODS FOR PARABOLIC
          EQUATIONS. SIAM J. NUMER.ANAL.7, 575-626, 1971.

625 DOUGLAS J., AND DUPONT T., A FINITE ELEMENT COLLOCATION METHOD
          FOR THE HEAT EQUATION. SYMPOSIA MATHEMATICA 10,
          MONOGRAF, BOLOGNA, 1972.

626 DOUGLAS J. AND DUPONT T., A FINITE ELEMENT COLLOCATION METHOD FOR
          QUASILINEAR PARABOLIC EQUATIONS. MATH. COMP.27, 17-28, 1973

627 DOUGLAS J. AND DUPONT T., GALERKIN METHODS FOR PARABOLIC EQUATIONS
          WITH NONLINEAR BOUNDARY CONDITIONS. NUMER.MATH.20, 213-237,1973.

628 DOUGLAS J., AND DUPONT T., A FINITE ELEMENT COLLOCATION METHOD FOR
          QUASILINEAR PARABOLIC EQUATIONS. MATH.COMP.27, 17-28,
          1973.

629 DOUGLAS J. AND DUPONT T., SOME SUPERCONVERGENCE RESULTS FOR
          GALERKIN METHODS FOR THE APPROXIMATE SOLUTION OF TWO-POINT
          BOUNDARY VALUE PROBLEMS. PP.89-92 OF J.J.H. MILLER (ED.),
          TOPICS IN NUMERICAL ANALYSIS. ACADEMIC PRESS, LONDON, 1973.

630 DOUGLAS J., AND DUPONT T., COLLOCATION METHODS FOR PARABOLIC
          EQUATIONS IN A SINGLE SPACE VARIABLE. LECTURE NOTES IN
          MATHEMATICS NO. 385, SPRINGER-VERLAG, BERLIN, 1974.

631 DOUGLAS J., AND DUPONT T., GALERKIN APPROXIMATIONS FOR THE TWO
          POINT BOUNDARY PROBLEM USING CONTINUOUS PIECEWISE
          POLYNOMIAL SPACES. NUMER.MATH.22, 99-110, 1974.

632 DOUGLAS J., DUPONT T. AND HENDERSON G.E., SIMULATION OF GAS WELL
          PERFORMANCE BY VARIATIONAL METHODS. SPE 2891, 1971.

633 DOUGLAS J., DUPONT T. AND RACHFORD H.H., THE APPLICATION OF VARIATIONAL
          METHODS TO WATERFLOODING PROBLEMS. J. CANADIAN PETROLEUM
          TECHNOLOGY 8, 79 - 85, 1969.

634 DOUGLAS J., DUPONT T., AND WAHLBIN L., OPTIMAL LINFINITY ERROR
          ESTIMATES FOR GALERKIN APPROXIMATIONS TO SOLUTIONS OF TWO
          POINT BOUNDARY VALUE PROBLEMS. (TO APPEAR).

635 DOUGLAS J., DUPONT T., AND WHEELER MARY F., SOME SUPERCONVERGENCE
          RESULTS FOR AN H1-GALERKIN PROCEDURE FOR THE HEAT EQUATION.
          TECHNICAL REPORT 1382, MATHEMATICS RESEARCH CENTER,
          UNIVERSITY OF WISCONSIN, MADISON, 1973.

636 DUNCAN J.M., FINITE ELEMENT ANALYSES OF PORT ALLEN LOCK. J.SOIL MECH.
          AND FOUNDATIONS DIV. A.S.C.E. 97, SM8, 1053-1068, 1971.

637 DUNGAR R., VIBRATIONS OF PLATE AND SHELL STRUCTURES USING
          TRIANGULAR FINITE ELEMENTS. J. STRAIN ANALYSIS 2,
          73-83, 1967.

638 DUNGAR R. AND SEVERN R.T., TRIANGULAR FINITE-ELEMENTS OF
          VARIABLE THICKNESS AND THEIR APPLICATION TO PLATE AND
          SHELL PROBLEMS. J.STRAIN ANALYSIS 4, 10-21, 1971.

639 DUNHAM R.S., AND BECKER E.B., ORGANIZATION AND FUNCTIONAL
          PURPOSE OF A FINITE ELEMENT COMPUTER PROGRAM. PRESSURE
          VESSELS AND PIPING CONF., NUCLEAR AND MATERIALS DIV.
          A.S.M.E., MIAMI, 1974.

640   DUNHAM R.S. AND PISTER K.S.,   A FINITE ELEMENT APPLICATION OF THE
           HELLINGER-REISSNER VARIATIONAL THEOREM. PROC. 2ND CONF.
           MATRIX METHODS IN STRUCTURAL MECHANICS, WRIGHT-PATTERSON
           AFB, OHIO, AFFDL-TR-68-150, 1968.
641   DUNNE P.C.,   COMPLETE POLYNOMIAL DISPLACEMENT FIELDS FOR THE FINITE
           ELEMENT METHOD.   J. ROY. AERONAUT. SOC. 72, 245 - 246, 1968.
642   DUPONT T.,   A FACTORIZATION PROCEDURE FOR THE SOLUTION OF ELLIPTIC
           DIFFERENCE EQUATIONS. SIAM J.NUMER.ANAL.5, 559-574, 1968.
643   DUPONT T.,   ADI METHODS FOR FINITE ELEMENT SYSTEMS. IN B.E.
           HUBBARD (ED.), NUMERICAL SOLUTION OF PARTIAL DIFFERENTIAL
           EQUATIONS - II, SYNSPADE 1970. ACADEMIC PRESS, NEW YORK,
           1971.
644   DUPONT T.,   SOME L2 ERROR ESTIMATES FOR PARABOLIC GALERKIN
           METHODS. PP. 491-504 OF A.K. AZIZ (ED.),   THE MATHEMATICAL
           FOUNDATIONS OF THE FINITE ELEMENT METHOD WITH APPLICATIONS
           TO PARTIAL DIFFERENTIAL EQUATIONS. ACADEMIC PRESS, NEW
           YORK, 1972.
645   DUPONT T.,   L2-ESTIMATES FOR GALERKIN METHODS FOR SECOND ORDER
           HYPERBOLIC EQUATIONS. SIAM J. NUMER.ANAL.10, 880-889,1973.
646   DUPONT T.,   GALERKIN METHODS FOR FIRST ORDER HYPERBOLICS: AN EXAMPLE.
           SIAM J. NUMER.ANAL.10, 890-899,1973.
647   DUPONT T., FAIRWEATHER G., AND JOHNSON J.P.,   THREE LEVEL GALERKIN METHODS
           FOR PARABOLIC EQUATIONS. SIAM J. NUMER.ANAL.11, 392-410, 1974.
648   DUPONT T., AND WAHLBIN L.,   L2 OPTIMALITY AND WEIGHTED H1
           PROJECTIONS INTO PIECEWISE POLYNOMIAL SPACES. (TO APPEAR).
649   DUPUIS G., AND GOEL J.J.,   A CURVED ELEMENT FOR THIN ELASTIC
           SHELLS. TECHNICAL REPORT, BROION UNIVERSITY, 1969.
650   DUPUIS G., AND GOEL J.J.,   FINITE ELEMENT WITH HIGH DEGREE OF
           REGULARITY. INT.J.NUMER.METHODS IN ENG.2, 563-577, 1970.
651   DUPUIS G., AND GOEL J.J.,   ELEMENTS FINIS RAFFINES EN ELASTICITE
           BIDIMENSIONNELLE. ZAMP 20, 858- 881, 1969.
652   DUPUIS G.A., PFAFFINGER D.D., AND MARCAL P.V.,   EFFECTIVE USE OF THE
           INCREMENTAL STIFFNESS MATRICES IN NONLINEAR GEOMETRIC ANALYSIS.
           PROC.SYMP.HIGH SPEED COMPUTING OF ELASTIC STRUCTURES, UNIVERSITY OF
           LIEGE, 1971.
653   DUVAUT G. AND LIONS J.L.,   LES INEQUATIONS EN MECANIQUE ET EN
           PHYSIQUE. DUNOD, PARIS, 1972.
654   DZISKARIANI A.V.,   ON THE RATE OF CONVERGENCE OF THE BUBNOV-
           GALERKIN METHOD. ZH.VYCH.MAT.4, 343-348,1964.
655   EAGLE J. AND BECKETT V.,   ANALYSIS OF LOADING AND THERMAL EFFECTS
           ON FUELLED GRAPHITE BRICKS FOR A HIGH-TEMPERATURE GAS
           COOLED REACTOR. PAPER D2/4, PROC.1ST STRUCTURAL MECH. IN
           REACTOR TECH. CONF., BERLIN, 1973.
656   ERNER A.M., AND UCCIFERRO J.J.,   A THEORETICAL AND NUMERICAL COMPARISON
           OF ELASTIC NONLINEAR FINITE ELEMENT METHODS. NATIONAL SYMP. ON
           COMPUTERISED STRUCT.ANAL. AND DESIGN. GEORGE WASHINGTON
           UNIVERSITY, 1972.
657   ECKHARDT U.,   QUADRATIC PROGRAMMING FOR SUCCESSIVE OVERRELAXATION.
           REPORT JUL-1064-MA, ZENTRAL INSTITUT FUR ANGEWANDTE
           MATHEMATIK, K.F.A., JULICH, 1974.
658   EGEBERG J.L.,   MESHGEN, A COMPUTER CODE FOR AUTOMATIC FINITE
           ELEMENT MESH GENERATION. REPORT SCL-DR-69-49, SANDIA
           LABORATORIES, ALBUQUERQUE, 1969.
659   EGELAND O., AND ARALDSEN PER O.,   SESAM-69, A GENERAL PURPOSE
           FINITE ELEMENT METHOD PROGRAM. COMPUTERS AND STRUCTURES
           4, 41-68, 1974.
660   EIDSON H.D. AND SCHUMAKER L.L.,   COMPUTATION OF G-SPLINES VIA A
           FACTORIZATION METHOD. TECH.RPT.60, CENTER FOR NUMERICAL
           ANALYSIS, UNIVERSITY OF TEXAS, AUSTIN, TEXAS, 1973.
661   EISENBERG M.A. AND MALVERN L.E.,   ON FINITE ELEMENT INTEGRATION IN
           NATURAL CO-ORDINATES. INT.J.NUMER.METH.ENG.7, 574-575, 1973.
662   EISENSTAT S.C., AND SCHULTZ M.H.,   COMPUTATIONAL ASPECTS OF THE
           FINITE ELEMENT METHOD. PP. 505-524 OF A.K. AZIZ (ED.),
           THE MATHEMATICAL FOUNDATIONS OF THE FINITE ELEMENT METHOD
           WITH APPLICATIONS TO PARTIAL DIFFERENTIAL EQUATIONS.
           ACADEMIC PRESS, NEW YORK, 1972.

663 EISENSTAT S.C. AND SCHULTZ M.H.,   COMPUTATIONAL ASPECTS OF THE FINITE
           ELEMENT METHOD. TECH.REPT.72-1,DEPT.OFCOMPUTER SCIENCE,YALE
           UNIVERSITY,CONNECTICUT, 1972.
664 ELIAS Z.M.,   DUALITY IN FINITE ELEMENT METHODS. J. ENG. MECH. DIV.
           A.S.C.E. 94, FM4, 931-946, 1968.
665 ELIAS Z.M.,   DYNAMIC ANALYSIS OF FRAME STRUCTURES BY THE FORCE METHOD.
           PP.275-297 OF J.T. ODEN, R.W. CLOUGH, AND Y.YAMOMOTO (EDS.),
           ADVANCES IN COMPUTATIONAL METHODS IN STRUCTURAL MECHANICS AND
           DESIGN. UNIVERSITY OF ALABAMA PRESS, HUNTSVILLE, 1972.
666 ELIAS Z.M.,   MIXED FINITE ELEMENT METHOD FOR AXISYMMETRIC SHELLS.
           INT.J.NUMER.METH.ENG.4, 261-277, 1972.
667 ELIAS Z.M.,   MIXED VARIATIONAL PRINCIPLES FOR SHELLS. INT. CONF. VARIATIONAL
           METHODS IN ENG., SOUTHAMPTON UNIVERSITY, 1972.
668 ELLINGTON J.P.,   A SIMPLE TRIANGULAR MEMBRANE FINITE ELEMENT
           REPORT TRG-2504, U.K.A.E.A. REACTOR GROUP, RISLEY, 1974.
669 ELTER C.,   STRESS ANALYSIS OF BWR PRESSURE VESSEL WITH BERSAFE
           AND FLHE. PP.221-238, PROC. 2ND INT.CONF.PRESSURE VESSEL
           TECHNOLOGY, A.S.M.E., NEW YORK, 1973.
670 ELTER C., WALDNER K., AND REYMEN J.,   STRESS ANALYSIS OF BWR PRESSURE
           VESSEL WITH BERSAFE AND FLHE. PP.221-238, PROC.2ND INT.CONF.
           PRESSURE VESSEL TECHNOLOGY, AMERICAN SOC.MECH.ENG., NEW YORK,
           1973.
671 EMERY A.F.,   THE USE OF SINGULARITY PROGRAMMING IN FINITE-DIFFERENCE AND
           FINITE-ELEMENT COMPUTATIONS OF TEMPERATURE. TRANS.ASME,J.HEAT
           TRANSFER   344-351,1973.
672 EMERY A.F., AND CARSON W.W.,   AN EVALUATION OF THE USE OF THE FINITE
           ELEMENT METHOD IN THE COMPUTATION OF TEMPERATURE.
           TRANS ASME., J.HEAT TRANSFER, 136-145, 1971.
673 EMERY A.F., AND SEGEDIN C.M.,   A NUMERICAL TECHNIQUE FOR
           DETERMINING THE EFFECT OF SINGULARITIFS IN FINITE
           DIFFERENCE SOLUTIONS ILLUSTRATED BY APPLICATION TO
           PLANE ELASTIC PROBLEMS. INT.J.NUMER.METH.ENG.6, 367-380,
           1973.
674 ENGELS H.,   ZUR ANWENDUNG KUBISCHER SPLINES AUF DIE RICHARDSON -
           EXTRAPOLATION. J.DEUTCH.MATH.VEREIN 74, 66-83, 1972.
675 ENGELS H.,   UBER ALLGEMEINE GAUSSCHE QUADRATUREN.
           COMPUTING 10, 83-95,1972.
676 ENGELS H.,   NUMERICAL QUADRATURE AND CUBATURE. VOLUME IN SERIES
           ON COMPUTATIONAL MATHEMATICS AND APPLICATIONS. ACADEMIC
           PRESS, LONDON, (TO APPEAR).
677 ENGELS H., AND ECKHARDT U.,   SYMMETRISCHE WILF'SCHE
           QUADRATURFORMELN. REPORT JUL-1094-MA, ZENTRALINSTITUT
           FUR ANGEWANDTE MATHEMATIK, KFA, JULICH, 1974.
678 ENGLISH W.J.,   A COMPUTER-IMPLEMENTED VECTOR VARIATIONAL SOLUTION
           OF LOADED RECTANGULAR WAVEGUIDES. SIAM J.APPL.MATH.
           21, 461-468, 1971.
679 EPSTEIN B.,   PARTIAL DIFFERENTIAL EQUATIONS. MCGRAW HILL,NEW YORK,1962.
680 EPSTEIN M.P., AND HAMMING R.W.,   NONINTERPOLATORY QUADRATURE
           FORMULAS. SIAM J.NUMER.ANAL.9, 464-475, 1972.
681 ERGATOUDIS J., IRONS B.M., AND ZIENKIEWICZ O.C.,   CURVED ISOPARAMETRIC
           QUADRILATERAL ELEMENTS IN FINITE ELEMENT ANALYSIS. INT. J.
           SOLIDS AND STRUCTURES 4, 31-42, 1968.
682 ERGATOUDIS J., IRONS B., AND ZIENKIEWICZ O.C.,   THREE-DIMENSIONAL
           ANALYSIS OF ARCH DAMS AND THEIR FOUNDATIONS. PROC.SYMP.ON
           ARCH DAMS. INST.CIV.ENG., LONDON, 1968.
683 EVANS D.J.,   THE ANALYSIS AND APPLICATION OF SPARSE MATRIX
           ALGORITHMS IN THE FINITE ELEMENT METHOD. PP. 427-447 OF
           J.R. WHITEMAN (ED.), THE MATHEMATICS OF FINITE ELEMENTS
           AND APPLICATIONS. ACADEMIC PRESS, LONDON, 1973.
684 EVANS D.J. (ED.),   SOFTWARE FOR NUMERICAL MATHEMATICS. ACADEMIC PRESS,
           LONDON, 1974.
685 EWING D.J.F., FAWKES A.J., AND GRIFFITHS J.,   RULES GOVERNING THE
           NUMBER OF NODES AND ELEMENTS IN A FINITE ELEMENT MESH.
           INT. J.NUMER.METH.ENG.2, 597-601, 1970.
686 FACCIOLI E., AND VITIELLO E.,   A FINITE ELEMENT LINEAR
           PROGRAMMING METHOD FOR THE LIMIT ANALYSIS OF THIN PLATES.

INT.J.NUMER.METH.ENG.5, 311-325, 1973.

687   FAIRWEATHER G.,   A SURVEY OF DISCRETE GALERKIN METHODS FOR
          PARABOLIC EQUATIONS IN ONE SPACE VARIABLE.
          MATH.COLLOQ.U.C.T. 7, 1971-72.

688   FAIRWEATHER G.,   GALERKIN METHODS FOR VIBRATION PROBLEMS IN TWO SPACE
          VARIABLES. SIAM J. NUMER. ANAL.4, 702-714, 1972.

689   FAIRWEATHER G. AND JOHNSON J.P.,   RICHARDSON EXTRAPOLATION FOR
          PARABOLIC GALERKIN METHODS.   PP.767-768 OF K.AZIZ (ED.),
          THE MATHEMATICAL FOUNDATIONS OF THE FINITE ELEMENT METHOD
          WITH APPLICATIONS TO PARTIAL DIFFERENTIAL EQUATIONS.
          ACADEMIC PRESS, NEW YORK, 1972.

690   FARHOOMAND I.,   NONLINEAR DYNAMIC STRESS ANALYSIS OF TWO-
          DIMENSIONAL SOLIDS.   PH.D.THESIS, UNIVERSITY OF CALIFORNIA,
          BERKELEY, 1970.

691   FARRELL J.J. AND DAI P.K.,   EFFECTS OF TIME AND DAMPING ON FINITE
          ELEMENT ANALYSIS OF RESPONSE OF STRUCTURES. PROC. 3RD CONF.
          MATRIX METHODS IN STRUCTURAL MECHANICS, WRIGHT-PATTERSON
          AFB., OHIO, 1971.

692   FELIPPA A.C.,   REFINED FINITE ELEMENT ANALYSIS OF LINEAR AND NONLINEAR
          TWO-DIMENSIONAL STRUCTURES.   SESM REPORT NO.66-22, UNIVERSITY
          OF CALIFORNIA, BERKELEY, 1967.

693   FELIPPA C.A.,   AN ALPHANUMERIC FINITE ELEMENT MESH PLOTTER.   INT.
          J.NUMER.METH.ENG.5, 217-236, 1972.

694   FELIPPA C.A.,   BASIS FOR FORMULATION OF FINITE ELEMENT MODELS.   PP. 127-132
          OF J.T. ODEN, R.W. CLOUGH, AND Y. YAMOMOTO (EDS.),   ADVANCES IN
          COMPUTATIONAL METHODS IN STRUCTURAL MECHANICS AND DESIGN.
          UNIVERSITY OF ALABAMA PRESS, HUNTSVILLE, 1972.

695   FELIPPA C.A.,   FINITE ELEMENT ANALYSIS OF THREE-DIMENSIONAL CABLE
          STRUCTURES.   PROC.CONF.COMP.METHODS IN NONLINEAR MECH.,
          AUSTIN, TEXAS, 1974.

696   FELIPPA C.A. AND CLOUGH R.W.,   THE FINITE ELEMENT METHOD IN SOLID
          MECHANICS. PP.210-252 OF G. BIRKHOFF AND R.S. VARGA (EDS.),
          NUMERICAL SOLUTION OF FIELD PROBLEMS IN CONTINUUM
          MECHANICS. SIAM-AMS PROCEEDINGS 2, AMERICAN MATHEMATICAL
          SOCIETY, PROVIDENCE, 1969.

697   FENVES S.J., PERRONE N., ROBINSON A.R.,   AND SCHNOBRICH W.C.,
          NUMERICAL AND COMPUTER METHODS IN STRUCTURAL MECHANICS.
          ACADEMIC PRESS, NEW YORK, 1973.

698   FIALA J.,   AN ALGORITHM FOR HERMITE-BIRKHOFF INTERPOLATION.
          APLIKACE MATEMATIKY 18, 167-175, 1973.

699   FINLAYSON B.A.,   CONVERGENCE OF THE GALERKIN METHOD FOR NONLINEAR
          PROBLEMS INVOLVING CHEMICAL REACTION.   SIAM J.NUMER.ANAL.8,
          316-325,1971.

700   FINLAYSON B.A., AND SCRIVEN L.E.,   THE METHOD OF WEIGHTED RESIDUALS AND
          ITS RELATION TO CERTAIN VARIATIONAL PRINCIPLES FOR THE ANALYSIS
          OF TRANSPORT PROCESSES.   CHEM.ENG.SCI.20, 395-404, 1965.

701   FINLAYSON B.A., AND SCRIVEN L.E.,   THE METHOD OF WEIGHTED RESIDUALS-
          A REVIEW.   APPLIED. MECH. REVIEWS 19, 735-748, 1966.

702   FINLAYSON B.A., AND SCRIVEN L.E.,   ON THE SEARCH FOR VARIATIONAL
          PRINCIPLES.   INT.J.HEAT MASS TRANSFER 10, 799-821, 1967.

703   FINN W.D.L.,   FINITE ELEMENT ANALYSIS OF SEEPAGE THROUGH DAMS.   J.SOIL.
          MECH. AND FOUNDATIONS DIV. A.S.C.E.93, SM6, PROC.PAPER 5552,
          41-48, 1967.

704   FIX G.,   ORDERS OF CONVERGENCE OF THE RAYLEIGH-RITZ AND WEINSTEIN-BAZLEY
          METHODS.   PROC.NAT.ACAD.SCI., U.S.A. 61,1219-1223, 1968.

705   FIX G.,   BOUNDS AND APPROXIMATIONS FOR EIGENVALUES OF SELF-ADJOINT
          BOUNDARY VALUE PROBLEMS.   TECHNICAL REPORT, HARVARD UNIVERSITY,
          MATH.DEPT. JAN.1968.

706   FIX G.,   HIGHER-ORDER RAYLEIGH-RITZ APPROXIMATIONS. J.MATH.MECH.18,
          645-657, 1969.

707   FIX G.J.,   EFFECTS OF QUADRATURE ERRORS IN FINITE ELEMENT
          APPROXIMATION OF STEADY STATE, EIGENVALUE, AND PARABOLIC
          PROBLEMS.   PP. 525-556 OF A.K. AZIZ (ED.),   THE MATHEMATICAL
          FOUNDATIONS OF THE FINITE ELEMENT METHOD WITH
          APPLICATIONS TO PARTIAL DIFFERENTIAL EQUATIONS.   ACADEMIC
          PRESS, NEW YORK, 1972.

708   FIX G., ON THE EFFECTS OF QUADRATURE ERRORS IN THE FINITE ELEMENT
       METHOD. PP. 55-68 OF J.T. ODEN, R.W. CLOUGH, AND Y. YAMOMOTO
       (EDS.) ADVANCES IN COMPUTATIONAL METHODS IN STRUCTURAL MECHANICS
       AND DESIGN. UNIVERSITY OF ALABAMA PRESS, HUNTSVILLE, 1972.

709   FIX G.J., FINITE ELEMENT METHOD FOR NONLINEAR HYPERBOLIC
       EQUATIONS. PROC.CONF.COMP.METHODS IN NONLINEAR MECH.,
       AUSTIN, TEXAS, 1974.

710   FIX G.J., GULATI S. AND WAKOFF G.I., ON THE USE OF SINGULAR FUNCTIONS
       WITH FINITE ELEMENT APPROXIMATION. J.COMP.PHYS.13, 209-228,1973.

711   FIX G.J., AND LARSEN K., ON THE CONVERGENCE OF SOR ITERATIONS FOR
       FINITE ELEMENT APPROXIMATIONS TO ELLIPTIC BOUNDARY VALUE
       PROBLEMS. SIAM J.NUMER.ANAL.8, 536-547, 1971.

712   FIX G. AND NASSIF N., ON FINITE ELEMENT APPROXIMATIONS TO TIME
       DEPENDENT PROBLEMS. NUMER.MATH.19, 127-135, 1972.

713   FIX G., AND STRANG G., FOURIER ANALYSIS OF THE FINITE ELEMENT
       METHOD IN RITZ-GALERKIN THEORY. STUDIES IN APPLIED MATH.,
       48, 265-273, 1969.

714   FIX G., AND WAKOFF G.I., EIGENVALUES OF RECTANGULAR POLYGONAL
       MEMBRANES. APENDIX C OF BIRKHOFF'S PAPER IN SCHOENBERG
       (ED.), APPROXIMATIONS WITH SPECIAL EMPHASIS ON SPLINE
       FUNCTIONS. ACADEMIC PRESS, NEW YORK, 1969.

715   FLEMING W., FUNCTIONS OF SEVERAL VARIABLES. ADDISON-WESLEY,
       MASSACHUSETTS, 1965.

716   FLETCHER R. (ED.), OPTIMIZATION. ACADEMIC PRESS, LONDON, 1969.

717   FOCHT A.S., SOME INEQUALITIES FOR THE SOLUTION AND ITS DERIVATIVES
       OF AN EQUATION OF ELLIPTIC TYPE IN L NORM (RUSSIAN).
       PROC.OF THE STEKLOV.INST.OF MATH., 77, 160-191, 1965.

718   FORSYTHE G., AND MOLER C.B., COMPUTER SOLUTION OF LINEAR ALGEBRAIC
       SYSTEMS. PRENTICE HALL, NEW JERSEY,1967.

719   FORSYTHE G., AND WASOW W., FINITE DIFFERENCE METHODS FOR PARTIAL
       DIFFERENTIAL EQUATIONS. WILEY, NEW YORK, 1960.

720   FOSTER K., AND ANDERSON R., DUAL AND COMPLEMENTARY VARIATIONAL
       PRINCIPLES FOR TWO DIMENSIONAL ELECTROSTATIC PROBLEMS.
       J.INST.MATH.APPLICS., 8,221-224,1971.

721   FORSTER P., DIE DISKRETE GREENSCHE FUNKTION UND FEHLERABSCHATZUNGEN
       ZUM GALERKIN-VERFAHREN.
       NUMER.MATH.19, 407-418,1972.

722   FOX L., (ED.), NUMERICAL SOLUTION OF ORDINARY AND PARTIAL DIFFERENTIAL
       EQUATIONS. PERGAMON PRESS, OXFORD, 1962.

723   FOX R.L. AND STANTON E.L., DEVELOPMENTS IN STRUCTURAL ANALYSIS
       BY DIRECT ENERGY MINIMIZATION. AIAA J.6, 1036-1042, 1968.

724   FRAEIJS DE VEUBEKE B., UPPER AND LOWER BOUNDS IN MATRIX STRUCTURAL
       ANALYSIS.. IN F.DE VEUBEKE (ED.), MATRIX METHODS OF STRUCTURAL
       ANALYSIS, PERGAMON PRESS, OXFORD, 1964.

725   FRAEIJS DE VEUBEKE B. (ED.), MATRIX METHODS OF STRUCTURAL ANALYSIS.
       PERGAMON PRESS, OXFORD, 1964.

726   FRAEIJS DE VEUBEKE B., BENDING AND STRETCHING OF PLATES - SPECIAL
       MODELS FOR UPPER AND LOWER BOUNDS. PROC.1ST CONF.MATRIX
       METHODS IN STRUCTURAL MECHANICS, WRIGHT-PATTERSON AFB, OHIO,
       AFFDL TR66-80,1965.

727   FRAEIJS DE VEUBEKE B., DISPLACEMENT AND EQUILIBRIUM MODELS IN THE FINITE
       ELEMENT METHOD. PP.145-197 OF O.C. ZIENKIEWICZ AND G.HOLISTER
       (EDS,)STRESS ANALYSIS. WILEY. NEW YORK. 1965.

728   FRAEIJS DE VEUBECKE B., A CONFORMING FINITE ELEMENT FOR PLATE BEINDING.
       INT.J.SOLIDS STRUCTURES 4, 95-108, 1968.

729   FRAEIJS DE VEUBEKE B., AN EQUILIBRIUM MODEL FOR PLATE BENDING.
       INT. J. SOLIDS STRUCTURES 4, 447-468, 1968.

730   FRAEIJS DE VEUBEKE B., A NEW VARIATIONAL PRINCIPLE FOR FINITE
       ELASTIC DISPLACEMENTS. INT.J.ENG.SCI.10, 745-765, 1972.

731   FRAEIJS DE VEUBEKE B., DUALITY IN STRUCTURAL ANALYSIS BY FINITE
       ELEMENTS. PP.323-356 OF J.T. ODEN AND E.R. DE A. OLIVEIRA (EDS.),
       LECTURES ON FINITE ELEMENT METHODS IN CONTINUUM MECHANICS.
       UNIVERSITY OF ALABAMA PRESS, HUNTSVILLE, 1973.

732   FRAEIJS DE VEUBEKE B., STATIC-GEOMETRIC ANALOGIES AND FINITE ELEMENT
       MODELS. PP. 299-322 OF J.T. ODEN AND E.R. DE A. OLIVEIRA (EDS.),
       LECTURES ON FINITE ELEMENT METHODS IN CONTINUUM MECHANICS.

UNIVERSITY OF ALABAMA PRESS, HUNTSVILLE, 1973.

733 FRAEIJS DE VEUBEKE B., THE DUAL PRINCIPLES OF ELASTODYNAMICS FINITE
     ELEMENT APPLICATIONS. PP. 357-378 OF J.T. ODEN AND E.R. DE A.
     OLIVEIRA (EDS.), LECTURES ON FINITE ELEMENT METHODS IN CONTINUUM
     MECHANICS. UNIVERSITY OF ALABAMA PRESS, HUNTSVILLE, 1973.
734 FRAEIJS DE VEUBEKE B., VARIATIONAL PRINCIPLES AND THE PATCH TEST.
     INT.J.NUMER.METH.ENG.8, 783-801, 1974.
735 FRAEIJS DE VEUBEKE B.M., AND HOGGE M.A., DUAL ANALYSIS FOR HEAT
     CONDUCTION PROBLEMS BY FINITE ELEMENTS. INT.J.NUMER.METH.
     ENG.5, 65-82, 1972.
736 FRAEIJS DE VEUBEKE B., AND SANDER G., AN EQUILIBRIUM MODEL FOR PLATE
     BENDING. INT.J.SOLIDS AND STRUCTURES 4,447-468, 1968.
737 FRAEIJS DE VEUBEKE B. AND ZIENKIEWICZ O.C., STRAIN ENERGY BOUNDS
     IN FINITE ELEMENT ANALYSIS BY SLAB ANALYSIS, J.STRAIN
     ANALYSIS 2, 265-271, 1967.
738 FRANCE P.W., PAREKH C.J., AND PETERS J.C., NUMERICAL ANALYSIS OF FREE
     SURFACE SEEPAGE PROBLEMS. J.IRRIGATION AND DRAINAGE DIV. A.S.C.E. 97,
     IR1, 165-179, 1971.
739 FRANCE P.W., PAREKH C.J., PETERS J.C. AND TAYLOR C., NUMERICAL ANALYSIS
     OF FREE SURFACE SEEPAGE PROBLEMS. PROC.A.S.C.E., MARCH 1971.
740 FREDERICK C.O., WONG Y.C., AND EDGE F.W., TWO DIMENSIONAL AUTOMATIC
     MESH GENERATION FOR STRUCTURAL ANALYSIS. INT.J.NUMER.METH.
     ENG.2, 145-150, 1970.
741 FREDERICKSON P.O., GENERALIZED TRIANGULAR SPLINES. TECHNICAL
     REPORT 7, MATHEMATICS DEPARTMENT, LAKEHEAD UNIVERSITY,
     CANADA, 1971.
742 FREHSE J., ZUR KNOVERGENZORDNUNG IN DER METHODE BEI ELLIPTISCHEN
     RANDVERTPROBLEM. PAPER AT DEUTCHE MATH.V.MEETING, HANNOVER,
     1974.
743 FREMOND M., FORMULATIONS DUALES DES ENERGIES POTENTIELLES ET
     COMPLEMENTAIRES. APPLICATION A LA METHODE DES ELEMENTS FINIS.
     C.R.ACAD. SC.PARIS SERIES A, 273, 775-777, 1971.
744 FREMOND M., LA METHODE DES ELEMENTS FINIS. REPORT, ECOLE NATIONALE
     DES PONTS ET CHAUSSEES, PARIS, 1972.
745 FREMOND M., DUAL FORMULATIONS FOR POTENTIAL AND COMPLEMENTARY
     ENERGIES. UNILATERAL BOUNDARY CONDITIONS. APPLICATIONS
     TO THE FINITE ELEMENT METHOD. PP. 175-188 OF J.R. WHITEMAN
     (ED.), THE MATHEMATICS OF FINITE ELEMENTS AND APPLICATIONS.
     ACADEMIC PRESS, LONDON, 1973.
746 FREMOND M., VARIATIONAL FORMULATION OF THE STEFAN PROBLEM,
     COUPLED STEFAN PROBLEM, FROST PROPAGATION IN POROUS MEDIA.
     PROC.CONF.COMP.METHODS IN NONLINEAR MECH., AUSTIN, TEXAS,
     1974.
747 FRIED I., FINITE ELEMENT METHOD IN FLUID DYNAMICS AND HEAT TRANSFER.
     REPORT 38, INSTITUT FUR STATIK UND DYNAMIK DER LUFT-UND
     RAUMFAHRKONSTRUKTIONEN, UNIVERSITAT STUTTGART,1967.
748 FRIED I., FINITE ELEMENT ANALYSIS OF TIME DEPENDENT PHENOMENA. AIAA
     J.7, 1170-1173,1969.
749 FRIED I., MORE ON GRADIENT ITERATIVE METHODS IN FINITE ELEMENT ANALYSIS.
     AIAA J. 7, 565-567, 1969.
750 FRIED I., GRADIENT METHODS FOR FINITE ELEMENT EIGENPROBLEMS. AIAA J.7,
     739 - 741, 1969.
751 FRIED I., SOME ASPECTS OF THE NATURAL CO-ORDINATE SYSTEM IN THE
     FINITE-ELEMENT METHOD. AIAA J. 7, 1366-1368, 1969.
752 FRIED I., A GRADIENT COMPUTATIONAL PROCEDURE FOR THE SOLUTION OF LARGE
     PROBLEMS ARISING FROM THE FINITE ELEMENT DISCRETIZATION METHOD.
     INT.J.NUMER.METH.ENG. 2, 477-494, 1970.
753 FRIED I., BASIC COMPUTATIONAL PROBLEMS IN THE FINITE ELEMENT ANALYSIS
     OF SHELLS. INT.J.SOLIDS STRUCTURES 7, 1705-1715, 1971.
754 FRIED I., CONDITION OF FINITE ELEMENT MATRICES GENERATED FROM
     NONUNIFORM MESHES. AIAA J.10, 219-221, 1971.
755 FRIED I., DISCRETIZATION AND COMPUTATIONAL ERRORS IN HIGHER-
     ORDER FINITE ELEMENTS. AIAA J.9, 2071-2072, 1971.
756 FRIED I., ACCURACY OF COMPLEX FINITE ELEMENTS. AIAA J. 10, 347-349,
     1972.
757 FRIED I., BOUNDARY AND INTERIOR APPROXIMATION ERRORS IN THE

FINITE ELEMENT METHOD. TRANS.A.S.M.E. SER.E 40, 1113-1117, 1973.

758 FRIED I., L2 AND L INFINITY CONDITION NUMBERS OF THE FINITE ELEMENT STIFFNESS AND MASS MATRICES, AND THE POINTWISE CONVERGENCE OF THE METHOD. PP.163-174 OF J.R. WHITEMAN (ED.), THE MATHEMATICS OF FINITE ELEMENTS AND APPLICATIONS. ACADEMIC PRESS, LONDON, 1973.

759 FRIED I., SHEAR IN C0 AND C1 BENDING FINITE ELEMENTS. INT.J.SOLIDS. STRUCTURES 9, 449-460,1973.

760 FRIED I., FINITE ELEMENT METHOD; ACCURACY AT A POINT Q. APPL. MATH.32, 149-162, 1974.

761 FRIED I., RESIDUAL ENERGY BALANCING TECHNIQUE IN THE GENERATING OF PLATE BENDING FINITE ELEMENTS. COMPUTERS AND STRUCTURES (TO APPEAR).

762 FRIED I. AND YANG S.K., BEST FINITE ELEMENTS DISTRIBUTION AROUND A SINGULARITY. AIAA J.10, 1244-1246, 1972.

763 FRIEDMAN A., PARTIAL DIFFERENTIAL EQUATIONS OF PARABOLIC TYPE. PRENTICE HALL, NEW JERSEY, 1964.

764 FRIEDMAN A., PARTIAL DIFFERENTIAL EQUATIONS HOLT,RHINEHART AND WINSTON, NEW YORK,1969.

765 FRIEDMAN A., REGULARITY THEOREMS FOR VARIATIONAL INEQUALITIES IN UNBOUNDED DOMAINS AND APPLICATIONS TO STOPPING TIME PROBLEMS. ARCH.RAT.MECH.ANAL.52, 134-160, 1973.

766 FRIEDMAN E., DIRECT ITERATION METHOD FOR THE INCORPORATION OF PHASE CHANGE IN FINITE ELEMENT HEAT CONDUCTION PROGRAMS. REPORT WAPD-TM-1133, BETTIS ATOMIC POWER LABORATORIES, PITTSBURG, 1974.

767 FRIEDRICHS K., DIE RANDWERT UND EIGENWERT/PROBLEM AUS DER THEORIE DER ELASTISCHEN PLATEN. MATH.ANN.98, 205-247, 1928.

768 FRIEDRICHS K.O., AND KELLER H.B., A FINITE-DIFFERENCE SCHEME FOR GENERALISED NEUMANN PROBLEMS. IN BRAMBLE (ED.), NUMERICAL SOLUTION OF PARTIAL DIFFERENTIAL EQUATIONS, ACADEMIC PRESS, NEW YORK, 1966.

769 FRITSCH F.N., ON SELF-CONTAINED NUMERICAL INTEGRATION FORMULAS FOR SYMMETRIC REGIONS. SIAM J.NUMER.ANAL.8, 213-221, 1971.

770 FROMM J.E., NEW DEVELOPMENTS IN NUMERICAL ANALYSIS OF THREE-DIMENSIONAL INCOMPRESIBLE VISCOUS FLOW. PROC.CONF.COMP. METHODS IN NONLINEAR MECH., AUSTIN, TEXAS, 1974.

771 FU C.C., A METHOD FOR THE NUMERICAL INTEGRATION OF EQUATIONS OF MOTION ARISING FROM A FINITE ELEMENT ANALYSIS. J.APPL.MECH 37, 599 - 605, 1970.

772 FU C.C., ON THE STABILITY OF EXPLICIT METHODS FOR THE NUMERICAL INTEGRATION OF THE EQUATIONS OF MOTION IN FINITE ELEMENT METHODS. INT.J. NUMER. METH. ENG.4, 95-107, 1972.

773 FUHRING H., A DISCUSSION ON THE DENSITY OF FINITE ELEMENT MATRICES. INT.J.NUMER.METH.ENG.8, 432-433, 1974.

774 FUJII H., FINITE ELEMENT SCHEMES: STABILITY AND CONVERGENCE. PP.201-218 OF J.T. ODEN, R.W. CLOUGH, AND Y. YAMOMOTO (EDS.), ADVANCES IN COMPUTATIONAL METHODS IN STRUCTURAL MECHANICS AND DESIGN. UNIVERSITY OF ALABAMA PRESS, HUNTSVILLE, 1972.

775 FUJINO T. AND OHSAKA K., HEAT CONDUCTION AND THERMAL STRESS ANALYSIS BY THE FINITE ELEMENT METHOD. PROC. 2ND CONF. MATRIX METHODS IN STRUCTURAL MECHANICS, WRIGHT-PATTERSON AFB., OHIO, AFFDL-TR-68-150,1968.

776 FULLARD K., THE COMPUTATION OF TEMPERATURE DISTRIBUTIONS AND THERMAL STRESSES USING FINITE ELEMENTS TECHNIQUES. PAPER M5/3 PROCEEDINGS 1ST STRUCTURAL MECH. IN REACTOR TECH. CONF., BERLIN, 1971.

777 GALERKIN B.G., RODS AND PLATES. SERIES OCCURRING IN VARIOUS QUESTIONS CONCERNING THE ELASTIC EQUILIBRIUM OF RODS AND PLATES. ENG.BULL. (VEST.INZH.) 19, 897-908, 1915.

778 GALLAGHER R.H., A CORRELATION STUDY OF METHODS OF MATRIX STRUCTURAL ANALYSIS. AGARDOGRAPH 69, PERGAMON PRESS, 1962.

779 GALLAGHER R.H., TECHNIQUES FOR THE DERIVATIVATION OF ELEMENT STIFFNESS MATRICES. AIAA J.1, 1431-1432, 1963.

780 GALLAGHER R.H., COMMENTS ON DERIVATION OF ELEMENT STIFFNESS MATRICES BY
        T.H.H. PIAN. AIAA J.3, 186-187, 1965.
781 GALLAGHER R.H., STIFFNESS MATRIX FOR SHALLOW RECTANGULAR SHELL
        ELEMENT. J.ENG.MECH.DIV. A.S.C.E.94, EM2, 708-709, 1968.
782 GALLAGHER R.H., ANALYSIS OF PLATE AND SHELL STRUCTURES.
        PROC.CONF. ON APPLICATION OF FINITE ELEMENT METHOD IN CIVIL
        ENGINEERING. VANDERBILT UNIVERSITY, 1969.
783 GALLAGHER R.H., SURVEY AND EVALUATION OF THE FINITE ELEMENT
        METHOD IN LINEAR FRACTURE MECHANICS ANALYSIS. PAPER L6/9,
        PROCEEDINGS 1ST STRUCTURAL MECH. IN REACTOR TECH. CONF.,
        BERLIN, 1971.
784 GALLAGHER R.H., TRENDS AND DIRECTIONS IN THE APPLICATIONS OF NUMERICAL
        ANALYSIS. ONR SYMPOSIUM ON NUMERICAL AND COMPUTER METHODS IN
        STRUCTURAL MECHANICS, UNIVERSITY OF ILLINOIS, 1971.
785 GALLAGHER R.H., APPLICATIONS OF FINITE ELEMENT ANALYSIS. PP.641-678
        OF J.T. ODEN, R.W. CLOUGH, AND Y. YAMOMOTO (EDS.), ADVANCES IN
        COMPUTATIONAL METHODS IN STRUCTURAL MECHANICS AND DESIGN,
        UNIVERSITY OF ALABAMA PRESS, HUNTSVILLE, 1972.
786 GALLAGHER R.H., THE FINITE ELEMENT METHOD OF THIN-SHELL STABILITY
        ANALYSIS. NATIONAL SYMP. COMPUTERISED STRUCT. ANALYSIS AND DESIGN,
        GEORGE WASHINGTON UNIVERSITY, 1972.
787 GALLAGHER R.H., COMPUTATIONAL METHODS IN NUCLEAR REACTOR STRUCTURAL DESIGN
        FOR HIGH TEMPERATURE APPLICATIONS: AN INTERPRETIVE REPORT. REPORT
        ORNL-4756, OAK RIDGE NATIONAL LABORATORY, TENNESSEE, 1973.
788 GALLAGHER R.H., GENERAL POTENTIAL ENERGY AND COMPLEMENTARY ENERGY
        MODELS BASED ON STRESS PARAMETERS. LECTURES AT ICCAD
        COURSE ON ADVANCED TOPICS IN FINITE ELEMENT ANALYSIS,
        SANTA MARGHERITA, ITALY, 1974.
789 GALLAGHER R.H., PERTURBATION PROCEDURES IN NONLINEAR FINITE
        ELEMENT ANALYSIS. PROC.CONF.COMP.METHODS IN NONLINEAR
        MECHANICS, AUSTIN, TEXAS, 1974.
790 GALLAGHER R.H., FINITE ELEMENT ANALYSIS: FUNDAMENTALS. (TO APPEAR).
791 GALLAGHER R.H. AND CHAN S.T.K., HIGHER-ORDER FINITE ELEMENT
        ANALYSIS OF LAKE CIRCULATION. INT.J.COMPUTERS AND FLUIDS
        1, 119-132, 1973.
792 GALLAGHER R.H. AND DHALLA A.K., DIRECT FLEXIBILITY FINITE ELEMENT
        ELASTOPLASTIC ANALYSIS. PAPER M6/9, PROCEEDINGS 1ST
        STRUCTURAL MECH. IN REACTOR TECH. CONF., BERLIN, 1971.
793 GALLAGHER R.H., GELLATLY R.A., PADLOG J., AND MALLETT R.H., A
        DISCRETE ELEMENT FOR THIN-SHELL INSTABILITY ANALYSIS.
        AIAA J.5, 138-145, 1967
794 GALLAGHER R.H., AND LEE C-H., MATRIX DYNAMIC AND INSTABILITY ANALYSIS
        WITH NON-UNIFORM ELEMENTS. INT.J.NUMER.METH.ENG.2, 265-276,
        1970.
795 GALLAGHER R.H., LIEN S. AND MAU S.T., A PROCEDURE FOR FINITE
        ELEMENT PLATE AND SHELL PRE- AND POST-BUCKLING ANALYSIS.
        PROC. 3RD CONF. MATRIX METHODS IN STRUCTURAL MECHANICS,
        WRIGHT-PATTERSON AFB, OHIO, 1971.
796 GALLAGHER R.H., AND PADLOG J., DISCRETE ELEMENT APPROACH TO STRUCTURAL
        INSTABILITY ANALYSIS. AIAA J.1, 1437-1439, 1963.
797 GALLAGHER R.H., PADLOG J., AND BIJLARD P.P., STRESS ANALYSIS
        IN HEATED COMPLEX SHAPES. J. AMERICAN ROCKET SOC. 32,
        700-707, 1962.
798 GALLAGHER R.H., YAMADA Y., AND ODEN J.T., (EDS.), RECENT ADVANCES
        IN MATRIX METHODS OF STRUCTURAL ANALYSIS AND DESIGN.
        UNIVERSITY OF ALABAMA PRESS, TUSCALOOSA, 1970.
799 GALLAGHER R.H., AND YOUNG H.T.V., ELASTIC INSTABILITY PREDICTIONS
        FOR DOUBLY CONNECTED REGIONS. PROC.2ND CONF.MATRIX METHODS
        IN STRUCTURAL MECHANICS, WRIGHT-PATTERSON AFB, OHIO, 1968.
800 GALLAGHER R.H. AND ZIENKIEWICZ O.C. (EDS.), OPTIMUM STRUCTURAL DESIGN.
        THEORY AND APPLICATIONS. WILEY,LONDON,1973.
801 GAMBOLATI G., DIAGONALLY DOMINANT MATRICES FOR THE FINITE
        ELEMENT METHOD IN HYDROLOGY. INT.J.NUMER.METH.ENG.6,
        587-608, 1973.
802 GANGAL M.D., DIRECT FINITE ELEMENT ANALYSIS OF ELASTIC CONTACT
        PROBLEMS. INT.J.NUMER.METH.ENG.5, 145-147, 1972.

803 GARABEDIAN H.L., (ED.), APPROXIMATION OF FUNCTIONS. ELSEVIER,
        AMSTERDAM, 1965.
804 GARABEDIAN P.R., PARTIAL DIFFERENTIAL EQUATIONS. WILEY, NEW YORK, 1964.
805 GARTLING D.K., AND BECKER E.B., COMPUTATIONALLY EFFICIENT FINITE
        ELEMENT ANALYSIS OF VISCOUS FLOW PROBLEMS. PROC.CONF.COMP.
        METHODS IN NONLINEAR MECHANICS, AUSTIN, TEXAS, 1974.
806 GELFAND I.M., AND FOMIN S.V., CALCULUS OF VARIATIONS. PRENTICE HALL,
        NEW JERSEY, 1963.
807 GELFAND I.M. AND SHILOV G.E., GENERALIZED FUNCTIONS. ACADEMIC
        PRESS, NEW YORK, 1964.
808 GEORGE J.A., COMPUTER IMPLEMENTATION OF THE FINITE ELEMENT METHOD.
        REPORT CS 208, COMPUTER SCIENCE DEPARTMENT, STANFORD UNIVERSITY,
        CALIFORNIA, 1971.
809 GEORGE J.A., ON THE DENSITY OF FINITE ELEMENT MATRICES. INT.J.
        NUMER.METH.ENG. 5, 297-300, 1972.
810 GEORGE J.A., ON BLOCK ELIMINATION OF SPARSE LINEAR SYSTEMS.
        SIAM J. NUMER.ANAL.11, 585-603, 1974.
811 GERIJ J., THERMO-ELASTIC-PLASTIC CYCLIC ANALYSIS BY FINITE
        ELEMENT METHOD. PAPER L7/7, PROC.2ND STRUCTURAL MECH.
        IN REACTOR TECH. CONF., BERLIN, 1973.
812 GHABOUSSI J., AND WILSON E.L., VARIATIONAL FORMULATION OF DYNAMICS
        OF FLUID SATURATED POROUS ELASTIC SOLIDS. J. ENG. MECH. DIV.
        A.S.C.E. 98, EM4, 947-963, 1972.
813 GHABOUSSI J., AND WILSON E.L., FLOW OF COMPRESSIBLE FLUID IN
        POROUS ELASTIC MEDIA. INT.J.NUMER.METH.ENG.5, 419-442,
        1973.
814 GHABOUSSI J., WILSON E.L. AND ISENBERG J., FINITE ELEMENT FOR ROCK JOINTS
        AND INTERFACES. PROC. A.S.C.E.99, SM10, 833-848, 1973.
815 GHERI G. AND CELLA A., SPLINE-BLENDED APPROXIMATION OF HARTMAN'S FLOW.
        REPORT, DIVISIONE DI FISICA, CAMEN,PISA,OCTOBER 1973.
816 GHERI G., AND CELLA A., SPLINE BLENDED APPROXIMATION OF HARTMANN'S FLOW,
        INT.J.NUMER.METH.ENG.8, 529-536, 1974.
817 GHISTA D.N., AND RAJU I.S., FINITE ELEMENT ANALYSES IN BIOMECHANICS,
        INT CONF. VARIATIONAL METHODS IN ENG., SOUTHAMPTON UNIVERSITY,
        1972.
818 GIANNINI M., AND MILES G.A., A CURVED ELEMENT APPROXIMATION IN THE
        ANALYSIS OF AXI-SYMMETRIC THIN SHELLS. INT. J.NUMER.
        METH.ENG.2, 459-476, 1970.
819 GIENCKE E., SIMPLE FINITE ELEMENT METHOD FOR STRESS CONCENTRATION
        PROBLEMS IN PRESSURE VESSELS. PP. 215-237, VOL.4, PROCEEDINGS
        1ST STRUCTURAL MECH. IN REACTOR TECH. CONF., BERLIN, 1971.
820 GIENCKE E., A SIMPLE MIXED METHOD FOR PLATE AND SHELL PROBLEMS.
        PAPER M2/5, PROCEEDINGS 2ND STRUCTURAL MECH. IN REACTOR
        TECH.CONF., BERLIN, 1973.
821 GILES G.L. AND BLACKBURN C.L., PROCEDURE FOR EFFICIENTLY
        GENERATING, CHECKING AND DISPLAYING NASTRAN INPUT AND
        OUTPUT DATA FOR ANALYSIS OF AEROSPACE VEHICLE STRUCTURES.
        NASTRAN USERS EXPERIENCES. N.A.S.A. TECHNICAL REPORT
        TM-X-2378, 1971.
822 GIRAULT V., THEORY OF A FINITE DIFFERENCE METHOD ON IRREGULAR NETWORKS.
        SIAM J. NUMER.ANAL. 11, 260-282, 1974.
823 GLADWELL G.M.L., A FINITE ELEMENT METHOD FOR ACOUSTICS.
        CONGRESS INTERNATIONAL D'ACOUSTIQUE, LIEGE, 1965.
824 GLADWELL G.L.M. (ED.), PROCEEDINGS SYMPOSIUM ON COMPUTER AIDED
        ENGINEERING. UNIVERSITY OF WATERLOO, 1971.
825 GLADWELL G.M.L. AND MASON V., VARIATIONAL FINITE ELEMENT
        CALCULATION OF THE ACOUSTIC RESPONSE OF A RECTANGULAR
        PANEL. J. SOUND AND VIBRATION 14, 115-136, 1971.
826 GLASS J.M., SMOOTH-CURVE INTERPOLATION: A GENERALIZED SPLINE FIT
        PROCEDURE. B.I.T. 6, 277-293, 1966.
827 GLOWINSKI R., SUR UNE METHODE D'APPROXIMATION EXTERNE PAR ELEMENTS
        FINIS D'ORDRE TWO. C.R. ACAD.SC.PARIS A 275, 201-204, AND
        333-335, 1972.
828 GLOWINSKI R., APPROXIMATIONS EXTERNES, PAR ELEMENTS FINIS DE LAGRANGE
        D'ORDRE UN ET DEUX, DU PROBLEME DE DIRICHLET POUR L'OPERATEUR
        BIHARMONIQUE. METHODE ITERATIVE DE RESOLUTION DES PROBLEMES

APPROCHES. PP.123-171 OF J.J.H. MILLER (ED.), TOPICS IN NUMERICAL
ANALYSIS. ACADEMIC PRESS, LONDON,1973.

829   GLOWINSKI R. AND MARROCCO A.,   ANALYSE NUMERIQUE DU CHAMP MAGNETIQUE D'UN
ALTERNATEUR PAR ELEMENTS FINIS ET SUR-RELAXATION PONCTUELLE NON
LINEAIRE. COMP.METH.APPL.MECH.ENG.3, 55-85,1974

830   GODBOLE P.N., AND MEEK J.L.,   APPLICATION OF FINITE ELEMENT
METHOD TO PLANE STRAIN EXTRUSION PROCESSES. PROC.CONF.
COMP.METHODS IN NONLINEAR MECHANICS, AUSTIN, TEXAS 1974.

831   GODFREY D.A., AND MOUSSA M.M.,   DYNAMIC ANALYSIS OF AXISYMMETRIC SHELLS
UNDER ARBITRARY TRANSIENT PRESSURES. NUCL. ENG.DESIGN 23,
187-194, 1972.

832   GOEL J-J.,   CONSTRUCTION OF BASIC FUNCTIONS FOR NUMERICAL UTILISATION
OF RITZ'S METHOD. NUMERISCHE MATH.12, 435-447, 1968.

833   GOEL J-J.,   LIST OF BASIC FUNCTIONS FOR NUMERICAL UTILISATION OF
RITZ'S METHOD. APPLICATION TO THE PROBLEM OF THE PLATE.
ECOLE POLYTECHNIQUE FEDERALE, LAUSANNE, 1969.

834   GOLDBERG J.E., BALUCH M.H., KORMAN T., AND KOH S.I.,   FINITE
ELEMENT APPROACH TO BENDING OF MICROPOLAR PLATES. INT.
J.NUMER.METH.ENG.8, 311-322, 1974.

835   GOLDSTEIN S., AND HOFFMANN A.,   PROGRAMMES DE CALCUL ELASTOPLASTIQUE
DE STRUCTURES EN USAGE AU COMMISSARIAT A L'ENERGIE
ATOMIQUE. FRENCH CONGRESS ON MECHANICS. POITERS, 1973.

836   GOLOMB M.,   APPROXIMATION BY PERIODIC SPLINE INTERPOLATION ON UNIFORM
MESHES. J.APPROX. THEORY 1, 26-65, 1968.

837   GOODIER J.H., AND HOFF H.J.(EDS.),   PROCEEDINGS 1ST SYMPOSIUM ON
NAVAL STRUCTURAL MECHANICS. PERGAMON PRESS, OXFORD, 1960.

838   GOODMAN R.E., TAYLOR R.L., AND BREKKE T.,   A MODEL FOR THE MECHANICS
OF JOINTED ROCK. J.SOIL MECH.FOUND.DIV.PROC. A.S.C.E.94,
637-659, 1968.

839   GORDON W.J.,   DISTRIBUTIVE LATTICES AND APPROXIMATION OF MULTIVARIATE
FUNCTIONS. PP.223-277 OF I.J. SCHOENBERG (ED.), APPROXIMATIONS
WITH SPECIAL EMPHASIS ON SPLINE FUNCTIONS, ACADEMIC PRESS, NEW
YORK, 1969.

840   GORDON W.J.,   SPLINE-BLENDED INTERPOLATION THROUGH CURVE NETWORKS.
J.MATH.MECH. 18, 931-952, 1969.

841   GORDON W.J.,   BLENDING FUNCTION METHODS FOR BIVARIATE AND MULTIVARIATE
APPROXIMATION. SIAM J. NUMER.ANAL.8, 158-177, 1971.

842   GORDON W.J., AND HALL C.A.,   GEOMETRIC ASPECTS OF THE FINITE ELEMENT
METHOD. PP.769-783 OF K.AZIZ (ED.), THE MATHEMATICAL
FOUNDATIONS OF THE FINITE ELEMENT MATHOD. ACADEMIC
PRESS,NEW YORK,1972.

843   GORDON W.J. AND HALL C.A.,   CONSTRUCTION OF CURVILINEAR CO-ORDINATE
SYSTEMS AND APPLICATION TO MESH GENERATION. INT.J.NUMER,METH.ENG.
461-478,1973.

844   GORDON W.J. AND HALL C.A.,   TRANSFINITE ELEMENT METHODS: BLENDING FUNCTION
INTERPOLATION OVER ARBITRARY CURVED ELEMENT DOMAINS.
NUMER. MATH. 21, 109-129, 1973.

845   GOTOH M.,   A FINITE ELEMENT ANALYSIS OF GENERAL DEFORMATION OF
SHEET METALS. INT.J.NUMER.METH.ENG.8, 731-741, 1974.

846   GOUDREAU G.L. AND TAYLOR R.L.,   EVALUATION OF NUMERICAL INTEGRATION
METHODS IN ELASTODYNAMICS. COMP.METH.APPL.MECH.ENG.2, 69-97,1973

847   GOULD L.,   FINITE ELEMENT ANALYSIS OF SHELLS OF REVOLUTION BY
MINIMIZATION OF THE POTENTIAL ENERGY FUNCTIONAL. PP.
279-308 OF APPLICATION OF FINITE ELEMENT METHODS IN CIVIL
ENGINEERING, A.S.C.E. 1969.

848   GOULD P.L. AND SEN S.K.,   REFINED MIXED METHOD FINITE ELEMENTS
FOR SHELLS OF REVOLUTION. PROC. 3RD CONF. MATRIX METHODS
IN STRUCTURAL MECHANICS, WRIGHT-PATTERSON AFB.,
OHIO, 1971.

849   GOULD S.H.,   VARIATIONAL METHODS FOR EIGENVALUE PROBLEMS.
UNIVERSITY OF TORONTO PRESS, TORONTO, 1957.

850   GOURLAY A.R.,   SOME RECENT METHODS FOR THE NUMERICAL SOLUTION OF TIME
DEPENDENT PARTIAL DIFFERENTIAL EQUATIONS. PROC.ROY.SOC.LONDON A,
323, 219-235, 1971.

851   GRAFTON P.E., AND STROME D.R.,   ANALYSIS OF AXISYMMETRIC SHELLS BY
THE DIRECT STIFFNESS METHOD. AIAA J.1, 2342-2347, 1963.

852   GRAM J. (ED.),   NUMERICAL SOLUTION OF PARTIAL DIFFERENTIAL
          EQUATIONS. D.REIDEL, AMSTERDAM, 1973.
853   GREENBAUM G.A., AND RUBINSTEIN M.F.,   CREEP ANALYSIS OF AXISYMMETRIC
          BODIES USING FINITE ELEMENTS. NUCLEAR ENG.AND DESIGN 10,
          379-397, 1968.
854   GREENBAUM G.A., HOFMEISTER L.D., AND EVENSEN D.A.,   PURE MOMENT
          LOADING OF AXISYMMETRIC FINITE ELEMENT MODELS. INT.J.NUMER.
          METH.ENG.5, 459-463, 1973.
855   GREENE B.E., JONES R.E., MCLAY R.W. AND STROME D.R.,   GENERALISED
          VARIATIONAL PRINCIPLES IN THE FINITE ELEMENT METHOD.
          AIAA J. 7, 1254-1260, 1969.
856   GREENE B.E., JONES R.E., MCLAY R.W. AND STROME D.R.,   DYNAMIC
          ANALYSIS OF SHELLS USING DOUBLY-CURVED FINITE ELEMENTS.
          PROC. 2ND CONF. MATRIX METHODS IN STRUCTURAL MECHANICS,
          WRIGHT-PATTERSON AFB., OHIO, AFFDL-TR-68-150, 1968.
857   GREENSPAN D.,   ON APPROXIMATING EXTREMALS OF FUNCTIONS I. I.C.C.
          BULLETIN 4, 99-120, 1965.
858   GREENSPAN D.,   ON APPROXIMATING EXTREMALS OF FUNCTIONALS - II,
          INT. J.ENG.SCI.5, 571, 1967.
859   GREGORY J.A.,   SMOOTH INTERPOLATION WITHOUT TWIST CONSTRAINTS.
          IN R.E. BARNHILL AND R.F. RIESENFELD (EDS.), COMPUTER
          AIDED GEOMETRIC DESIGN. ACADEMIC PRESS, NEW YORK, 1975.
860   GREGORY J.A.,   SYMMETRIC SMOOTH INTERPOLATION ON TRIANGLES.
          REPORT TR/34, DEPARTMENT OF MATHEMATICS, BRUNEL UNIVERSITY,
          1973.
861   GREGORY J.A.,   SMOOTH INTERPOLATION WITHOUT TWIST CONSTRAINTS.
          TECHNICAL REPORT TR/47, DEPARTMENT OF MATHEMATICS, BRUNEL
          UNIVERSITY, 1974.
862   GREGORY J.A.,   PIECEWISE INTERPOLATION THEORY. PH.D.THESIS,
          DEPARTMENT OF MATHEMATICS, BRUNEL UNIVERSITY, 1975.
863   GREGORY J.A., AND WHITEMAN J.R.,   LOCAL MESH REFINEMENT WITH
          FINITE ELEMENTS FOR ELLIPTIC PROBLEMS. TECHNICAL REPORT
          TR/24, DEPARTMENT OF MATHEMATICS, BRUNEL UNIVERSITY, 1974.
864   GREVILLE T.N.E.,   INTERPOLATION BY GENERALIZED SPLINE FUNCTIONS
          TECHNICAL SUMMARY REPORT NO.476, MATHEMATICS RESEARCH
          CENTER, UNIVERSITY OF WISCONSIN, MADISON, 1964.
865   GREVILLE T.N.E. (ED.),   THEORY AND APPLICATION OF SPLINE FUNCTIONS.
          ACADEMIC PRESS, NEW YORK, 1969.
866   GRIEGER I.,   INGA INTERACTIVE GRAPHISCHE ANALYSE BENUTZERHANDBUCH.
          INSTITUT FUR STATIK UND DYNAMIK DER LUFT UND
          ROUMFARTKONSTRUKTIONEN, UNIVERSITAT STUTTGART, 1973.
867   GRIEGER I.,   INGA AN INTERATICE GRAPHIC SYSTEM FOR STRUCTURAL
          ANALYSIS. PROCEEDINGS CAD 74, CONF. ON COMPUTERS IN
          ENGINEERING AND BUILDING DESIGN, IMPERIAL COLLEGE,
          LONDON, 1974.
868   GRIEMANN L.F.,   FINITE ELEMENT ANALYSIS OF PLATE BENDING WITH TRANSVERSE
          SHEAR DEFORMATION. NUCL.ENG.DESIGN.14, 223-230, 1970.
869   GRIFFIN D.S., AND KELLOGG R.B.,   A NUMERICAL SOLUTION OF AXIALLY
          SYMMETRICAL AND PLANE ELASTICITY PROBLEMS.
          INT.J.SOLIDS AND STRUCTURES 3, 781-794, 1967.
870   GROSS H.,   THERMOVISCOELASTIC AXISYMMETRIC STRESS ANALYSIS BY
          FINITE ELEMENT AND FINITE DIFFERENCE COMPUTER TECHNIQUES.
          PAPER M5/5, PROCEEDINGS 1ST STRUCTURAL MECH. IN REACTOR
          TECH. CONF., BERLIN, 1971.
871   GROTKOP G.,   FINITE ELEMENT ANALYSIS OF LONG-PERIOD WATER WAVES. COMP.
          METH.IN APPL.MECH.ENG.2, 153-146,1973.
872   GUDERLEY K.G. AND HSU C.-C.,   A SPECIAL FORM OF GALERKIN'S METHOD APPLIED
          TO HEAT TRANSFER IN COUETTE-POISEUILLE FLOWS. J.COMP.PHYS.11,
          70-89,1973.
873   GUILINGER W.H.,   FINITE ELEMENT METHOD. TRANS.AMER.MATH.SOC,14,
          199-XXX, 1971.
874   GUNDERSON R.H. AND CETINER A.,   ELEMENT STIFFNESS MATRIX
          GENERATION. J. STRUCTURAL DIV. A.S.C.E. 97, ST1,
          363-375,1971.
875   GUPTA A.K., AND MOHRAZ B.,   A METHOD OF COMPUTING NUMERICALLY
          INTEGRATED STIFFNESS MATRICES. INT.J.NUMER.METH.ENG.5,

83-89, 1972.

876 GUPTA K.K., VIBRATION OF FRAMES AND OTHER STRUCTURES WITH BANDED
        STIFFNESS MATRIX. INT.J.NUMER.METH.ENG.2, 221-228, 1970.

877 GUPTA K.K., RECENT ADVANCES IN NUMERICAL ANALYSIS OF STRUCTURAL
        EIGENVALUE PROBLEMS. PP.249-271 OF THEORY AND PRACTICE IN FINITE
        ELEMENT STRUCTURAL ANALYSIS. UNIVERSITY OF TOKYO PRESS, 1973.

878 GUPTA K.K., SOLUTION OF QUADRATIC MATRIX EQUATIONS FOR FREE
        VIBRATION ANALYSIS OF STRUCTURES. INT.J.NUMER.METH.ENG.
        6, 129-135, 1973.

879 GURTIN M.E., VARIATIONAL PRINCIPLES FOR LINEAR INITIAL VALUE PROBLEMS.
        QUART.APPL.MATH. 22, 252-256, 1964.

880 GUYAN R.J., DISTRIBUTED MASS MATRIX FOR PLATE ELEMENTS IN
        BENDING. AIAA J.3, 567-568, 1965.

881 GUYAN R.J., REDUCTION OF STIFFNESS AND MASS MATRICES. AIAA J.3,
        380, 1965.

882 GUYAN R.J., UJIHARA B.H. AND WELCH P.W., HYDROELASTIC ANALYSIS OF
        AXISYMMETRIC SYSTEMS BY A FINITE ELEMENT METHOD. PROC.
        2ND CONF. MATRIX METHODS IN STRUCTURAL MECHANICS,
        WRIGHT-PATTERSON AFB., OHIO, AFFDL-TR-68-150, 1968.

883 GUYMON G.L. AND SCOTT V.H., APPLICATION OF FINITE ELEMENT METHOD TO
        GENERAL NUMERICAL SOLUTION OF TWO DIMENSIONAL DIFFUSION-CONVECTION
        EQUATION. TRANS AM. GEOPHYSICAL UNION 51, 282-   , 1970.

884 HAFTKA R.T., HALLETT P.H. AND NACHBAR W., ADAPTATION OF
        KOITER'S METHOD TO FINITE ELEMENT ANALYSIS OF SNAP
        THROUGH BUCKLING BEHAVIOUR. J. SOLIDS AND STRUCTURES 7,
        1427-1446, 1971.

885 HAISLER W., AND STRICKLIN J.A., RIGID - BODY DISPLACEMENTS OF
        CURVED ELEMENTS IN THE ANALYSIS OF SHELLS BY THE MATRIX
        DISPLACEMENT METHOD. AIAA J.5, 1525-1527, 1967.

886 HAISLER W., AND STRICKLIN J.A., NONLINEAR FINITE ELEMENT ANALYSIS
        INCLUDING HIGHER-ORDER STRAIN ENERGY TERMS. AIAA J. 8,
        1158-1159, 1970.

887 HAISLER W.E., AND STRICKLIN J.A., COMPUTATIONAL METHODS FOR
        SOLVING NONLINEAR STRUCTURE PROBLEMS. PROC.CONF.COMP.
        METHODS IN NONLINEAR MECH., AUSTIN, TEXAS, 1974.

888 HAISLER W.E., STRICKLIN J.A., AND VON REISEMANN W.A., DYNAPLAS A FINITE
        ELEMENT PROGRAM FOR THE DYNAMIC, LARGE DEFLECTION, ELASTIC-PLASTIC
        ANALYSIS OF STIFFENED SHELLS OF REVOLUTION. REPORT TEES-72-27,
        DEPARTMENT OF AEROSPACE ENGINERING, TEXAS A+M. UNIVERSITY, 1972.

889 HAISLER W.E., STRICKLIN J.A., AND VON RIESEMANN W.A., SAMMSOR 3 A FINITE
        ELEMENT PROGRAM TO DETERMINE STIFFNESS AND MASS MATRICES OF RING
        STIFFENED SHELLS OF REVOLUTION. REPORT TEES-72-26, TEXAS A+M
        UNIVERSITY, 1972.

890 HALL C.A., ERROR BOUNDS FOR SPLINE INTERPOLATION. J. APPROX. THEORY 1,
        209-218, 1968.

891 HALL C.A., BICUBIC INTERPOLATION OVER TRIANGLES.
        J.MATH.MECH. 19, 1-11, 1969.

892 HALL C.A., NATURAL CUBIC AND BICUBIC SPLINE INTERPOLATION, SIAM.J.
        NUMER.ANAL.10, 1055-1060,1973.

893 HALL C.A. AND KENNEDY J., FOURTH ORDER CONVERGENCE OF A RITZ
        APPROXIMATION TO A PLANE STRAIN ELASTICITY PROBLEM. TO APPEAR.

894 HALL C.A., AND KENNEDY J. CONVERGENCE OF A RITZ APPROXIMATION FOR THE
        STEADY STATE HEAT FLOW PROBLEM. J. INST. MATH. APPLICS.12,
        187-196,1973.

895 HALMOS P.R., INTRODUCTION TO HILBERT SPACE AND THEORY OF SPECTRAL
        MULTIPLICITY. CHELSEA, NEW YORK, 1951.

896 HALMOS P.R., FINITE DIMENSIONAL VECTOR SPACES. PRINCETON UNIVERSITY
        PRESS, 1955.

897 HANNA M.S., AND SMITH K.T., SOME REMARKS ON THE DIRICHLET
        PROBLEM IN PIECEWISE SMOOTH DOMAINS. COMM.PURE APPL.
        MATH.20, 575-593, 1967.

898 HANSEN H.R., SOME EXAMPLES OF THE APPLICATION OF THE FINITE
        ELEMENT METHOD TO SKIP STRUCTURES. COMPUTERS AND STRUCTURES
        4, 205-212, 1974.

899 HANSEN K.F., DIRECT METHODS FOR MULTIDIMENSIONAL KINETICS PROBLEMS. TRANS,
        AMERICAN NUCL. SOC.16, 291-292, 1973.

900   HANSTEEN O.E., A CONICAL ELEMENT FOR DISPLACEMENT ANALYSIS OF
              AXI-SYMMETRIC SHELLS. FINITE ELEMENT METHODS, TAPIR,
              TRONDHEIM, 1969.
901   HANSTEEN O.E., FINITE ELEMENT METHODS AS APPLICATIONS OF VARIATIONAL
              FINITE ELEMENT METHODS IN STRESS ANALYSIS, TAPIR, TRONDHEIM,1969.
902   HARDY C., BARONET C.N., AND TORDION G.V., THE ELASTO-PLASTIC
              INDENTATION OF A HALF-SPACE BY A RIGID SPHERE. INT.J.NUMER.
              METH.ENG.3, 451-463, 1971.
903   HARKEGARD G., A FINITE ELEMENT ANALYSIS OF ELASTIC-PLASTIC
              PLATES CONTAINING CAVITIES AND INCLUSIONS WITH REFERENCE
              TO FATIGUE CRACK INITIATION. INT.J.FRACTURE 9, 437-447,
              1973.
904   HARRIS H.G., AND PIFKO A.B., ELASTIC-PLASTIC BUCKLING OF STIFFENED
              RECTANGULAR PLATES. PP.207-254 OF APPLICATION OF FINITE
              ELEMENT METHODS IN CIVIL ENGINEERING, A.S.C.E., 1969.
905   HARRIS S., THE BOLTZMANN EQUATION. HOLT, RHINEHART AND WINSTON, NEW
              YORK, 1971.
906   HARRISON D.G. AND CHEUNG Y.K., A HIGHER-ORDER TRIANGULAR FINITE
              ELEMENT FOR THE SOLUTION OF FIELD PROBLEMS IN ORTHOTROPIC MEDIA.
              INT.J.NUMER.METH.ENG.7, 287-295,1973.
907   HARTIG D., HERMITEAN RECTANGLE FINITE ELEMENT FAMILY. PAPER M2/4,
              PROC.2ND STRUCTURAL MECH. IN REACTOR TECH.CONF., BERLIN,
              1973.
908   HARTMAN P. AND STAMPACCHIA G., ON SOME NONLINEAR ELLIPTIC DIFFERENTIAL
              FUNCTIONAL EQUATIONS. ACTA MATHEMATICS 115, 271-310, 1966.
909   HARTZ B.J., FINITE ELEMENT CONTRIBUTION TO FRACTURE MECHANICS.
              PP.457-471 OF J.T. ODEN (FD.), ADVANCES IN COMPUTATIONAL METHODS
              IN STRUCTURAL MECHANICS AND DESIGN. UNIVERSITY OF ALABAMA
              PRESS, HUNTSVILLE, 1972.
910   HARTZ B.J., AND CHOPRA P.S., A FINITE ELEMENT CONTRIBUTION TO FRACTURE
              MECHANICS. PP. 457-472 OF J.T. ODEN, R.W. CLOUGH AND Y. YAMOMOTO
              (EDS.), ADVANCES IN COMPUTATIONAL METHODS IN STRUCTURAL MECHANICS
              AND DESIGN. UNIVERSITY OF ALABAMA PRESS, HUNTSVILLE, 1972.
911   HARTZMAN M., STATIC STRESS ANALYSIS OF AXISYMMETRIC SOLIDS WITH
              MATERIAL AND GEOMETRIC NONLINEARITIES BY THE FINITE
              ELEMENT METHOD. REPORT UCRL-51390, LAWRENCE LIVERMORE
              LABORATORY, UNIVERSITY OF CALIFORNIA, LIVERMORE, 1973.
912   HARUNG H.S., MILLAR M.A., AND BROTTON D.M., IMPERFECTIONS IN
              AXIALLY LOADED PLANE FRAMES. INT.J.NUMER.METH.ENG.8, 847-
              864, 1974.
913   HARVEY J.W., AND KELSEY S., TRIANGULAR PLATE BENDING ELEMENTS
              WITH ENFORCED COMPATIBILITY. AIAA J.9, 1023-1026, 1971.
914   HARTIG D., THE HERMITIAN RECTANGLE FINITE ELEMENT FAMILY. PAPER
              M2/4, PROCEEDINGS 2ND STRUCTURAL MECH. IN REACTOR TECH.
              CONF., BERLIN, 1973.
915   HARTUNG R.F., AN ASSESSMENT OF CURRENT CAPABILITY FOR COMPUTER
              ANALYSIS OF SHELL STRUCTURES. COMPUTERS AND STRUCTURES,
              1, 3-32, 1971.
916   HARTZMAN M. AND HUTCHINSON J.R., NONLINEAR DYNAMICS OF SOLIDS BY
              THE FINITE ELEMENT METHOD. COMPUTERS AND STRUCTURES 2,
              47-72, 1972.
917   HARZMAN M., NONLINEAR DYNAMIC ANALYSIS OF AXISYMMETRIC SOLIDS
              BY THE FINITE ELEMENT METHOD. REPORT UCRL-74978,
              LAWRENCE LIVERMORE LABORATORY, UNIVERSITY OF CALIFORNIA,
              1974.
918   HASSELIN G., HELIOT J. AND VOUILLON C., USE OF THE TITUS FINITE
              ELEMENT CODE FOR THE NUCLEAR STRUCTURES AND INTERNAL STRESS-
              ANALYSIS. PAPER M2/3, PROCEEDINGS 2ND STRUCTURAL MECH.
              IN REACTOR TECH.CONF., BERLIN, 1973.
919   HAYES D.J. AND MARCAL P.V., DETERMINATION OF UPPER BOUNDS FOR
              PROBLEMS IN PLANE STRESS USING FINITE ELEMENT TECHNIQUES,
              INT.J.MECH.SCI.9, 245-252, 1967.
920   HAYS D.F., VARIATIONAL FORMULATION OF THE HEAT EQUATION. CHAPTER 2,
              OF R.J. DONNELLY (ED.), NON-EQUILIBRIUM THERMODYNAMICS,
              VARIATIONAL TECHNIQUES AND STABILITY. UNIVERSITY OF CHICAGO
              PRESS, 1966.

921   HEAD J.L. AND JEZERNIK A., FINITE ELEMENT ANALYSIS OF STRESSES
           IN FUEL ELEMENTS OF A HIGH-TEMPERATURE GAS COOLED REACTOR.
           PAPER D2/2, PROCEEDINGS 1ST STRUCTURAL MECH. IN REACTOR
           TECH. CONF., BERLIN, 1971.

922   HEDSTROM G.W. AND VARGA R.S., APPLICATION OF BESOV SPACES TO SPLINE
           APPROXIMATION. J. APPROX.THEORY 4, 295-327,1971.

923   HEISE U., COMPILED APPLICATION OF FINITE ELEMENT METHODS AND
           RICHARDSON EXTRAPOLATION TO THE TORSION PROBLEM.
           PP.225-237 OF J.R.WHITEMAN (ED.), THE MATHEMATICS OF FINITE
           ELEMENTS AND APPLICATIONS, ACADEMIC PRESS, LONDON, 1973.

924   HEISE U., A FINITE ELEMENT ANALYSIS IN POLAR CO-ORDINATES OF THE
           SAINT VENANT TORSION PROBLEM. INT.J.NUMER.METH.ENG.8,
           713-729, 1974.

925   HELFRICH H-P., FEHLERABSCHATZUNGEN FUR DAS GALERKINVERFAHREN ZUR
           LOSUNG VON EVOLUTIONSGLEICHUNGEN. MANUSCRIPTA MATHEMATICA
           (TO APPEAR).

926   HELLAN K., ON THE UNITY OF THE CONSTANT STRAIN/CONSTANT MOMENT FINITE
           ELEMENT METHODS. INT.J.NUMER.METH.ENG.6, 191-200, 1973.

927   HELLEN T.K., ECONOMICAL COMPUTER TECHNIQUES FOR SOLID FINITE
           ELEMENTS. REPORT RD/B/N-1217, CENTRAL ELECTRICITY
           GENERATING BOARD, BERKELEY NUCLEAR LABORATORY, 1969.

928   HELLEN T.K., SOLID WEDGE FINITE ELEMENTS FOR STRESS ANALYSIS
           REPORT R.D/B/N-1305, CENTRAL ELECTRICITY GENERATING BOARD,
           BERKELEY NUCLEAR LABORATORY, 1969.

929   HELLEN T.K., THE APPLICATION OF THE BERSAFE FINITE ELEMENT SYSTEM
           TO NUCLEAR DESIGN PROBLEMS. PROCEEDINGS 1ST STRUCTURAL
           MECH. IN REACTOR TECH.CONF., BERLIN, 1971.

930   HELLEN T.K., FINITE ELEMENT CALCULATION OF STRESS INTENSITY
           FACTORS USING ENERGY TECHNIQUES. PAPER 65/3, PROC.2ND
           STRUCTURAL MECH. IN REACTOR TECH.CONF., BERLIN, 1973.

931   HELLEN T.K., AND MONEY H.A., THE APPLICATION OF THREE DIMENSIONAL
           FINITE ELEMENTS TO A CYLINDER-CYLINDER INTERSECTION.
           INT. J.NUMER.METH.ENG.ENG.2,415-418, 1970.

932   HELLWIG G., DIFFERENTIAL OPERATORS OF MATHEMATICAL PHYSICS. ADDISON
           WESLEY, MASSACHUSETTS, 1964.

933   HELLWIG G., PARTIAL DIFFERENTIAL EQUATIONS. BLAISDELL, NEW YORK, 1964.

934   HELMHOLTZ H., ZUR THEORIE DER STATIONAREN STROME IN REIBENDEN FLUSSIGKEITEN,
           COLLECTED WORKS, VOL.1, PAPER XII, BARTH, LEIPZIG, 1882.

935   HENDRY J.A. AND DELVES L.M., VARIATIONAL SOLUTION OF TWO DIMENSIONAL
           FLUID FLOW PROBLEMS. INTERNAL REPORT CSS-73-4, DEPARTMENT OF
           COMPUTATIONAL AND STATISTICAL SCIENCE, UNIVERSITY OF LIVERPOOL,
           1973.

936   HENNART J.P., COMPARISON OF EXTRAPOLATION TECHNIQUES WITH HIGHER
           ORDER FINITE ELEMENT MODELS FOR DIFFUSION EQUATIONS WITH
           PIECEWISE CONTINUOUS MATERIAL PROPERTIES. PP.95-119,
           MATHEMATICAL MODELS AND COMPUTATIONAL TECHNIQUES FOR ANALYSIS
           OF NUCLEAR SYSTEMS. AMERICAN NUCL.SOC., 1973.

937   HENNART J.P., SINGULARITIES IN THE FINITE ELEMENT APPROXIMATION
           OF TWO DIMENSIONAL DIFFUSION PROBLEMS. TRANS.AMER.NUCL.
           SOC.17, 239-240, 1973.

938   HENSHELL R.D., PAFEC 70+, THE MANUAL OF THE PAFEC-SYSTEM. DEPARTMENT
           OF MECHANICAL ENGINEERING, UNIVERSITY OF NOTTINGHAM, 1972.

939   HENSHELL R.D., ON HYBRID FINITE ELEMENTS. PP299-312 OF WHITEMAN (ED.), THE
           MATHEMATICS OF FINITE ELEMENTS AND APPLICATIONS, ACADEMIC
           PRESS, LONDON, 1973.

940   HENSHELL R.D. NEALE B.K. AND WARBURTON G.B., A NEW HYBRID
           CYLINDRICAL SHELL FINITE ELEMENT. J.SOUND AND VIBRATION
           16, 519-532, 1971.

941   HENSHELL R.D. AND SHAW K.G., CRACK TIP FINITE ELEMENTS ARE UNNECESSARY.
           TECH.RPT., DEPARTMENT OF MECHANICAL ENGINEERING,
           UNIVERSITY OF NOTTINGHAM, 1973.

942   HERBOLD R.J., SCHULTZ M.H., AND VARGA R.S., QUADRATURE SCHEMES
           FOR THE NUMERICAL SOLUTION OF BOUNDARY VALUE PROBLEMS BY
           VARIATIONAL TECHNIQUES. AEQUAT. MATH. 3, 96-119, 1969.

943   HERBOLD R.J. AND VARGA R.S., THE EFFECT OF QUADRATURE ERRORS
           IN THE NUMERICAL SOLUTION OF TWO-DIMENSIONAL BOUNDARY

VALUE PROBLEMS BY VARIATIONAL TECHNIQUES. AEQUAT.MATH.
7, 36-58, 1972.

944 HERRERA I., AND BIELAK J., A SIMPLIFIED VERSION OF GURTIN'S
VARIATIONAL PRINCIPLE. ARCH. RAT. MECH. ANAL. 53, 131-149,
1974.

945 HERRMANN L.R., A BENDING ANALYSIS OF PLATES. PROC.1ST CONF.MATRIX
METHODS IN STRUCTURAL MECHANICS, WRIGHT-PATTERSON AFB.,
OHIO, AFFDL TR66-80, 1965.

946 HERRMANN L.R., ELASTICITY EQUATIONS FOR INCOMPRESSIBILE AND
NEARLY INCOMPRESSIBLE MATERIALS BY A VARIATIONAL THEOREM,
AIAA J.3, 1896-1900, 1965.

947 HERRMANN L., ELASTIC AND TORSIONAL ANALYSIS OF IRREGULAR SHAPES,
J.ENG.MECH.DIV.PROC.ASCE 91,EM6, 11-19, 1965.

948 HERRMANN L.R., FINITE ELEMENT BENDING ANALYSES OF PLATES.
J.ENG.MECH.DIV. ASCE,EM5,13-25,1968.

949 HERRMANN L.R. AND CAMPBELL D.M., A FINITE ELEMENT ANALYSIS FOR
THIN SHELLS. AIAA J.6, 1842-1846, 1968.

950 HIBBITT H.D., ELASTIC-PLASTIC AND CREEP ANALYSIS OF PIPELINES
BY FINITE ELEMENTS. PP.239-251, PROC.2ND INT.CONF.
PRESSURE VESSEL TECHNOLOGY, A.S.M.E., NEW YORK, 1973.

951 HIBBITT H.D. MARCAL P.V. AND RICE J.R., A FINITE ELEMENT
FORMULATION FOR PROBLEMS OF LARGE STRAIN AND LARGE
DISPLACEMENT. INT.J.SOLIDS STRUCTURES 6, 1069-1086, 1970.

952 HIBBITT H.D., SORENSEN E.P., AND MARCAL P.V., ELASTIC-PLASTIC AND CREEP
ANALYSIS OF PIPELINES BY FINITE ELEMENTS. PP.239-251 PROC. 2ND
INT. CONF. PRESSURE VESSEL TECHNOLOGY, AMERICAN SOC. MECH. ENG.,
NEW YORK, 1973.

953 HICKS G.W., FINITE ELEMENT ELASTIC BUCKLING ANALYSES. J.STRUCT.
DIV. A.S.C.E. 93, ST6,71-86, 1967.

954 HILBERT S., NUMERICAL METHODS FOR ELLIPTIC BOUNDARY PROBLEMS,
PH.D. THESIS, UNIVERSITY OF MARYLAND, 1969.

955 HILTON P.D., AND SIH G.C., APPLICATIONS OF THE FINITE ELEMENT
METHOD TO THE CALCULATIONS OF STRESS INTENSITY FACTORS.
PP. 426-483 OF G.C. SIH (ED.), MECHANICS OF FRACTURE:
METHODS OF ANALYSIS AND SOLUTIONS OF CRACK PROBLEMS.
NOORDHOFF, LEYDEN, 1972.

956 HINATA M., SHIMASAKI M. AND KIYONO T., NUMERICAL SOLUTION OF
PLATEAU'S PROBLEM BY A FINITE ELEMENT METHOD. MATH.COMP,
28, 45-60, 1974.

957 HINTON E., AND CAMPBELL J.S., LOCAL GLOBAL SMOOTHING OF DISCONTINUOUS FINITE
ELEMENT FUNCTIONS USING A LEAST SQUARES METHOD. INT.J.NUMER.METH.ENG.
8, 461-480, 1974.

958 HINTON E., OWEN D.R.J., AND IRONS B.M., FINITE ELEMENT PROGRAMMING,
VOLUME IN SERIES COMPUTATIONAL MATHEMATICS AND
APPLICATIONS. ACADEMIC PRESS, LONDON, (TO APPEAR).

959 HLAVACEK I., ON A SEMI-VARIATIONAL METHOD FOR PARABOLIC EQUATIONS I,
APLIKACE MATHEMATIKY 17,327-351,1972.

960 HLAVACEK I., ON A SEMI-VARIATIONAL METHOD FOR PARABOLIC EQUATIONS II,
APLIKACE MATEMATIKY 18,43-64,1973.

961 HLAVACEK I., ON A CONJUGATE SEMI-VARIATIONAL METHOD FOR PARABOLIC
EQUATIONS. APLIKACE MATEMATIKY 18, 434-444, 1973.

962 HODGE P.G., NUMERICAL APPLICATIONS OF MINIMUM PRINCIPLES IN
PLASTICITY. ENGINEERING PLASTICITY. CAMBRIDGE UNIVERSITY
PRESS, 1968.

963 HODGE P.G., AND HCMAHON A.A., A SIMPLE FINITE ELEMENT MODEL FOR ELASTIC
PLASTIC PLATE BENDING. NATIONAL SYMP. COMPUTERISED STRUCTURAL
ANALYSIS AND DESIGN. GEORGE WASHINGTON UNIVERSITY, 1972.

964 HOEG K., FINITE ELEMENT ANALYSIS OF STRAIN SOFTENING CLAY. J.SOIL.MECH.
AND FOUNDATIONS DIV. A.S.C.E. 98, SM10, 43-58, 1972.

965 HOFF N.J. (ED.), CREEP IN STRUCTURES. SPRINGER VERLAG, BERLIN,
1962.

966 HOFFMANN A., LIVOLANT M., AND ROCHE R., ANALYSIS OF SHELLS OF ANY
FORM IN THE PLASTIC REGION BY THE FINITE ELEMENT METHOD.
PROC.2ND STRUCTURAL MECH. IN REACTOR TECH.CONF., BERLIN,
1973.

967 HOFFMANN A., LIVOLANT M., AND ROCHE R., PLASTIC ANALYSIS OF

SHELLS BY FINITE ELEMENT METHODS. PAPER L6/2, PROC.
2ND STRUCTURAL MECH. IN REACTOR TECH. CONF., BERLIN, 1973.

968 HOFMANN H.H., LOAD LIMIT ANALYSIS, COMBINATION OF STATISTICAL METHODS
       (MONTE CARLO METHOD) AND FINITE ELEMENT METHOD FOR FRACTURE
       CALCULATIONS. PP.253-272, VOL.4, PROCEEDINGS 1ST STRUCTURAL
       MECH. IN REACTOR TECH. CONF., BERLIN, 1971.

960 HOFMANN H.H., PRESTRESSED CONCRETE REACTOR PRESSURE VESSEL
       FAILURE ANALYSIS. THE INCORPORATION OF STATISTICAL
       METHODS IN FINITE ELEMENT CALCULATIONS. PAPER H3/6,
       PROCEEDINGS 1ST STRUCTURAL MECH. IN REACTOR TECH. CONF.,
       BERLIN, 1971.

970 HOFMEISTER L.D., AND EVENSEN D.A., VIBRATION PROBLEMS USING
       ISOPARAMETRIC SHELL ELEMENTS. INT.J.NUMER.METH.ENG.
       5, 142-145, 1972.

971 HOFMEISTER L.D., GREENBAUM G.A. AND EVENSEN D.A., LARGE STRAIN
       ELASTO-PLASTIC FINITE ELEMENT ANALYSIS. AIAA J. 9, 1248-
       1254, 1971.

972 HOHN W., UBER DER NUMERISCHE BEHANDLUNG VON VARIATIONSPROBLEM
       MIT NATURLICHEN RANDBEDINGUNGEN IN ZWEI DIMENSIONEN.
       REPORT 161, FACHBEREICH MATHEMATIK, TECHNISCHE HOCHSCHULE,

973 HOLAND I., STIFFNESS MATRICES FOR PLATE BENDING ELEMENTS. FINITE
       ELEMENT METHODS. TAPIR, TRONDHEIM, 1969.

974 HOLAND I., THE FINITE ELEMENT METHOD IN PLANE STRESS ANALYSIS.
       FINITE ELEMENT METHODS. TAPIR, TRONDHEIM, 1969.

975 HOLAND I., FUNDAMENTALS OF THE FINITE ELEMENT METHOD. COMPUTERS
       AND STRUCTURES 4, 3-16, 1974.

976 HOLAND I., AND BELL K., (EDS.), THE FINITE ELEMENT METHOD IN STRESS
       ANALYSIS. TAPIR, TRONDHEIM, 1969.

977 HOLAND I., AND BERGAN P.G., HIGHER ORDER FINITE ELEMENT FOR PLANE
       STRESS. PROC. A.S.C.E., EM2, 698-702, 1968.

978 HOLAND I., AND MOAN T., THE FINITE ELEMENT METHOD IN PLATE BUCKLING.
       FINITE ELEMENT METHODS.. TPAIR, TRONDHEIM, 1969.

970 HOLUSA L., KRATOCHVIL J., ZLAMAL M., AND ZENISEK A., CALCULATION OF A
       PLATE OF CONSTANT THICKNESS BY THE FINITE ELEMENT METHOD.
       STAVEBNICKY CASOPIS 17, 779-783, 1969.

980 HOLZLOHNER U., A FINITE ELEMENT ANALYSIS FOR TIME-DEPENDENT PROBLEMS.
       INT.J.NUMER.METH.ENG.8,55-69,1974

981 HOOD P., FINITE ELEMENT FORMULATION WITH REFERENCE TO FLUID
       DYNAMICS. PROC.CONF.COMP.METHODS IN NONLINEAR MECHANICS.
       AUSTIN, TEXAS, 1974.

982 HOPF E., A REMARK ON LINEAR ELLIPTIC DIFFERENTIAL EQUATIONS OF SECOND
       ORDER. PROC.AMERICAN MATH. SOC.3, 791-793, 1952.

983 HOPPE V., FINITE ELEMENTS WITH HARMONIC INTERPOLATION FUNCTIONS,
       PP.131-142 OF J.R.WHITEMAN (ED.), THE MATHEMATICS OF FINITE
       ELEMENTS AND APPLICATIONS. ACADEMIC PRESS, LONDON, 1973.

984 HORMANDER L., LINEAR PARTIAL DIFFERENTIAL OPERATORS.
       SPRINGER-VERLAG, BERLIN, 1963.

985 HOUSEHOLDER A.S., KWIC INDEX FOR NUMERICAL ALGEBRA. TECHNICAL
       REPORT ORNL-4778, OAK RIDGE NATIONAL LABORATORY,
       TENNESSEE, 1972.

986 HRENIKOFF A., A SOLUTION OF PROBLEMS IN ELASTICITY BY THE
       FRAMEWORK METHOD. J.APPL.MECH. 8A, 169-175, 1941.

987 HSU T.R., BERTELS A.W.M., ARYA B., AND BANERJEE S., APPLICATION
       OF THE FINITE ELEMENT METHOD TO THE NONLINEAR ANALYSIS OF
       NUCLEAR REACTOR FUEL BEHAVIOUR. PROC.CONF.COMP.METHODS
       IN NONLINEAR MECH., AUSTIN, TEXAS, 1974.

988 HU HAI-CHANG., ON SOME VARIATIONAL PRINCIPLES IN THE THEORY OF
       ELASTICITY AND THE THEORY OF PLASTICITY. SCIENTIA SINICA
       4, 33-35, 1955.

980 HUANG Y.H., FINITE ELEMENT ANALYSIS OF NONLINEAR SOIL MEDIA.
       PP. 662-689 OF APPLICATION OF FINITE ELEMENT METHODS IN
       CIVIL ENGINEERING, A.S.C.E., 1969.

990 HUBBARD B.E. (ED.), NUMERICAL SOLUTION OF PARTIAL DIFFERENTIAL
       EQUATIONS - II, SYNSPADE 1970. ACADEMIC PRESS, NEW YORK,
       1971.

991 HUEBNER K.H., APPLICATION OF FINITE ELEMENT METHODS TO

THERMOHYDRODYNAMIC LUBRICATION. INT.J.NUMER.METH.ENG.8, 139-165, 1974.

992 HUGHES T.J.R., AND ALLIK H., FINITE ELEMENTS FOR COMPRESSIBLE AND INCOMPRESSIBLE CONTINUA. PP. 27-62 OF APPLICATION OF FINITE ELEMENT METHODS IN CIVIL ENGINEERING. A.S.C.E., 1969.

993 HULME B.L., INTERPOLATION BY RITZ APPROXIMATION. J.MATH.MECH.18, 337-342, 1968.

994 HULME B.L., PIECEWISE BICUBIC METHODS FOR PLATE BENDING PROBLEMS. PH.D. THESIS, HARVARD UNIVERSITY, 1969.

995 HULME B.L., A NEW BICUBIC INTERPOLATION OVER RIGHT TRIANGLES. J.APPROX.THEORY 5, 66-73, 1972.

996 HULME B.L., DISCRETE GALERKIN AND RELATED ONE-STEP METHODS FOR ORDINARY DIFFERENTIAL EQUATIONS. MATH.COMP.26, 881-891, 1972.

997 HULME B.L., ONE-STEP PIECEWISE POLYNOMIAL GALERKIN METHODS FOR INITIAL VALUE PROBLEMS. MATH.COMP.26, 415-426,1972.

998 HUNG H.D., DUALITY IN THE ANALYSIS OF SHELLS BY THE FINITE ELEMENT METHOD. INT.J.SOLIDS AND STRUCTURES 7, 281-300, 1971.

999 HUNT J.T., KNITTEL M.R., AND BARACH O., FINITE ELEMENT APPROACH TO ACOUSTIC RADIATION FROM ELASTIC STRUCTURES. J.ACCOUST. SOC.AMERICA 55, 269-280, 1974.

1000 HURLEY F.X., AND LINBACK R.K., A SELF-ADJUSTING GRID FOR FINITE-DIFFERENCE PROGRAMS. INT.J.NUMER.METH.ENG.5, 585-587, 1973.

1001 HURR R.T., MODELLING GROUND WATER FLOW BY THE FINITE ELEMENT METHOD. INT. CONF. VARIATIONAL METHODS IN ENG., SOUTHAMPTON UNIVERSITY, 1972.

1002 HUSSEY M.J.L., THATCHER R.W., AND BERNAL M.J.M., ON THE CONSTRUCTION AND USE OF FINITE ELEMENTS. J.INST.MATH.APPLICS., 6, 263-282, 1970.

1003 HUTCHINS G.J., AND SOLER A.I., APPROXIMATE ELASTICITY SOLUTION FOR MODERATELY THICK SHELLS OF REVOLUTION. TRANS. A.S.M.E. SER.E 40, 955-960, 1973.

1004 HUTT J.M., AND SALAM A.E., DYNAMIC STABILITY OF PLATES BY FINITE ELEMENTS. J.ENG.MECH.DIV. A.S.C.E. 97, EM3, 879-899, 1971.

1005 HUTTON S.G., AND ANDERSON D.L., FINITE ELEMENT METHOD; A GALERKIN APPROACH. J.ENG.MECH.DIV. A.S.C.E. 97, EM5, 1503-1520, 1971.

1006 HUTULA D.N., AND WIANCKO B.E., MATUS A THREE-DIMENSIONAL FINITE ELEMENT PROGRAM FOR SMALL STRAIN ELASTIC ANALYSIS. REPORT WAPD-TM-1081, BETTIS ATOMIC POWER LABORATORY, PITTSBURG, 1973.

1007 HUTULA D.N., WIANCKO B.E., AND ZEILER S.M., APACHE A THREE-DIMENSIONAL FINITE ELEMENT PROGRAM FOR STEADY-STATE OR TRANSIENT HEAT CONDUCTION ANALYSIS. REPORT WAPD-TM-1080, BETTIS ATOMIC POWER LABORATORY, PITTSBURG, 1973.

1008 HUTULA D.N., AND ZEILER S.M., MESH-3D A THREE-DIMENSIONAL FINITE ELEMENT MESH GENERATOR PROGRAM FOR EIGHT NODE ISOPARAMETRIC ELEMENTS. REPORT WAPD-TM-1079, BETTIS ATOMIC POWER LABORATORY, PITTSBURG, 1973.

1009 HUANG C.T., HO M.K. AND WILSON N.E., FINITE ELEMENT ANALYSIS OF SOIL DEFORMATIONS. PP. 729-746 OF APPLICATION OF FINITE ELEMENT METHODS IN CIVIL ENGINEERING. A.S.C.E., 1969.

1010 HUANG C., AND PI W.S., NONLINEAR ACOUSTIC RESPONSE ANALYSIS OF PLATES USING THE FINITE ELEMENT METHOD. AIAA J.10, 276-281, 1972.

1011 IDING R.H., PISTER K.S., AND TAYLOR R.L., IDENTIFICATION OF NONLINEAR ELASTIC SOLIDS BY A FINITE ELEMENT METHOD. COMP.METH.APPL.MECH.ENG.4, 121-142, 1974.

1012 IKEBE Y., THE GALERKIN METHOD FOR THE NUMERICAL SOLUTION OF FREDHOLM INTEGRAL EQUATIONS OF THE SECOND KIND. SIAM REVIEW 14, 465-491, 1972.

1013 IKEGAWA M. AND WASHIZU K., FINITE ELEMENT METHOD APPLIED TO ANALYSIS OF FLOW OVER A SPILLWAY CREST. INT.J.NUMER.METH.ENG.6,179-189,1973.

1014 IRONS B.M., STRUCTURAL EIGENVALUE PROBLEMS; ELIMINATION OF UNWANTED VARIABLES. AIAA J.3, 961-962, 1965.

1015 IRONS B.M., DISTRIBUTED MASS MATRIX FOR PLATE ELEMENT BENDING.

                AIAA J.4, 189, 1966.

1016   IRONS B.M., ENGINEERING APPLICATION OF NUMERICAL INTEGRATION IN
          STIFFNESS METHOD. AIAA J. 4, 2035-2037, 1966.

1017   IRONS B.M., ROUNDOFF CRITERIA IN DIRECT STIFFNESS SOLUTIONS,
          AIAA J.6, 1308-1312, 1968.

1018   IRONS B.M., A CONFORMING QUARTIC TRIANGULAR ELEMENT FOR PLATE
          BENDING. INT.J.NUMER.METH.ENG.1, 29-46, 1969.

1019   IRONS B.M., ECONOMICAL COMPUTER TECHNIQUES FOR NUMERICALLY INTEGRATED
          FINITE ELEMENTS. INT.J.NUMER.METH.ENG.1, 201-204, 1969.

1020   IRONS B.M., A FRONTAL SOLUTION PROGRAM FOR FINITE ELEMENT ANALYSIS,
          INT.J.NUMER.METH.ENG.2, 5-32, 1970.

1021   IRONS B.M., A BOUND THEOREM IN EIGENVALUES AND ITS PARACTICAL
          APPLICATIONS. PROC. 3RD CONF. MATRIX METHODS IN STRUCTURAL
          MECHANICS, WRIGHT-PATTERSON AFB, OHIO, 1971.

1022   IRONS B.M., QUADRATURE RULES FOR BRICK BASED FINITE ELEMENTS.
          INT.J.NUM.METH.ENG. 3, 293-294, 1971.

1023   IRONS B.M., A NEW FORMULATION FOR PLATE BENDING ELEMENTS. INT.CONF.
          VARIATIONAL METHODS IN ENG., SOUTHAMPTON UNIVERSITY, 1972.

1024   IRONS B.M., A TECHNIQUE FOR DEGENERATING BRICK-TYPE ISOPARAMETRIC
          ELEMENTS USING HIERARCHICAL MIDSIDE NODES. INT.J.NUMER.METH.ENG.
          8,203-209,1974.

1025   IRONS B.M., AND DRAPER K.J., INADEQUACY OF NODAL CONNECTIONS IN
          A STIFFNESS SOLUTION FOR PLATE BENDING. AIAA J.5, 961,
          1965.

1026   IRONS B.M., ERGATOUDIS J.G., AND ZIENKIEWICZ O.C., COMPLETE
          POLYNOMIAL DISPLACEMENT FIELDS FOR FINITE ELEMENT METHODS.
          TRANS. ROYAL AERO.SOC. 72, 709-711, 1968.

1027   IRONS B.M., OLIVEIRA E.R. DE A., AND ZIENKIEWICZ O.C., COMMENTS
          ON THE PAPER: THEORETICAL FOUNDATIONS OF THE FINITE
          ELEMENT METHOD. INT.J.SOLIDS STRUCTURES 6, 695-697, 1970.

1028   IRONS B.M., AND RAZZAQUE A., EXPERIENCE WITH THE PATCH TEST FOR
          CONVERGENCE OF FINITE ELEMENTS. PP. 557-587 OF A.K. AZIZ
          (ED.), THE MATHEMATICAL FOUNDATIONS OF THE FINITE ELEMENT
          METHOD WITH APPLICATIONS TO PARTIAL DIFFERENTIAL EQUATIONS,
          ACADEMIC PRESS, NEW YORK, 1972.

1029   IRONS B.M., AND RAZZAQUE A., SHAPE FUNCTION FORMULATIONS FOR ELEMENTS
          OTHER THAN DISPLACEMENT MODELS. PROC. SYMP. VARIATIONAL METHODS,
          UNIVERSITY OF SOUTHAMPTON, 1972.

1030   IRONS B.M., AND RAZZAQUE A., A FURTHER MODIFICATION TO AHMAD'S
          SHELL ELEMENT. INT.J.NUMER.METH.ENG.5, 588-589, 1973.

1031   IRONS B.M., AND ZIENKIEWICZ O.C., THE ISOPARAMETRIC FINITE ELEMENT
          SYSTEM - A NEW CONCEPT IN FINITE ELEMENT ANALYSIS. RECENT
          ADVANCES IN STRESS ANALYSIS. ROYAL AERO.SOC., 1969.

1032   IRONS B.M., ZIENKIEWICZ O.C. AND OLIVEIRA E.R. DE A., COMMENTS ON
          THE PAPER: THEORETICAL FOUNDATIONS OF THE FINITE ELEMENT
          METHOD. INT.J.SOLIDS STRUCTURES 6, 695-697,1970.

1033   IRVING J., AND CARMICHAEL G.D.T., COMPARISON OF CALCULATED AND MEASURED
          STRAINS IN OLDBURY CONCRETE PRESSURE VESSELS. J.BRIT.NUCL.ENERGY
          SOC.8, 60-64, 1969.

1034   ISAACS L.T., A CURVED CUBIC TRIANGULAR FINITE ELEMENT FOR
          POTENTIAL FLOW PROBLEMS. INT.J.NUMER.METH.ENG.7, 537-544,
          1973.

1035   ISAKSON G., DISCRETE ELEMENT PLASTIC ANALYSIS OF STRUCTURES IN A
          STATE OF MODIFIED PLANE STRAIN. AIAA.J.7, 545-546, 1971.

1036   ISAKSON G., AND LEVY A., FINITE ELEMENT ANALYSIS OF INTERLAMINAR
          SHEAR FIBROUS COMPOSITES. J. COMPOSITE MATERIALS 5, 273-276,
          1971.

1037   ISENBERG J., AND ADHAM S.A., SEISMIC INTERACTION OF SOIL AND POWER
          PLANTS. J.POWER DIV.A.S.C.E. 98, PO2, 273-279, 1972.

1038   ITO F., A COLLOCATION METHOD FOR BOUNDARY VALUE PROBLEMS USING SPLINE
          FUNCTIONS. PH.D. THESIS, BROWN UNIVERSITY, 1972.

1039   IVERSEN P.A., SOME ASPECTS OF THE FINITE ELEMENT METHOD IN TWO
          DIMENSIONAL PROBLEMS. TAPIR, TRONDHEIM, 1969.

1040   IWASAKI T., AND SHIMIZU T., ANALYSIS OF BRANCH PIPE STRESS (JAPANESE),
          HITACHI HYDRON 54, 1053-1058, 1972.

1041   JACQUIN J.C., AND HANSEN K.F., FINITE ELEMENT SOLUTIONS FOR

MULTIREGION PROBLEMS.   TRANS.AMER.NUCL.SOC.17, 238-239,
1973.

1042   JAEGER T.A. (ED.),   STRUCTURAL MECHANICS IN REACTOR TECHNOLOGY.   VOL.5:
STRUCTURAL ANALYSIS AND DESIGN, PART M, DESIGN, RELIABILITY,
COMPUTATIONAL METHODS.   E.E.C. LUXEMBURG, 1971.

1043   JAMET P.,   ON THE CONVERGENCE OF FINITE-DIFFERENCE APPROXIMATIONS
TO ONE-DIMENSIONAL SINGULAR BOUNDARY VALUE PROBLEMS.
NUMER.MATH.14, 355-378, 1969.

1044   JAMET P.,   ESTIMATIONS D'ERREUR POUR DES ELEMENTS FINIS DROITS
PRESQUE DEGENERES.   REPORT CRM-447, CENTRE D'ETUDES DE
LIMEIL, VILLENEUVE SAINT-GEORGES, 1974.

1045   JAMET P. AND RAVIART P.A.,   NUMERICAL SOLUTION OF THE STATIONARY
NAVIER-STOKES EQUATIONS BY FINITE ELEMENT METHODS.

1046   JANENKO N.N. AND KVASOV B.I., AN ITERATIVE METHOD FOR THE CONSTRUCTION OF
POLYCUBIC SPLINE FUNCTIONS.
SOVIET MATH.DOKL.11,1643-1645,1970.

1047   JANOVSKY V.,   ELLIPTIC BOUNDARY VALUE PROBLEMS WITH NONVARIATIONAL
PERTURBATION AND THE FINITE ELEMENT METHOD.   APLIKACE
MATEMATIKY 18, 422-433, 1973.

1048   JASWON M.A., AND SYMM G.,   INTEGRAL EQUATION METHODS IN POTENTIAL
THEORY AND ELASTOSTATICS. VOLUME IN SERIES ON COMPUTATIONAL
MATHEMATICS AND APPLICATIONS, ACADEMIC PRESS, LONDON.
(TO APPEAR).

1049   JAVANDEL I. AND WITHERSPOON P.A.,   APPLICATION OF THE FINITE ELEMENT
METHOD TO TRANSIENT FLOW IN POROUS MEDIA.   J.SOC.PETROLIUM
ENG. 241-252, 1968.

1050   JENKINS W.M.,   MATRIX AND DIGITAL COMPUTER METHODS IN STRUCTURAL
ANALYSIS.   MCGRAW HILL, NEW YORK, 1968.

1051   JEROME J.,   ON N-WIDTH IN SOBOLEV SPACES AND APPLICATIONS TO ELLIPTIC
BOUNDARY VALUE PROBLEMS.   MATHEMATICS RESEARCH CENTER,
TECHNICAL REPORT NO.927, UNIVERSITY OF WISCONSIN, MADISON,1968.

1052   JEROME J.W.   MINIMIZATION PROBLEMS AND LINEAR AND NONLINEAR SPLINE
FUNCTIONS. I:EXISTENCE,II:CONVERGENCE.
SIAM J.NUMER.ANAL.10, 808-819 AND 820-830, 1973.

1053   JEZERNIK A., ALUJEVIC A., AND HEAD J.L.,   FINITE ELEMENT MATRIX DISPLACEMENT
METHOD FOR TIME DEPENDENT STRESS ANALYSIS OF REACTOR MATERIALS
IN TWO- AND THREE-DIMENSIONS.   PAPER M5/8, PROCEEDINGS 1ST
STRUCTURAL MECH. IN REACTOR TECH. CONF., BERLIN, 1971.

1054   JEZERNIK A., AND LEECH A.,   THE COMPARISON OF ITERATIVE AND DIRECT
SOLUTION TECHNIQUES IN THE ANALYSIS OF TIME-DEPENDENT
STRESS PROBLEMS, INCLUDING CREEP, BY THE FINITE ELEMENT
METHOD. PP.449-462 OF J.R. WHITEMAN (ED.), THE MATHEMATICS
OF FINITE ELEMENTS AND APPLICATIONS. ACADEMIC PRESS,
LONDON, 1973.

1055   JEZERNIK A., AND MILLER M.C.,   LARGE USER-ORIENTATED SYSTEMS OF
PROGRAMS FOR STRUCTURAL ANALYSIS AND DESIGN. PAPER M1/2,
PROC.2ND STRUCTURAL MECH. IN REACTOR TECH. CONF., BERLIN,
1973.

1056   JEZERNIK A., AND MILLER M.C.,   LARGE USER ORIENTED SYSTEMS OF
PROGRAMS FOR STRUCTURAL ANALYSIS AND DESIGN. NUCL.ENG.
AND DESIGN 27, 238-273, 1974.

1057   JOFRIET J.C., AND MCNEICE G.M.,   FINITE ELEMENT ANALYSIS OF
REINFORCED CONCRETE BEAMS. J.STRUCTURAL DIV.A.S.C.E. 97,
ST3, 785-806, 1971.

1058   JOHNSON C.,   ON THE CONVERGENCE OF A MIXED FINITE ELEMENT FOR PLATE
BENDING PROBLEMS. NUMER.MATH.21, 43-62, 1973.

1059   JOHNSON W.H., AND MCLAY R.W.,   CONVERGENCE OF THE FINITE ELEMENT
METHOD IN THE THEORY OF ELASTICITY. J.APPL.MECH.TRANS.
ASME, 274-278, 1968.

1060   JONES D.S.,   GENERALISED FUNCTIONS.   MCGRAW HILL, LONDON, 1966.

1061   JONES R.E.,   A GENERALIZATION OF THE DIRECT STIFFNESS METHOD OF
STRUCTURAL ANALYSIS. AIAA J.2, 821-826, 1964.

1062   JONES R.E.,   A SELF-ORGANIZING MESH GENERATION PROGRAM.   PRESSURE
VESSELS AND PIPING CONF., NUCLEAR AND MATERIALS DIV. A.S.
M.E., MIAMI, 1974.

1063   JONES R.E., AND STROME D.R.,   A SURVEY OF THE ANALYSIS OF SHELLS

BY THE DISPLACEMENT METHOD. PROC. 1ST CONF.MATRIX METHODS
IN STRUCTURAL-MECHANICS. WRIGHT-PATERSON AFB., OHIO,
AFFDL TR66-80, 1965.

1064 JONES R.E., AND STROME D.R., DIRECT STIFFNESS METHOD OF ANALYSIS
OF SHELLS OF REVOLUTION UTILIZING CURVED ELEMENTS. AIAA,
J.4, 1519-1525, 1966.

1065 JORDAN W.B., IMPROVED CALCULATION OF ELASTO-PLASTIC SYSTEMS BY
THE FINITE ELEMENT METHOD. TECHNICAL REPORT KAPL-M-7108,
KNOLLS ATOMIC POWER LABORATORY SCHENECTADY, NEW YORK, 1970.

1066 JORDAN W.B., A.E.C., R AND D REPORT KAPL-M-7112, 1970.

1067 KACHANOV L.M., VARIATIONAL PRINCIPLES FOR ELASTIC-PLASTIC SOLIDS.
PRIKL.MATH.I MEKH.6, 1942.

1068 KALKANI E.C., MESH GENERATION PROGRAM FOR HIGHWAY EXCAVATION
CUTS. INT.J.NUMER.METH.ENG.8, 369-394, 1974.

1069 KAMEL H.A., AND EISENSTEIN H.K., AUTOMATIC MESH GENERATION IN TWO
AND THREE DIMENSIONAL INTERCONNECTED DOMAINS. PAPER AT
SYMP. HIGH SPEED COMPUTING OF ELASTIC STRUCTURES,
IUTAM,LEIGE, BELGIUM, 1970.

1070 KAMEL H.A., AND LIU D., APPLICATION OF THE FINITE ELEMENT METHOD
TO SHIP STRUCTURES. COMPUTERS AND STRUCTURES, 1, 103-130,
1971.

1071 KAMMERER W.J., REDDIEN G.W., AND VARGA R.S., QUADRATIC
INTERPOLATING SPLINES: THEORY AND APPLICATIONS. N.A.T.O.
ADV.STUDY INST., NUMERICAL SOLUTION OF PARTIAL DIFFERENTIAL
EQUATIONS, KJELLER, NORWAY, AUGUST 1973.

1072 KAMMERER W.J. AND VARGA R.S., ON ASYMPTOTICALLY BEST NORMS FOR POWERS
OF OPERATORS. NUMER. MATH. 20, 93-98, 1972.

1073 KAN D., MESH AND CONTOUR PLOT FOR TRIANGLE AND ISOPARAMETRIC ELEMENTS.
COMPUTER REPORT CNME/CR/39, DEPARTMENT OF CIVIL ENGINEERING,
UNIVERSITY OF WALES,SWANSEA,1970.

1074 KANG, C.M., AND HANSEN K.F., FINITE ELEMENT METHODS FOR SPACE-TIME
REACTOR ANALYSIS. REPORT MITNE-135, DEPARTMENT OF NUCLEAR ENG.,
MASSACHUSETTS INSTITUTE OF TECHNOLOGY, CAMBRIDGE, 1971.

1075 KANG C.M., AND HANSEN K.F., FINITE ELEMENT METHODS FOR REACTOR
ANALYSIS. NUCLEAR SCIENCE AND ENG.51, 456-495, 1973.

1076 KANTOROVICH L.V., AND AKILOV G.P., FUNCTIONAL ANALYSIS IN NORMED
SPACES, PERGAMON PRESS, OXFORD, 1964.

1077 KANTOROVICH L.V. AND KRYLOV V.I., APPROXIMATE METHODS OF HIGHER
ANALYSIS. NOORDHOFF, GRONINGEN, 1964.

1078 KAPER H.G., LEAK G.K., AND LINDEMAN, A.J., APPLICATION OF
FINITE ELEMENT METHODS IN REACTOR MATHEMATICS.
NUMERICAL SOLUTION OF THE NEUTRON DIFFUSION EQUATION,
REPORT ANL-7925, ARGONNE NATIONAL LABORATORY, ARGONNE,
ILLINOIS.

1079 KAPER H.G., LEAF G.K., AND LINDEMAN A.J., APPLICATIONS OF FINITE ELEMENT
METHODS IN REACTOR MATHEMATICS; NUMERICAL SOLUTION OF THE NEUTRON
DIFFUSION EQUATION. REPORT ANL-7925, ARGONNE NATIONAL LABORATORY,
ILLINOIS, 1972.

1080 KAPER H.G., LEAF G.K., AND LINDEMAN A.J., TIMING COMPARISON STUDY FOR SOME
HIGH ORDER FINITE ELEMENT PROCEDURES AND A LOW ORDER FINITE
DIFFERENCE APPROXIMATION PROCEDURE FOR THE NUMERICAL SOLUTION OF
THE MULTIGROUP NEUTRON DIFFUSION EQUATION. NUCL.SCI.ENG,49,
27-48, 1972.

1081 KAPUR K.K., PREDICTION OF PLATE VIBRATIONS USING A CONSISTENT
MASS MATRIX. AIAA.J.4, 565-566, 1966.

1082 KAPUR K.K., STABILITY OF PLATES USING THE FINITE ELEMENT METHOD, J.ENG.
MECH.DIV.A.S.C.E. 92, EMX, 177-195, 1966.

1083 KAPUR K.K., VIBRATIONS OF A TIMOSHENKO BEAM USING A FINITE ELEMENT
APPROACH. J. AMERICAN ACOUSTICAL SOC. 40, 1058-1063, 1966.

1084 KAPUR K.K., AND HARTZ B.J., STABILITY OF PLATES USING THE FINITE
ELEMENT METHOD. J.ENG.MECH.DIV.PROC. A.S.C.E.92,177-195,1966.

1085 KARCHER H.J., UBER DIE ANWENDUNG EINES KONJUGIERTEN-GRADIENTEN-
VERFAHRENS AUF DIE FINIT-ELEMENT-BERECHNUNG NICHTLINEARER
STRUCTUREN. Z.A.M.M. 54, 85-86, 1974.

1086 KARLIN S., TOTAL POSITIVITY VOL.1. STANFORD UNIVERSITY PRESS, 1968.

1087 KATO T., PERTURBATION THEORY FOR LINEAR OPERATORS. SPRINGER

VERLAG, BERLIN, 1966.

1088   KAUFMAN S., AND HALL D.B.,   BENDING ELEMENTS FOR PLATE AND SHELL
       NETWORKS.   AIAA J.5, 402-405, 1967.

1089   KAVANAGH K.T.,   ORTHOGONAL MODES FOR THE SOLUTION TO STATIC NONLINEAR
       PROBLEMS.   REPORT SC-DC-722277,SANDIA LABORATORIES, ALBUQUERQUE,
       1972.

1090   KAVANAGH K.T., AND CLOUGH R.W.,   FINITE ELEMENT APPLICATIONS IN
       THE CHARACTERIZATION OF ELASTIC SOLIDS. J. SOLIDS STRUCTURES 7,
       11-24, 1971.

1091   KAWAHARA M.,   A NUMERICAL ANALYSIS ON VISCOPLASTIC STRUCTURES
       BY THE FINITE ELEMENT METHOD.   BULL.FACUL.SCI. AND ENG.
       CHUO UNIVERSITY 15, 75-91,1972.

1092   KAWAHARA M.,   LARGE STRAIN, VISCOELASTIC NUMERICAL ANALYSIS BY MEANS OF
       FINITE ELEMENT METHOD.   PROC.J.S.C.E. 204, 141-149,1972.

1093   KAWAHARA M., AND KAMEMURA K.,   ELASTOVISCOPLASTIC FINITE ELEMENT
       ANALYSIS BY PERTURBATION METHOD.   PROC.CONF.COMP.METHODS IN
       NONLINEAR MECH., AUSTIN, TEXAS, 1974.

1094   KAWAHARA M., YOSHIMURA N., NAKAGAWA K. AND OHSAKA H.,   STEADY FLOW
       ANALYSIS OF INCOMPRESSIBLE VISCOUS FLUID BY THE FINITE ELEMENT
       METHOD.   THEORY AND PRACTICE IN FINITE ELEMENT STRUCTURAL
       ANALYSIS, UNIVERSITY OF TOKYO PRESS,1973.

1095   KAWAI T.,   FINITE ELEMENT ANALYSIS OF GEOMETRICALLY NONLINEAR
       PROBLEMS. PROC., JAPAN-U.S. SEMINAR ON MATRIX METHODS IN
       STRUCTURAL ANALYSIS AND DESIGN, TOKYO, 1969.

1096   KAWAI T.,   APPLICATION OF THE FINITE ELEMENT METHOD TO PRESSURE VESSEL
       DESIGN (JAPANESE).   KOATSURYOKU 8, 1877-1885, 1970.

1097   KAWAI T., MURAKI T., AND TOYOSHI T.,   FINITE ELEMENT ANALYSIS OF
       THIN-WALLED STRUCTURES WITH APPLICATION TO STRUCTURAL
       PROBLEMS IN REACTOR TECHNOLOGY.   PAPER K7/3, PROCEEDINGS
       2ND STRUCTURAL MECHANICS IN REACTOR TECH.CONF., BERLIN,
       1973.

1098   KAWAI T., AND TADA Y.,   FINITE ELEMENT ANALYSIS OF A WING STRUCTURE,
       PP. 727-744 OF J.T. ODEN, R.W. CLOUGH, AND Y. YAMOMOTO (EDS.),
       ADVANCES IN COMPUTATIONAL METHODS IN STRUCTURAL MECHANICS AND
       DESIGN.   UNIVERSITY OF ALABAMA PRESS, HUNTSVILLE, 1972.

1099   KAWAI T.,   THE APPLICATION OF FINITE ELEMENT METHODS TO SHIP
       STRUCTURES. NAT. SYMP.COMPUTERISED STRUCT. ANAL. AND DESIGN,
       GEORGE WASHINGTON UNIVERSITY, 1972.

1100   KAWAI T., MURAKI T., TANAKA N., AND TWAKI T.,   FINITE ELEMENT
       ANALYSIS OF THIN-WALLED STRUCTURES BASED ON THE MODERN
       ENGINEERING THEORY OF BEAMS. PROC.3RD CONF.MATRIX MATHODS IN
       STRUCTURAL MECHANICS, WRIGHT-PATTERSON AFB., OHIO, 1971.

1101   KAWAI T. AND OHTSUBO H.,   A METHOD OF SOLUTION FOR THE COMPLICATED
       BUCKLING PROBLEMS OF ELASTIC PLATES WITH COMBINED USE OF
       RAYLEIGH-RITZ PROCEDURE IN THE FINITE ELEMENT METHOD.
       PROC.2ND CONF.MATRIX METHODS IN STRUCTURAL MECHANICS,
       WRIGHT-PATTERSON AFB., OHIO, AFFDL-TR-68-150,1968.

1102   KAWAI T., AND OHTSUBO H.,   ON THE STATES OF STRESS AND DEFORMATION
       OF CYLINDRICAL SPECIMENS OF BRITTLE MATERIAL UNDER UNIAXIAL
       COMPRESSION.   PROC., JAPAN-U.S. SEMINAR ON MATRIX METHODS
       IN STRUCTURAL ANALYSIS AND DESIGN, TOKYO, 1969.

1103   KAWAI T., AND YOSHIMURA N.,   ANALYSIS OF LARGE DEFLECTION OF PLATES
       BY THE FINITE ELEMENT METHOD.   INT.J.NUMER.METH.ENG 1,
       123-134, 1969.

1104   KEEGSTRA P.N.R., HEAD J.L., AND ALUJEVIC A.,   FINITE ELEMENT
       STRESS ANALYSIS OF INTERACTING FUEL PELLET AND CANNING
       UNDER FAST REACTOR CONDITION.   PAPER C3/3, PROCEEDINGS
       2ND STRUCTURAL MECH. IN REACTOR TECH.CONF., BERLIN,
       1973.

1105   KELLOGG O.D.,   FOUNDATIONS OF POTENTIAL THEORY.   SPRINGER-VERLAG, BERLIN,
       1929.

1106   KELLOGG R.B.,   DIFFERENCE EQUATIONS ON A MESH ARISING FROM A GENERAL
       TRIANGULATION.   MATH.COMP. 18, 203-210, 1964.

1107   KELLOGG R.B.,   RITZ DIFFERENCE EQUATIONS ON A TRIANGULATION.   (TO APPEAR).

1108   KELLOGG R.B.,   HIGH ORDER SINGULARITIES FOR INTERFACE PROBLEMS.
       PP.589-602 OF A.K. AZIZ (ED.),   THE MATHEMATICAL

FOUNDATIONS OF THE FINITE ELEMENT METHOD WITH APPLICATIONS
TO PARTIAL DIFFERENTIAL EQUATIONS. ACADEMIC PRESS, NEW
YORK, 1972.

1109 KELLOGG R.G., INTERPOLATION BETWEEN SUBSPACES OF A HILBERT
SPACE. TECHNICAL NOTE BN-719, INSTITUTE FOR FLUID
DYNAMICS AND APPLIED MATHEMATICS, UNIVERSITY OF MARYLAND,
1972.

1110 KELLOGG R.B., SOME MATHEMATICAL ASPECTS OF FINITE ELEMENT CALCULATIONS.
TRANS. AMERICAN NUCL. SOC.16, 131-XXX, 1973.

1111 KELSEY S., FINITE ELEMENT METHODS IN CIVIL ENGINEERING. PP. 775-789
OF J.T. ODEN, R. CLOUGH, AND Y. YAMOMOTO (EDS.), ADVANCES IN
COMPUTATIONAL METHODS IN STRUCTURAL MECHANICS AND DESIGN,
UNIVERSITY OF ALABAMA PRESS, HUNTSVILLE, 1972.

1112 KELSEY S.J., THE APPLICATION OF QUASILINEARIZATION TO NON-LINEAR SYSTEMS.
INT.J.NUMER.METH.ENG.8, 589-611, 1974.

1113 KERR A.D., AN EXTENSION OF THE KANTOROVICH METHOD. Q.APPL.MATH.26,
219-228,1968

1114 KERR A.D., AN EXTENDED KANTOROVICH METHOD FOR THE SOLUTION OF
EIGENVALUE PROBLEMS. INT.J.SOLIDS STRUCTURES 5, 559-572,
1969.

1115 KERR A.D., A NEW ITERATIVE SCHEME FOR THE SOLUTION OF PARTIAL
DIFFERENTIAL EQUATIONS. (MANUSCRIPT)

1116 KERR A.D. AND ALEXANDER H., AN APPLICATION OF THE EXTENDED KANTOROVICH
METHOD TO THE STRESS ANALYSIS OF A CLAMPED RECTANGULAR PLATE.
ACTA MECHANICA 6, 180-196, 1968.

1117 KESLER C.E. (ED.), CONCRETE FOR NUCLEAR REACTORS, VOL.I. AMERICAN
CONCRETE INSTITUTE, DETROIT, 1972.

1118 KEY J.E., COMPUTER PROGRAM FOR SOLUTION OF LARGE, SPARSE,
UNSYMMETRIC SYSTEMS OF LINEAR EQUATIONS. INT. J.NUMER.
METH.ENG.6, 497-509,1973.

1119 KEY S.W., A CONVERGENCE INVESTIGATION OF THE DIRECT STIFFNESS METHOD.
PH.D. DISSERTATION, UNIVERSITY OF WASHINGTON, 1969.

1120 KEY S.W., THE TRANSIENT DYNAMIC ANALYSIS OF THIN SHELLS BY THE FINITE
ELEMENT METHOD. REPORT SC-DC-713889, SANDIA LABORATORIES,
ALBUQUERQUE, 1971.

1121 KEY S.W., A FINITE ELEMENT PROCEDURE FOR THE LARGE DEFORMATION
DYNAMIC RESPONSE OF AXISYMMETRIC SOLIDS. COMP.METH.APPL.
MECH.ENG.4, 195-218, 1974.

1122 KEY S.W. AND BEISINGER Z., THE ANALYSIS OF THIN SHELLS WITH
TRANSVERSE SHEAR STRAINS BY THE FINITE ELEMENT METHOD.
PROC.2ND CONF.MATRIX METHODS IN STRUCTURAL MECHANICS,
WRIGHT-PATTERSON AFB, OHIO, AFFDL-TR-80-150, 1968.

1123 KEY S.W., AND BEISINGER Z.E., SLADE. A COMPUTER PROGRAM FOR THE
STATIC ANALYSIS OF THIN SHELLS. REPORT SC-RR-69-369,
SANDIA LABORATORIES, ALBUQUERQUE, 1970.

1124 KEY S.W. AND BEISINGER Z.E., THE ANALYSIS OF THIN SHELLS BY THE
FINITE ELEMENT METHOD. PROC. IUTAM. SYMP. HIGH SPEED
COMPUTING OF ELASTIC STRUCTURES, LEIGE, 1970.

1125 KEY S.W., AND BEISINGER Z.E., THE TRANSIENT DYNAMIC ANALYSIS OF
THIN SHELLS IN THE FINITE ELEMENT METHOD. PROC. 3RD CONF.
MATRIX METHODS IN STRUCTURAL MECHANICS, WRIGHT-PATTERSON
AFB., OHIO, 1971.

1126 KEY S.W., AND BEISINGER Z.E., SLADE D., A COMPUTER PROGRAM FOR THE DYNAMIC
ANALYSIS OF THIN SHELLS. REPORT SLA-73-79, SANDIA LABORATORIES,
ALBUQUERQUE, 1973.

1127 KEY S.W., AND KRIEG R.D., COMPARISON OF FINITE ELEMENT AND FINITE
DIFFERENCE METHODS. O.N.R. SYMP. NUMERICAL AND COMPUTER
METHODS IN STRUCTURAL MECHANICS, URBANA, 1971.

1128 KHANNA J., CRITERION FOR SELECTING STIFFNESS MATRICES. AIAA J.
3, 1976.,1965.

1129 KHANNA J., AND HOOLEY R.F., COMPARISON AND EVALUATION OF STIFFNESS
MATRICES. AIAA.J.4, 2105-2111, 1966.

1130 KHATUA T.P., AND CHEUNG Y.K., BENDING AND VIBRATION OF MULTILAYER
SANDWICH BEAMS AND PLATES. INT.J.NUMER.METH.ENG.6, 11-24,
1973.

1131 KIKUCHI F., A FINITE ELEMENT METHOD FOR NON SELF-ADJOINT

PROBLEMS. INT.J.NUMER.METH.ENG.6, 39-54, 1973.

1132 KIKUCHI F., AND ANDO Y., A FINITE ELEMENT METHOD FOR INITIAL VALUE
PROBLEMS. PROC. 3RD CONF. MATRIX METHODS IN STRUCTURAL
MECHANICS, WRIGHT-PATTERSON AFB., OHIO, 1971.

1133 KIKUCHI F., AND ANDO Y., A NEW VARIATIONAL PRINCIPLE FOR THE
FINITE ELEMENT METHOD AND ITS APPLICATION TO PLATE AND
SHELL PROBLEMS. NUCL.ENG. AND DESIGN 21, 95-113, 1972.

1134 KIKUCHI F., AND ANDO Y., FINITE ELEMENT SOLUTIONS FOR PLATE BENDING
PROBLEMS BY SIMPLIFIED HYBRID DISPLACEMENT METHOD. J. NUCL.
SCI. TECH. (TOKYO) 15, 155-178, 1972.

1135 KIKUCHI F., AND ANDO Y., LUMPED FINITE ELEMENT BASES FOR BEAMS AND
PLATES. J.NUCL.SCI.TECH. (TOKYO) 9, 749-751, 1972.

1136 KIKUCHI F., AND ANDO Y., SIMPLIFIED HYBRID DISPLACEMENT METHOD
APPLIED TO PLATE BUCKLING PROBLEMS. J.NUCL.SCI.TECH,
9, 497-499, 1972.

1137 KIKUCHI F., AND ANDO Y., SOME FINITE ELEMENT SOLUTIONS FOR
PLATE BENDING BY SIMPLIFIED HYBRID DISPLACEMENT METHOD.
NUCL.ENG.AND DESIGN 23, 155-178, 1972.

1138 KIKUCHI F. AND ANDO Y., APPLICATION OF SIMPLIFIED HYBRID
DISPLACEMENT METHOD TO PLATE AND SHELL PROBLEMS. PAPER
M5/5, PROCEEDINGS 2ND STRUCTURAL MECH. IN REACTOR TECH.
CONF., BERLIN, 1973.

1139 KIKUCHI F. AND ANDO Y., CONVERGENCE OF FINITE ELEMENT SCHEMES
BASED ON NONCONFORMING AND SIMPLIFIED HYBRID DISPLACEMENT
METHODS. PAPER M2/6, PROC.2ND STRUCTURAL MECH. IN REACTOR
TECH.CONF., BERLIN, 1973.

1140 KIKUCHI F. AND ANDO Y., CONVERGENCE OF LUMPED FINITE ELEMENT
SCHEMES FOR SELECTED INITIAL VALUE PROBLEMS. PAPER M2/7,
PROCEEDINGS 2ND STRUCTURAL MECH. IN REACTOR TECH. CONF.,
BERLIN, 1973.

1141 KIKUCHI F., AND ANDO Y., CONVERGENCE OF A MIXED FINITE ELEMENT SCHEME
FOR PLATE BENDING. NUCL. ENG. AND DESIGN 24, 357-373, 1973.

1142 KIKUCHI F. AND ANDO Y., ON THE CONVERGENCE OF FINITE ELEMENT
SCHEMES BASED ON NON-CONFORMING AND SIMPLIFIED HYBRID
DISPLACEMENT METHODS. PAPER M2/6, PROCEEDINGS 2ND
STRUCTURAL MECH. IN REACTOR TECH. CONF., BERLIN, 1973.

1143 KIKUCHI F., OHYA H., AND ANDO Y., APPLICATION OF FINITE ELEMENT
METHOD TO AXISYMMETRIC BUCKLING OF SHALLOW SPHERICAL
SHELLS UNDER EXTERNAL PRESSURE. J.NUCL.SCI.TECH.(TOKYO)
10, 339-347, 1973.

1144 KING I.P., TRIANGULAR MESH GENERATION PROGRAM. C.P.R.2, CENTRE FOR
NUMERICAL METHODS IN ENGINEERING, SWANSEA, 1967.

1145 KINSER W., AND DELLA TORRE E., AN ITERATIVE APPROACH TO THE
FINITE ELEMENT METHOD IN FIELD PROBLEMS. I.E.E.E. TRANS,
MICROWAVE THEORY AND TECH., MTT-22, 221-228, 1974.

1146 KIRIOKA K., AND HIRATA T., COMPUTING SYSTEM FOR STRUCTURAL ANALYSIS OF
CAR BODIES. PP. 529-550 OF J.T. ODEN, R.W. CLOUGH, AND Y. YAMOMOTO
(EDS.), ADVANCES IN COMPUTATIONAL METHODS IN STRUCTURAL MECHANICS
AND DESIGN. UNIVERSITY OF ALABAMA PRESS, HUNTSVILLE, 1972.

1147 KLEIN S., DISCUSSION ON FINITE ELEMENT SOLUTION FOR ASIXYMMETRICAL SHELLS.
J.ENG.MECH DIV.A.S.C.E. 91, EMX, 262-268, 1965.

1148 KLEIN S.A., A STATIC AND DYNAMIC FINITE ELEMENT SHELL ANALYSIS WITH
EXPERIMENTAL VERIFICATION. INT.J.NUMER.METH.ENG.3,
299-316, 1971.

1149 KLOTZLER R., MEHRDIMENSIONALE VARIATIONSRECHNUNG. BIRKHAUSER-VERLAG,
BASEL, 1970.

1150 KNUDSON W., AND NAGY D., DISCRETE DATA SMOOTHING BY SPLINE
INTERPOLATION WITH APPLICATION TO INITIAL GEOMETRY OF CABLE
NETS. COMP.METH.APPL.MECH.ENG.4, 321-348, 1974.

1151 KOBAYASHI A.S., WOO S.L.Y., LAWRENCE C., AND SCHLEGEL W.A.,
ANALYSIS OF THE CORNEO-SCLERAL SHELL BY THE METHOD OF DIRECT
STIFFNESS. J. BIOMECHANICS 4, 323-330, 1971.

1152 KOBZA W., MULTIANGULAR ELEMENTS IN PLATE BENDING. INT.J.NUMER.METH.
ENG. 7, 545-551, 1973.

1153 KOHLER W., AND PITTR J., CALCULATION OF TRANSIENT TEMPERATURE FIELDS WITH
FINITE ELEMENTS IN SPACE AND TIME DIMENSIONS. INT.J.NUMER.METH.

ENG.8, 625-631, 1974.

1154 KOHN W., IMPROVEMENT OF RAYLEIGH-RITZ EIGENFUNCTIONS. SIAM REVIEW 14, 399-419, 1972.

1155 KOLAR V., SINGULARITATEN EINER VERALLEMEINERTEN METHODE DER EUDLICHEN ELEMENTS UND IHRE ANWENDUNG BEI BERECHNUNG NICHT LINEAREU PLATTENAUFGABEU. Z.A.M.M. 47, T205-207, 1967.

1156 KOLAR V., THE INFLUENCE OF DIVISION ON THE RESULTS IN THE FINITE ELEMENT METHOD. Z.A.M.M., GAMM-TAGUNG, DELFT, 1970.

1157 KOLAR V., KRATOCHVIL J., LEITNER F. AND ZENISEK A., STRESS ANALYSIS OF TWO-AND THREE-DIMENSIONAL STRUCTURES BY THE FINITE ELEMENT METHOD. SPRINGER VERLAG, WIEN, 1972. (IN GERMAN).

1158 KOLAR V., KRATOCHVIL J., ZLAMAL M., AND ZENISEK A., TECHNICAL PHYSICAL AND MATHEMATICAL PRINCIPLES OF THE FINITE ELEMENT METHOD. ACADEMIA, CZECHOSLOVAK ACADEMY OF SCIENCES, PRAGUE, 1971.

1159 KOLAR V. AND NEMEC I., THE EFFICIENT FINITE ELEMENT ANALYSIS OF RECTANGULAR AND SKEW LAMINATED PLATES. INT.J.NUMER.METH.ENG.7, 309-323,1973.

1160 KONDRATEV V.A., BOUNDARY PROBLEMS FOR ELLIPTIC EQUATIONS WITH CONICAL OR ANGULAR POINTS. TRANS. MOSCOW. MATH. SOC. 16, 297-313, 1967.

1161 KONRAD A., LINEAR ACCELERATOR CAVITY FIELD CALCULATION BY THE FINITE ELEMENT METHOD. I.E.E.E. TRANS. NUCL.SCI.NS20,802-808, 1973.

1162 KONRAD A. AND SILVESTER P., SCALAR FINITE ELEMENT PACKAGE FOR TWO DIMENSIONAL FIELD PROBLEMS. TRANS I.E.E.E. (MICROWAVE THEORY AND TECHNIQUES) 952-955,1971.

1163 KONRAD A., AND SILVESTER P., A FINITE ELEMENT PROGRAM PACKAGE FOR AXISYMMETRIC SCALAR FIELD PROBLEMS. COMP. PHYS. COMMUNICATIONS 5, 437-455, 1973.

1164 KORMAN T., MORGHEN F.T., AND BALUCH M.H., BENDING OF MICROPOLAR PLATES. NUCL.ENG.DESIGN 26, 432-439, 1974.

1165 KORNEEV V., THE COMPARISON OF THE FINITE ELEMENT METHOD WITH THE VARIATIONAL-DIFFERENCE METHOD IN THE THEORY OF ELASTICITY. (RUSSIAN). IZVESTIA REPORTS OF THE WHOLE-UNION SCIENTIFIC INVESTIGATION OF THE HYDRO-TECHNICAL INSTITUTE 83, 286-307, 1967.

1166 KOSHLYAKOV N.S., SMIRNOV M.M. AND GLINER E.B., DIFFERENTIAL EQUATIONS OF MATHEMATICAL PHYSICS. NORTH HOLLAND, AMERSTERDAM, 1964.

1167 KOUKAL S., PIECEWISE POLYNOMIAL INTERPOLATIONS AND THEIR APPLICATIONS TO PARTIAL DIFFERENTIAL EQUATIONS. CZECH, SBORNIK VAAZ,BRNO, 29-38, 1970.

1168 KOUKAL S., PIECEWISE POLYNOMIAL INTERPOLATIONS IN THE FINITE ELEMENT METHOD. APLIKACE MATEMATIKY 18, 146-160, 1973.

1169 KRAHULA J.L., ANALYSIS OF BENT AND TWISTED BARS USING THE FINITE ELEMENT METHOD. AIAA J.5, 1194-1197, 1967.

1170 KRAHULA J.L., A FINITE ELEMENT SOLUTION FOR SAINT-VENANT TORSION. AIAA J.7, 2200-2203, 1969.

1171 KRUHULA J.L. AND POLHEMUS J.F., USE OF FOURIER SERIES IN THE FINITE ELEMENT METHOD. AIAA J. 6, 726-727, 1968.

1172 KRAJCINOVIC D., A CONSISTENT DISCRETE ELEMENT TECHNIQUE FOR THIN WALLED ASSEMBLAGES. INT. J. SOLIDS STRUCTURES 5, 639-662, 1969.

1173 KRASNOSELSKII M.A., THE CONVERGENCE OF THE GALERKIN METHOD FOR NONLINEAR EQUATIONS. DOKL.AKAD.NAUK.S.S.S.R. 73, 1121-1124, 1950.

1174 KRASNOSELSKII M.A. AND VAINIKKO G.M., APPROXIMATE SOLUTION OF OPERATOR EQUATIONS. (RUSSIAN). NAUKA, MOSCOW, 1969.

1175 KRATOCHVIL J., AND ZENISEK A., CALCULATION OF RECTANGULAR PLATES BY THE FINITE ELEMENT METHOD. STAVEBNICKY CASOPIS 17, 641-653, 1969.

1176 KRATOCHVIL J., ZENISEK A., AND ZLAMAL M., A SIMPLE ALGORITHM FOR THE STIFFNESS MATRIX OF TRIANGULAR PLATE BENDING ELEMENTS. INT.J.NUMER. METH.ENG. 3, 553-564,1971.

1177 KRIEG R.D., AND KEY S.W., TRANSIENT SHELL RESPONSE BY NUMERICAL TIME INTEGRATION. REPORT SC-DC-721392, SANDIA LABORATORIES, ALBUQUERQUE, 1972.

1178 KREIN S.G., AND PETUNIN Y.I., SCALES OF BANACH SPACES. RUSSIAN

MATHEMATICAL SURVEYS, 21,NO.2. 85-160, 1966.

1179 KRESS R.,   ON GENERAL HERMITE TRIGONOMETRIC INTERPOLATION.

1180 KRESS R.,   EINABLEITINGSFREIES RESTGLIED FUR DIE TRIGONOMETRISCHE
                INTERPOLATION PERIODISCHER ANALYTISCHER FUNCTIONEN.
                NUMER.MATH.16, 389-396,1971.

1181 KRIEG R.D., AND KEY S.W., TRANSIENT SHELL RESPONSE BY NUMERICAL TIME
                INTEGRATION. PP.237-258 OF J.T. ODEN, R.W. CLOUGH AND Y. YAMOMOTO
                (EDS.), ADVANCES IN COMPUTATIONAL METHODS IN STRUCTURAL MECHANICS
                AND DESIGN. UNIVERSITY OF ALABAMA PRESS, HUNTSVILLE, 1972.

1182 KRIEG R.D. AND KEY S.W.,   TRANSIENT SHELL RESPONSE BY NUMERICAL TIME
                INTEGRATION. INT.J.NUMER.METH.ENG.7, 273-286,1973.

1183 KRISHNA MURTY A.V., AND RAO G.V., ASSESSMENT OF ACCURACIES OF
                FINITE ELEMENT EIGENVALUES. PP.379-386 OF J.R. WHITEMAN
                (ED.), THE MATHEMATICS OF FINITE ELEMENTS AND APPLICATIONS.
                ACADEMIC PRESS, LONDON, 1973.

1184 KRISHNA MURTHY N., THREE DIMENSIONAL FINITE ELEMENT ANALYSIS OF
                THICK WALLED PIPE-NOZZLE JUNCTIONS WITH CURVED TRANSITIONS.
                PAPER G2/7, PROCEEDINGS 1ST STRUCTURAL MECH. IN REACTOR
                TECH.CONF.,   BERLIN, 1971.

1185 KRISHNAMURTHY N., THREE-DIMENSIONAL FINITE ELEMENT ANALYSIS OF THICK-
                WALLED VESSEL-NOZZLE JUNCTIONS WITH CURVED TRANSITIONS. REPORT
                ORNL-TM-3315, OAK RIDGE NATIONAL LABORATORY, TENNESSEE, 1971.

1186 KRYLOV V.I., APPROXIMATE CALCULATION OF INTEGRALS. (TRANSLATED
                BY A.H. STROUD). MACMILLAN, NEW YORK, 1962.

1187 KUCERA M. (ED.), THEORY OF NONLINEAR OPERATORS. ACADEMIC PRESS, NEW
                YORK, 1973.

1188 KUDRUAVCEV L.D., IMBEDDING THEOREMS FOR CLASSES OF FUNCTIONS
                DEFINED IN THE ENTIRE SPACE OR IN THE HALFSPACE. MATH.
                SB. 10, 616-639, AND 11, 3-35, 1966.

1189 KURODA T., MURAKAMI H., AND YAMAMOTO S., STRESS ANALYSIS OF A
                PRESTRESSED CONCRETE NUCLEAR PRESSURE VESSEL BY THE FINITE
                ELEMENT METHOD USING VARIABLE STRAIN ELEMENTS. PAPER H3/2,
                PROCEEDINGS 1ST STRUCTURAL MECH. IN REACTOR TECH. CONF.,
                BERLIN, 1971.

1190 KUTTLER J.R., FINITE DIFFERENCE APPROXIMATIONS FOR EIGENVALUES
                OF UNIFORMLY ELLIPTIC OPERATORS. SIAM J. NUMER.ANAL.7,
                206-232, 1970.

1191 LAASONEN P., ON THE SOLUTION OF POISSON'S DIFFERENTIAL EQUATION.
                J.ASSOC.COMP.MACH.5, 370-382, 1958.

1192 LAASONEN P., ON THE TRUNCATION ERROR OF DISCRETE APPROXIMATIONS
                TO THE SOLUTION OF DIRICHLET PROBLEMS IN A DOMAIN WITH
                CORNERS. J.ASSOC.COMP.MACH.5, 32-38, 1958.

1193 LAASONEN P., ON THE DISCRETIZATION ERROR OF THE DIRICHLET PROBLEM
                IN A PLANE REGION WITH CORNERS. ANNALE ACAD. SCIENT.
                FENN. SERIES A, MATHEMATICA 408, 1-16,1967.

1194 LAKSHMIKANTHAM C., AND TONG P., STRESSES AROUND HOLES IN
                STIFFENED COMPOSITE PANELS USING LAURENT-SERIES AND
                FINITE ELEMENT METHODS. 5TH INT.CONF.EXP.STRESS ANALYSIS,
                UDINE, 1974.

1195 LAKSHMINARAYANAN V., AND LANG T.E., DISCRETIZATION-DISPERSION IN
                FINITE ELEMENT MODELLING OF WAVE PROPAGATION IN SOLIDS.
                PROC.CONF.COMP.METHODS IN NONLINEAR MECHANICS, AUSTIN,
                TEXAS, 1974.

1196 LANCASTER P., INTERPOLATION IN A RECTANGLE AND FINITE ELEMENTS
                OF HIGHER DEGREE. MANUSCRIPT.

1197 LANCZOS C., APPLIED ANALYSIS, PRENTICE HALL, NEW JERSEY, 1956.

1198 LANCZOS C., LINEAR DIFFERENTIAL OPERATORS. WILEY, NEW YORK, 1961.

1199 LANDAU L., AND LIFCHITZ E., THEORIE D'ELASTICITE. MIR, MOSCOW,
                1967.

1200 LANGBALLE M., AASEN E., AND MELLEM T., APPLICATION OF THE FINITE
                ELEMENT METHOD TO MACHINERY. COMPUTERS AND STRUCTURES
                4, 149-192, 1974.

1201 LANGER R.E., (ED.), ON NUMERICAL APPROXIMATION, UNIVERSITY OF
                WISCONSIN PRESS, MADISON, 1959.

1202 LARSSON S.G., AND HARKEGARD G., ON THE FINITE ELEMENT ANALYSIS
                OF CRACK AND INCLUSION PROBLEMS IN ELASTIC-PLASTIC

MATERIALS. COMPUTERS AND STRUCTURES 4, 293-306, 1974.

1203 LASCAUX P., APPLICATION OF THE FINITE ELEMENT METHOD IN TWO-DIMENSIONAL
HYDRODYNAMICS USING LAGRANGIAN VARIABLES. REPORT LA-TR-73-3,
CENTRE D'ETUDES, COMMISSARIAT A L'ENERGIE ATOMIQUE, LIMEIL-BREVANNES,
FRANCE, 1973.

1204 LATHROP K.D., TRANSPORT THEORY NUMERICAL METHODS. REPORT LA-UR-73-517,
LOS ALAMOS SCIENTIFIC LABORATORY, NEW MEXICO, 1972.

1205 LATTES R., AND LIONS J.L., THE METHOD OF QUASI-REVERSIBILITY:
APPLICATIONS TO PARTIAL DIFFERENTIAL EQUATIONS. AMERICAN ELSEVIER,
NEW YORK, 1969.

1206 LAUNAY P., CHARPENET G. AND VOUILLON C.C., THE TRIDIMENSIONAL
THERMOELASTIC COMPUTER CODE TITUS. PAPER M5/4, PROCEEDINGS
1ST STRUCTURAL MECH. IN REACTOR TECH. CONF., BERLIN, 1971.

1207 LAVRENTIEV M.M., SOME IMPROPERLY POSED PROBLEMS OF MATHEMATICAL
PHYSICS. SPRINGER-VERLAG, BERLIN, 1967.

1208 LAWRENCE K.L., TWISTED BEAM ELEMENT MATRICES FOR BENDING. AIAA.
J.8, 1160-1161, 1970.

1209 LEAF G.K., AND KAPER H.G., L-INFINITY BOUNDS FOR MULTIVARIATE LAGRANGE
APPROXIMATION. SIAM J.NUMER.ANAL.11, 363-381, 1974.

1210 LEAF G.K., LINDEMAN A.J., AND KAPER H.G., CONSTRUCTION OF A
FINITE ELEMENT APPROXIMATION WHICH CROSSES MATERIAL
INTERFACES. REPORT ANL-8052, ARGONNE NATIONAL LABORATORY,
ILLINOIS, 1974.

1211 LEBEDEV N.N., SPECIAL FUNCTIONS AND THEIR APPLICATIONS.
PRENTICE HALL, NEW JERSEY, 1965.

1212 LEBESGUE H., LECONS SUR L'INTEGRATION. CHELSEA PUBLISHING CO.,
NEW YORK, 1928-73.

1213 LEE C.H., FINITE ELEMENT METHOD FOR TRANSIENT LINEAR VISCOUS FLOW
PROBLEMS. · PROC.CONF.NUMERICAL METHODS IN FLUID DYNAMICS,
SOUTHAMPTON, 1973.

1214 LEE E.H., A SURVEY OF VARIATIONAL METHODS FOR ELASTIC PROPAGATION
ANALYSIS IN COMPOSITES WITH PERIODIC STRUCTURES. PP.122-139 OF
DYNAMICS OF COMPOSITE MEDIA. AMERICAN SOC.MECH.ENG., 1972.

1215 LEE E.H., AND FORSYTHE G.E., VARIATIONAL STUDY OF NONLINEAR SPLINE
CURVES. SIAM REVIEW 15, 120-133, 1973.

1216 LEE K.N., AND NEMAT-NASSER S., MIXED VARIATIONAL PRINCIPLES,
FINITE ELEMENTS AND FINITE ELASTICITY. PROC.CONF., COMP.
METHODS IN NONLINEAR MECHANICS, AUSTIN, TEXAS, 1974.

1217 LEE K.P., GENERALIZED STIFFNESS MATRIX OF A CURVED BEAM ELEMENT.
AIAA J.7, 2043-2045, 1969.

1218 LEECH A., BERDYNE (PHASE 1), A COMPUTER SYSTEM FOR THE DYNAMIC
ANALYSIS OF STRUCTURES. C.E.G.B. REPORT MS/C/P295,1973

1219 LEECH A., DYSAFE, A COMPUTER SYSTEM FOR THE SIMPLE DYNAMIC
RESPONSED OF STRUCTURES. C.E.G.B. REPORT MS/C/P223, 1973.

1220 LEHMAN R.S., DEVELOPMENTS AT AN ANALYTIC CORNER OF SOLUTIONS OF
ELLIPTIC PARTIAL DIFFERENTIAL EQUATIONS. J.MATH.MECH.,8,
727-760, 1959.

1221 LEONARD J.W., LINEARIZED COMPRESSIBLE FLOW BY THE FINITE ELEMENT METHOD.
BELL AEROSYSTEMS CO., TECH.NOTE TCTN-9500-920156, 1969.

1222 LEONARD J.W., GALERKIN FINITE ELEMENT FORMULATION FOR
INCOMPRESSIBLE FLOW. TECH.NOTE TCTN-9500-920181, BELL
AEROSYSTEMS CO., 1970.

1223 LEONARD J.W., FINITE ELEMENT ANALYSIS OF PERTURBED COMPRESSIBLE FLOW.
INT. J. NUMER. METH.ENG.4, 123-132, 1972.

1224 LEONARD J.W. AND BRAMLETTE T.T., FINITE ELEMENT SOLUTIONS OF
DIFFERENTIAL EQUATIONS J.ENG.MECH.DIV.A.S.C.E.96, EM6,
1277-1283, 1970.

1225 LEONARD J.W. AND MELFI D., 3-D FINITE ELEMENT MODEL FOR LAKE
CIRCULATION. PROC.3RD CONF. MATRIX METHODS IN STRUCTURAL
MECHANICS, WRIGHT-PATTERSON AFB., OHIO, 1971.

1226 LERAY J., AND LIONS J.L., QUELQUES RESULTATS DE VISIK SUR LES PROBLEMES
ELLIPTIQUES NON LINEAIRES PAR LES METHODES DE MINTY-BROWDER.
BULL.SOC.MATH.FRANCE 93, 97-107, 1965.

1227 LESAINT P., RAYLEIGH RITZ GALERKIN METHODS FOR SYMMETRIC POSITIVE
DIFFERENTIAL EQUATIONS APPLICATION TO FINITE ELEMENT
METHODS. REPORT, CENTRE D'ETUDES DE LIMEIL, VILLENEUVE-

SAINT-GEORGES, 1973.

1228 LESAINT P., FINITE ELEMENT METHODS FOR THE TRANSPORT EQUATION. RAIRO,
SERIE MATHEMATIQUES, (TO APPEAR).

1229 LESAINT P., FINITE ELEMENT METHODS FOR SYMMETRIC HYPERBOLIC EQUATIONS.
NUMER.MATH.21,244-255,1973.

1230 LESAINT P., AND GERIN-ROZE J., ISOPARAMETRIC FINITE ELEMENT METHODS
FOR THE NEUTRON TRANSPORT EQUATION. (TO APPEAR).

1231 LESAINT P., AND RAVIART P.A., ON A FINITE ELEMENT METHOD FOR SOLVING
THE NEUTRON TRANSPORT EQUATION. REPORT 74008, LABORATOIRE
ANALYSE NUMERIQUE, UNIVERSITY OF PARIS, 1974.

1232 LESTINGI J. AND BROWN S., COMPARISON OF THE NUMERICAL INTEGRATION
TECHNIQUE AND THE FINITE ELEMENT METHOD IN THE ANALYSIS
OF THIN-SHELL STRUCTURES. PAPER M5/6, PROCEEDINGS 2ND
STRUCTURAL MECH.IN REACTOR TECH. CONF., BERLIN, 1973.

1233 LEVY A., ARMEN H., AND WHITESIDE J., ELASTIC AND PLASTIC INTERLAMINAR
SHEAR DEFORMATION IN LAMINATED COMPOSITES UNDER GENERALIZED PLANE
STRESS. PROC.3RD CONF. MATRIX METHODS IN STRUCTURAL MECHANICS.
WRIGHT-PATTERSON AFB., OHIO, 1971.

1234 LEVY N.J., AND MARCAL P.V., THREE-DIMENSIONAL ELASTIC-PLASTIC
STRESS ANALYSIS FOR FRACTURE MECHANICS. PROC.5TH ANNUAL
INF.MEETING HEAVY SECTION STEEL TECH.PROGRAM, OAK RIDGE
NATIONAL LAB., TENNESSEE, 1971.

1235 LEVY N., MARCAL P.V., OSTERGREN W.J., AND RICE J.R., SMALL SCALE
YIELDING NEAR A CRACK IN PLAIN STRAIN: A FINITE ELEMENT
ANALYSIS. INT.J.FRACT.MECH.7, 143-156, 1971.

1236 LEWIS D.J., IRVING J., AND CARMICHAEL G.D.T., ADVANCES IN THE ANALYSIS
OF PRESTRESSED CONCRETE PRESSURE VESSELS. NUCL.ENG.DESIGN 20,
543-573, 1972.

1237 LEWIS R.W., AND BRUCH J.C., AN APPLICATION OF LEAST SQUARES TO ONE
DIMENSIONAL TRANSIENT PROBLEMS. INT.J.NUMER.METH.ENG.8, 633-647,
1974.

1238 LEWIS R.W., AND GARNDER R.W., A FINITE ELEMENT SOLUTION OF
COUPLED ELECTROKINETIC AND HYDRODYNAMIC FLOW IN POROUS
MEDIA. INT.J.NUMER.METH.ENG.5, 41-55, 1972.

1239 LEWIS E.E., AND MILLER W.F., FINITE ELEMENT INTEGRAL NEUTRON
TRANSPORT. TRANS.AMER.NUCL.SOC.17, 237-238, 1973.

1240 LEWY H., ON A VARIATIONAL PROBLEM WITH INEQUALITIES ON THE BOUNDARY.
J.MATH.MECH. 17, 861-884, 1968.

1241 LEWY H., ON A MINIMUM PROBLEM FOR SUPERHARMONIC FUNCTIONS. INTERNATIONAL
CONF. ON FUNCTIONAL ANALYSIS, TOKYO, 1969.

1242 LEWY H., AND STAMPACCHIA G., ON THE REGULARITY OF A SOLUTION OF A
VARIATIONAL INEQUALITY. COMM. PURE APPL.MATH. 22, 153-188, 1969.

1243 LEWY H., AND STAMPACCHIA G., ON THE REGULARITY OF CERTAIN SUPERHARMONIC
FUNCTIONS. J.D'ANALYSE MATH.23, 227-236, 1970.

1244 LIAM FINN W.D., FINITE ELEMENT ANALYSIS OF SEEPAGE THROUGH DAMS.
J.SOIL MECH.AND FOUNDATIONS DIV., PROC. A.S.C.E.SM6, 41-48,
1967.

1245 LIEBERSTEIN H.M., THEORY OF PARTIAL DIFFERENTIAL EQUATIONS. ACADEMIC
PRESS, NEW YORK, 1972.

1246 LINDBERG G.M., AND OLSON M.D., CONVERGENCE STUDIES OF EIGENVALUE
SOLUTIONS USING TWO FINITE PLATE BENDING ELEMENTS.
INT.J.NUMER.METH.ENG.2,99-116, 1970.

1247 LINDBERG G.M. AND OLSON M.D., A HIGH-PRECISION TRIANGULAR
CYLINDRICAL SHELL FINITE ELEMENT. AIAA J, 9, 530-532, 1972.

1248 LINDBERG G.M. AND OLSON M.D. AND COWPER G.R., NEW DEVELOPMENTS
IN THE FINITE ELEMENT ANALYSIS OF SHELLS. PROC 3RD CONF.
MATRIX METHODS IN STRUCTURAL MECHANICS. WRIGHT-PATTERSON
AFB., OHIO, 1971.

1249 LINDLEY P.B., PLANE STRESS ANALYSIS OF RUBBER AT HIGH STRAINS USING
FINITE ELEMENTS. J.STRAIN ANALYSIS 6, 45-52, 1971.

1250 LINK M., ZUR BERECHNUNG VON PLATTEN NACH DER THEORIE II.
ORDNUNG MIT HILFE EINES HYBRIDEN DEFORMATIONSMODELLS.
INGENIEUR ARCHIV 42, 381-394, 1973.

1251 LINZ P., AND KROPP T.E., A NOTE ON THE COMPUTATION OF INTEGRALS
INVOLVING PRODUCTS OF TRIGONOMETRIC AND BESSEL FUNCTIONS.
MATH.COMP.27, 871-872, 1973.

1252   LIONS J.L.,   ON THE NUMERICAL APPROXIMATION OF SOME EQUATIONS
              ARISING IN HYDRODYNAMICS. PP.11-23 OF G. BIRKHOFF AND
              R.S. VARGA (EDS.),   NUMERICAL SOLUTION OF FIELD PROBLEMS IN
              CONTINUUM MECHANICS. SIAM-AMS PROCEEDINGS II, AMERICAN
              MATHEMATICAL SOCIETY, PROVIDENCE, 1969.
1253   LIONS J.L.,   EQUATIONS DIFFERENTIELLES OPERATIONNELLES ET PROBLEME
              AUX LIMITES. SPRINGER VERLAG, BERLIN, 1971.
1254   LIONS J.L.,   QUELQUES METHODES DE RESOLUTION DES PROBLEMES AUX
              LIMITES NON LINEAIRES. DUNOD, PARIS.
1255   LIONS J.L.,   OPTIMAL CONTROL OF SYSTEMS GOVERNED BY PARTIAL
              DIFFERENTIAL EQUATIONS. SPRINGER VERLAG, BERLIN, 1971.
1256   LIONS J.L.,   PERTURBATIONS SINGULIERES DANS LES PROBLEMES AUX LIMITES
              ET EN CONTROLE OPTIMAL. LECTURE NOTES IN MATHEMATICS,
              NO. 323, SPRINGER-VERLAG, BERLIN, 1973.
1257   LIONS J.L.,   REMARKS ON SOME FREE BOUNDARY PROBLEMS. PROC.CONF.
              COMPUTATIONAL METHODS IN NONLINEAR MECH., AUSTIN, TEXAS,
              1974.
1258   LIONS J.L., AND MAGENES E.,   PROBLEMS AUX LIMITES NON HOMOGENES
              ET APPLICATIONS. DUNOD, PARIS, 1968.
1259   LIONS J. AND MAGENES E.,   NON-HOMOGENEOUS BOUNDARY VALUE PROBLEMS AND
              APPLICATIONS. VOLS I-III. (TRANSLATED FROM FRENCH BY P.KENNETH.)
              SPRINGER-VERLAG,NEW YORK,1972.
1260   LIONS J.L. AND STAMPACCHIA G.,   VARIATIONAL INEQUALITIES. COMM.
              PURE APPLIED MATH.20, 493-519, 1967.
1261   LIONS J.L., AND TEMAM R.,   UNE METHODE D'ECLATEMENT DES OPERATEURS ET
              DES CONSTRAINTES EN CALCUL DES VARIATIONS. COMPTES REND, ACAD.
              SCI.PARIS SER.A 263, 563-565, 1967.
1262   LIVESLEY R.K.,   MATRIX METHODS IN STRUCTURAL ANALYSIS, PERGAMON
              PRESS, OXFORD, 1964.
1263   LO K.S., AND SCORDELIS A.C.,   FINITE SEGMENT ANALYSIS OF FOLDED
              PLATES. J. STRUCT. DIV. A.S.C.E. 95, ST5, 831-852, 1969.
1264   LOBITZ D.W.,   MESH GENERATION CODE AND A CONTOUR PLOTTING ROUTINE.
              REPORT SC-TM-710557, SANDIA LABORATORIES, ALBUQUERQUE, 1972.
1265   LOCK A.C., AND SABIR A.B.,   ALGORITHM FOR THE LARGE DEFLECTION
              GEOMETRICALLY NONLINEAR PLANE AND CURVED STRUCTURES. PP.
              483-494 OF J.R. WHITEMAN (ED.), THE MATHEMATICS OF FINITE
              ELEMENTS AND APPLICATIONS. ACADEMIC PRESS, LONDON, 1973.
1266   LOGCHER R.D.,   THE DEVELOPMENT OF ICES STRUDL. PP. 591-606 OF J.T. ODEN,
              R.W. CLOUGH AND Y. YAMOMOTO (EDS.), ADVANCES IN COMPUTATIONAL
              METHODS IN STRUCTURAL MECHANICS AND DESIGN. UNIVERSITY OF
              ALABAMA PRESS, HUNTSVILLE, 1972.
1267   LOOK D.C., AND LOVE T.J.,   NUMERICAL QUADRATURE USED FOR RADIATIVE
              HEAT-TRANSFER COMPUTATIONS. INT.J.NUMER.METH.ENG.8,
              395-401, 1974.
1268   LOOV R.E.,   THE DETERMINATION OF STRESSED AND DEFORMATIONS OF REINFORCED
              CONCRETE AFTER CRACKING. PP. 1257-1261 OF M.TE'ENI (ED.),
              STRUCTURE, SOLID MECHANICS AND ENGINEERING DESIGN. WILEY-
              INTERSCIENCE, LONDON, 1971.
1269   LORENTZ G.G.,   APPROXIMATION OF FUNCTIONS, HOLT, RINEHART AND WINSTON,
              NEW YORK, 1966.
1270   LORENZ G.G. (ED.)   APPROXIMATION THEORY. ACADEMIC PRESS,NEW YORK,1973.
1271   LOSCALZO F.R., AND TALBOT T.D.,   SPLINE FUNCTION APPROXIMATIONS FOR
              SOLUTIONS OF ORDINARY DIFFERENTIAL EQUATIONS. SIAM J.NUMER.
              ANAL., 4, 433-445, 1967.
1272   LOVE A.E.H.,   MATHEMATICAL THEORY OF ELASTICITY. CAMBRIDGE UNIVERSITY
              PRESS, 1934.
1273   LUCAS T.R. AND REDDIEN G.W.,   A HIGH ORDER PROJECTION METHOD FOR NONLINEAR
              TWO POINT BOUNDARY VALUE PROBLEMS.
              NUMER.MATH.20,257-270,1973.
1274   LUFT R.W. ROESSET J.M. AND CONNOR J.J.,   AUTOMATIC GENERATION OF
              FINITE ELEMENT MATRICES. J.STRUCT.DIV.A.S.C.E. 97, ST1,
              349-362, 1971.
1275   LUKAS I.L.,   CURVED BOUNDARY ELEMENTS, GENERAL FORMS OF POLYNOMIAL
              MAPPINGS. PROC.CONF.COMP.METHODS IN NONLINEAR MECH.,
              AUSTIN, TEXAS, 1974.
1276   LUNDE JOHNSEN TH.,   ON THE COMPUTATION OF NATURAL MODES OF AN

UNSUPPORTED VIBRATING STRUCTURE BY SIMULTANEOUS ITERATION.
COMP.METH.APPL.MECH.ENG.2, 305-322, 1973.

1277 LUXMOORE A.R., GARDNER N.A. AND WYATT P.J., A COMPARISON OF FINITE
ELEMENT AND EXPERIMENTAL STUDIES ON THE DEFORMATION OF
ZIRCONIUM NOTCHED BEND SPECIMENS. J.MECH.PHYS.SOLIDS 9,
395-406,1971.

1278 LYCHE T. AND SCHUMAKER L.L., ALGOL PROCEDURES FOR COMPUTING SMOOTHING
AND INTERPOLATING NATURAL SPLINES.
TECH.RPT.31, CENTER FOR NUMERICAL ANALYSIS, UNIVERSITY OF TEXAS,
AUSTIN, TEXAS, 1973.

1279 LYCHE T. AND SCHUMAKER L.L., COMPUTATION OF SMOOTHING AND INTERPOLATING
NATURAL SPLINES VIA LOCAL BASES. SIAM.J.NUMER.ANAL.10,
1027-1038,1973.

1280 LYNCH F. DE S., A FINITE ELEMENT METHOD OF VISCOPLASTIC STRESS
ANALYSIS WITH APPLICATION TO ROLLING CONTACT PROBLEMS.
INT. J.NUMER.METH.ENG.1, 379-394, 1969.

1281 LYNN P.P., A LEAST SQUARES FINITE ELEMENT ANALYSIS OF NONLINEAR
BIOPHYSICAL DIFFUSION-KINETICS EQUATIONS. PROC.CONF.COMP.
METHODS IN NONLINEAR MECHANICS, AUSTIN, TEXAS, 1974.

1282 LYNN P.P., LEAST SQUARES FINITE ELEMENT ANALYSIS OF LAMINAR
BOUNDARY LAYER FLOWS. INT.J.NUMER.METH.ENG.8, 865-876,
1974.

1283 LYNN P.P., AND ARYA S.K., USE OF THE LEAST SQUARES CRITERION IN
THE FINITE ELEMENT FORMULATION. INT.J.NUMER.METH.ENG.
6, 75-88, 1973.

1284 LYNN P.P. AND ARYA S.K., FINITE ELEMENTS FORMULATED BY THE WEIGHTED
DISCRETE LEAST SQUARES METHOD. INT.J.NUMER.METH.ENG.8,
71-90,1974.

1285 LYNN P.P. AND RAMEY G.E., A METHOD TO GENERATE BOTH UPPER AND LOWER
BOUNDS TO PLATE EIGENVALUES BY CONFORMING DISPLACEMENT
FINITE ELEMENTS. PAPER M6/4, PROCEEDINGS 1ST STRUCTURAL
MECH. IN REACTOR TECH. CONF., BERLIN, 1971.

1286 LYNN P.P. AND QUILLON B.S., TRIANGULAR THICK PLATE BENDING ELEMENTS
PAPER M6/5, PROCEEDINGS 1ST STRUCTURAL MECH. IN REACTOR TECH.
CONF., BERLIN, 1971.

1287 LYNN P.P., AND ARYA S.K., USE OF THE LEAST SQUARES CRITERION IN THE
FINITE ELEMENT FORMULATION. INT.J.NUMER.METH.ENG.6, 75-88, 1973.

1288 MACCALLAM K.J., SURFACES FOR INTERACTIVE GRAPHICAL DESIGN. COMP.J.,
13,4,1970.

1289 MACDONALD D.A., SOLUTION OF THE INCOMPRESSIBLE BOUNDARY LAYER
EQUATIONS VIA THE GALERKIN KANTOROVICH TECHNIQUE.
J.INST.MATH.APPLICS, 6, 115-130,1970.

1290 MACNEAL R.H., A HYBRID METHOD OF COMPONENT MODE SYNTHESIS.
COMPUTERS AND STRUCTURES 1, 581-602, 1971.

1291 MACNEAL R.H., SOME ORGANIZATIONAL ASPECTS OF NASTRAN. PAPER M1/3,
PROC.2ND STRUCTURAL MECH. IN REACTOR TECH. CONF., BERLIN,
1973.

1292 MACNEAL R.H., AND MCCORMICK C.W., THE NASTRAN COMPUTER PROGRAM
FOR STRUCTURAL ANALYSIS. J.COMPUTERS STRUCTURES 1,
389-412, 1971.

1293 MADSEN N.K., A POSTERIORI ERROR BOUNDS FOR NUMERICAL SOLUTIONS
OF THE NEUTRON TRANSPORT EQUATION. MATH.COMP.27, 773-780,
1973.

1294 MAGNUS W., AND OBERHETTINGER F., SPECIAL FUNCTIONS OF MATHEMATICAL
PHYSICS. CHELSEA, NEW YORK, 1949.

1295 MAHATA P.C., AND MCNARY O., A NEW TRIANGULAR ELEMENT FOR FINITE DIFFERENCE
SOLUTION OF AXISYMMETRIC CONDUCTION PROBLEMS IN CYLINDRICAL
CO-ORDINATES. INT.J.NUMER.METH.ENG.8, 547-567, 1974.

1296 MAK C.K., AND KAO D.W., FINITE ELEMENT ANALYSIS OF BUCKLING AND POST-
BUCKLING BEHAVIOR OF STRUCTURES WITH GEOMETRIC IMPERFECTIONS.
NAT. SYMP. ON COMPUTERISED STRUCT. ANALYSIS AND DESIGN. GEORGE
WASHINGTON UNIVERSITY, 1972.

1297 MALCOLM M.A., NONLINEAR SPLINE FUNCTIONS. TECHNICAL REPORT STAN-CS-73
-372, COMPUTER SCIENCE DEPARTMENT, STANFORD UNIVERSITY, 1973.

1298 MALLETT R.H., AND BERKE L., AUTOMATED METHOD FOR THE FINITE
DISPLACEMENT ANALYSIS OF THREE DIMENSIONAL TRUSS AND FRAME

ASSEMBLIES. AFFDL-TR-102, 1966.

1299 MALLETT R.H., AND DISNEY R.K., CURRENT FINITE ELEMENT ANALYSIS PRACTICES. TRANS. AMERICAN NUCL. SOC. 16, 131-132, 1973.

1300 MALLETT R.H. AND HAFTKA R.T., PROGRESS IN NONLINEAR FINITE ELEMENT ANALYSIS USING ASYMPTOTIC SOLUTION TECHNIQUES. PP.357-374 OF J.T. ODEN, R.W. CLOUGH, AND Y. YAMOMOTO (EDS.), ADVANCES IN COMPUTATIONAL METHODS IN STRUCTURAL MECHANICS AND DESIGN, UNIVERSITY OF ALABAMA PRESS, HUNTSVILLE, 1972.

1301 MALLETT R.H., AND MARCAL P.V., FINITE ELEMENT ANALYSIS OF NONLINEAR STRUCTURES. J.STRUCT.DIV.A.S.C.E.94, ST9, 2081-2105, 1968.

1302 MALONE D.U., AND CONNOR J.J., FINITE ELEMENTS AND DYNAMIC VISCOELASTICITY J.ENG.MECH.DIV. A.S.C.E. 97, EM4, 1145-1158, 1971.

1303 MANGASARIAN O.L., NONLINEAR PROGRAMMING, MCGRAW HILL, NEW YORK, 1969.

1304 MANGASARIAN O.L., AND SCHUMAKER L.L., DISCRETE SPLINES VIA MATHEMATICAL PROGRAMMING. SIAM J. CONTROL 9, 174-183, 1971.

1305 MANSFIELD L., ON THE VARIATIONAL CHARACTERIZATION AND CONVERGENCE OF BIVARIATE SPLINES. NUMER. MATH.20, 99-114, 1972.

1306 MANSFIELD LOIS, HIGHER ORDER COMPATIBLE TRIANGULAR FINITE ELEMENTS. NUMER.MATH.22, 89-97, 1974.

1307 MARCAL P.V., COMPARATIVE STUDY OF NUMERICAL METHODS OF ELASTIC-PLASTIC ANALYSIS. AIAA J., 6,1,157-158, 1967.

1308 MARCAL P.V., EFFECT OF INITIAL DISPLACEMENT ON PROBLEM OF LARGE DEFLECTION AND STABILITY. TECH.REPT. ARPA E54, BROWN UNIVERSITY, 1967.

1309 MARCAL P.V., FINITE ELEMENT ANALYSIS OF COMBINED PROBLEMS OF MATERIAL AND GEOMETRIC BEHAVIOUR. TECH.REPT.1 ONY, BROWN UNIVERSITY, 1969.

1310 MARCAL P.V., ON GENERAL PURPOSE PROGRAMS FOR FINITE ELEMENT ANALYSIS, WITH SPECIAL REFERENCE TO GEOMETRIC AND MATERIAL NONLINEARITIES. IN HUBBARD (ED.), NUMERICAL SOLUTION OF PARTIAL DIFFERENTIAL EQUATIONS-II, SYNSPADE 1970. ACADEMIC PRESS, NEW YORK, 1971.

1311 MARCAL P.V., SURVEY OF GENERAL PURPOSE PROGRAMS FOR FINITE ELEMENT ANALYSIS. PP. 517-528 OF J.T. ODEN, R.W. CLOUGH AND Y. YAMOMOTO (EDS.), ADVANCES IN COMPUTATIONAL METHODS IN STRUCTURAL MECHANICS AND DESIGN. UNIVERSITY OF ALABAMA PRESS, HUNTSVILLE, 1972.

1312 MARCAL P.V. AND BETTIS R.S., ELASTIC-PLASTIC BEHAVIOUR OF A LONGITUDINAL SEMIELLIPTIC CRACK IN A THICK PRESSURE VESSEL. PAPER 16, PROC.6TH ANNUAL INFORMATION MEETING HEAVY SECTION STEEL TECHNOLOGY PROGRAM, OAK RIDGE NATIONAL LABORATORY, TENNESSEE, 1972.

1313 MARCAL P.V., AND KING I.P., ELASTIC-PLASTIC ANALYSIS OF TWO DIMENSIONAL STRESS SYSTEMS BY THE FINITE ELEMENT METHOD. INT.J.MECH,SCI,9, 143-155, 1967.

1314 MARCAL P.V., STUART P.M. AND BETTES R.S., ELASTIC-PLASTIC BEHAVIOUR OF A LONGITUDINAL SEMI-ELLIPTIC CRACK IN A THICK PRESSURE VESSEL. TECHNICAL REPORT 18252, OAK RIDGE NATIONAL LABORATORY, TENNESSEE, 1973.

1315 MARCHUK G.I., NUMERICAL METHODS FOR NUCLEAR REACTOR CALCULATIONS. CONSULTANTS BUREAU, NEW YORK, 1959.

1316 MAREK I., AND NEDOMA J., FINITE ELEMENT IN THE THEORY OF SH-WAVE PROPAGATION. PP. 31-37 OF NUMERISCHE METHODEN IN DER GEOPHYSIK, K.A.P.G. 52, CZECHOSLOVAK ACADEMY, PRAGUE, 1973.

1317 MARGUERRE K., SCHALK M., AND WOLFEL H., BERECHNUNG DER ERDBEBENSCHWINGUNGEN VON STRUKTUREN MIT DER FINITE ELEMENT METHODE. PAPER K2/7, PROCEEDINGS 1ST STRUCTURAL MECH. IN REACTOR TECH.CONF., BERLIN, 1971.

1318 MARGUERRE K., SCHALK M., AND WOELFEL H., CALCULATION OF VIBRATIONS OF STRUCTURES CAUSED BY EARTHQUAKES USING THE FINITE ELEMENT METHOD. PP.123-139 OF VOL.5 OF PROCEEDINGS 1ST STRUCTURAL MECH IN REACTOR TECH. CONF., BERLIN, 1971.

1319 MARSHALL J.A. AND MITCHELL A.R., AN EXACT BOUNDARY TECHNIQUE FOR IMPROVED ACCURACY IN THE FINITE ELEMENT METHOD. J.INST. MATH.APPLICS.12, 355-362, 1973.

1320 MARTIN H.C., DERIVATION OF STIFFNESS MATRICES FOR THE ANALYSIS OF LARGE DEFLECTION AND STABILITY PROBLEMS. PROC.1ST CONF. MATRIX METHODS IN STRUCTURAL MECHANICS, WRIGHT-PATTERSON AFB,

OHIO, AFFDL TR66-80, 1965.

1321 MARTIN H.C., INTRODUCTION TO MATRIX METHODS OF STRUCTURAL ANALYSIS,
MCGRAW HILL, NEW YORK, 1966.

1322 MARTIN H.C., STIFFNESS MATRIX FOR A TRIANGULAR SANDWICH
ELEMENT IN BENDING. TECH. REPT. 32-1158, JET PROPULSION
LAB., CALIFORNIA INSTITUTE OF TECHNOLOGY, PASADINA, 1967.

1323 MARTIN H.C., FINITE ELEMENT ANALYSIS OF FLUID FLOWS. PROC. 2ND CONF.
MATRIX METHODS IN STRUCTURAL MECHANICS. WRIGHT-PATTERSON AFB.,
OHIO, 1968.

1324 MARTIN H.C., FINITE ELEMENTS AND THE ANALYSIS OF GEOMETRICAL
NONLINEAR PROBLEMS. PP.343-381 OF R.H. GALLAGHER, Y.
YAMADA AND J.T. ODEN(EDS.), RECENT ADVANCES IN MATRIX
METHODS OF STRUCTURAL ANALYSIS AND DESIGN. UNIVERSITY OF
ALABAMA PRESS, HUNTSVILLE, 1971.

1325 MASON J.B., FINITE ELEMENT ANALYSIS OF COUPLED IRREVERSIBLE VECTOR
PROCESSES. PH.D. THESIS, UNIVERSITY OF MARYLAND, 1971.

1326 MASON W., RECTANGULAR FINITE ELEMENTS FOR ANALYSIS OF PLATE
VIBRATIONS. J. SOUND VIBRATION 7, 437-448, 1968.

1327 MASSINI G., PEPINO A., SERCHIA L., AND ZACCARELLI P., DESCRIPTION OF THE
IBM-360 VERSION OF THE FINITE ELEMENT COMPUTER PROGRAM SAFE 3D.
REPORT RT/ING(72)16, COMITATO NAZIONALE PER L'ENERGIA NUCLEARE,
ROME, 1972.

1328 MATSUMOTO K., VIBRATION OF AN ELASTIC BODY IMMERSED IN A FLUID. PP. 259-
274 OF J.T. ODEN, R.W. CLOUGH AND Y. YAMOMOTO (EDS.), ADVANCES IN
STRUCTURAL MECHANICS AND DESIGN. UNIVERSITY OF ALABAMA PRESS,
HUNTSVILLE, 1972.

1329 MAWENYA A.S., AND DAVIES J.D., FINITE ELEMENT BENDING ANALYSIS
OF MULTILAYER PLATES. INT.J.NUMER.METH.ENG.8, 215-225,
1974.

1330 MAWHIN J., DEGRE DE COINCIDENCE ET PROBLEMES AUX LIMITES POUR
DES EQUATIONS DIFFERENTIELLES ORDINAIRES ET
FONCTIONNELLES. REPORT 64, INSTITUT DE MATHEMATIQUE PURE
ET APPLIQUEE, UNIVERSITE CATHOLIQUE DE LOUVIN, 1973.

1331 MCCORMICK C.W., THE NASTRAN PROGRAM FOR STRUCTURAL ANALYSIS. PP. 551
-572 OF J.T. ODEN, R.W. CLOUGH AND Y. YAMOMOTO (EDS.), ADVANCES IN
COMPUTATIONAL METHODS IN STRUCTURAL MECHANICS AND DESIGN.
UNIVERSITY OF ALABAMA PRESS, HUNTSVILLE, 1972.

1332 MCGEORGE R., AND SWEC F., REFINED CRACKED CONCRETE ANALYSIS OF
CONCRETE CONTAINMENT STRUCTURES SUBJECT TO OPERATIONAL
AND ENVIRONMENTAL LOADINGS. PAPER J2/5, PROC.2ND
STRUCTURAL MECH. IN REACTOR TECH. CONF., BERLIN, 1973.

1333 MCLAURIN J.W., A GENERAL COUPLED EQUATION APPROACH FOR SOLVING THE
BIHARMONIC BOUNDARY VALUE PROBLEM. SIAM J. NUMER. ANAL. 11,
14-33, 1974.

1334 MCLAY R.W., AN INVESTIGATION INTO THE THEORY OF THE DISPLACEMENT METHOD
OF ANALYSIS FOR LINEAR ELASTICITY. PH.D. THESIS, UNIVERSITY OF
WISCONSIN, MADISON, 1963.

1335 MCLAY R.W., A SPECIAL VARIATIONAL PRINCIPLE FOR THE FINITE
ELEMENT METHOD. AIAA J.7, 533-539, 1969.

1336 MCLAY R.W., ON CERTAIN APPROXIMATIONS IN FINITE ELEMENT METHODS.
J.APPL.MECH., 58-61, 1971.

1337 MCLAY R.W., AND BULLIS S., CONVERGENCE OF THE FINITE ELEMENT
METHOD IN NONLINEAR HEAT CONDUCTION. PROC.CONF.COMP.
METHODS IN NONLINEAR MECH., AUSTIN, TEXAS, 1974.

1338 MCLAY R.W., AND BUTURLA E.M., ERROR BOUNDS IN AN OPTIMIZATION
PROBLEM USING THE FINITE ELEMENT METHOD. TRANS. ASME
SER.E. 40, 204-208, 1973.

1339 MCLEOD R.J.Y., BASIS FUNCTIONS FOR CURVED ELEMENTS IN THE
FINITE ELEMENT METHOD. PH.D.THESIS, UNIVERSITY OF DUNDEE,
1972.

1340 MCLEOD R.J.Y., AND MITCHELL A.R., THE USE OF PARABOLIC ARCS IN
MATCHING CURVED BOUNDARIES. J.INST.MATH.APPLICS. (TO APPEAR).

1341 MCLEOD R., AND MITCHELL A.R., THE CONSTRUCTION OF BASIS FUNCTIONS
FOR CURVED ELEMENTS IN THE FINITE ELEMENT METHOD. J.
INST.MATH.APPLICS.10, 382-393, 1972.

1342 MCLEOD I.A., NEW RECTANGULAR FINITE ELEMENT FOR SHEAR WALL

ANALYSIS. J. STRUCT. DIV. A.S.C.E. 95, 399-410, 1969.

1343 MCLOSH R.J., A STIFFNESS MATRIX FOR THE ANALYSIS OF THIN PLATES IN
        BENDING. J.AERO.SOC.28, 34-42, 1961.

1344 MCNEICE G.M. AND KEMP K.O. COMPARISON OF FINITE ELEMENT AND UNIQUE
        LIMIT ANALYSIS SOLUTIONS FOR CERTAIN REINFORCED CONCRETE
        SLABS. PROC.INST.CIV.ENGINEERS 43, 629-640, 1969.

1345 MCWHORTER L.B. AND HAISLER W.E., FAMSOR, A FINITE ELEMENT PROGRAM
        FOR THE FREQUENCIES AND MODE SHAPES OF SHELLS OF
        REVOLUTION. REPORT SC-CR-715125, SANDIA LABORATORIES,
        ALBUQUERQUE, 1970.

1346 MEBANE P.M., AND STRICKLIN J.A., IMPLICIT RIGID BODY MOTION IN
        CURVED FINITE ELEMENTS. AIAA J.9, 344-345, 1971

1347 MEEK J.L., THE FINITE ELEMENT METHOD IN STRUCTURAL MECHANICS.
        INST.ENG.AUSTRALIA CIV.ENG.TRANS.CE8, 166-178, 1966.

1348 MEGARD G., PLANAR AND CURVED SHELL ELEMENTS. IN
        FINITE ELEMENT METHODS, TAPIR, TRONDHEIM, 1969.

1349 MEHLUM E., A CURVE FITTING METHOD BASED ON A VARIATIONAL CRITERION.
        B.I.T. 10, 177-183, 1964.

1350 MEHROTRA B.L. AND MUFTI A.A., FINITE ELEMENT ANALYSIS OF THIN
        SHELLS. J.ENG.MECH.DIV.A.S.C.E. 95, EM4, 1021-1024,1969.

1351 MEI C., FREE VIBRATIONS OF CIRCULAR MEMBRANES UNDER ARBITRARY
        TENSION BY THE FINITE ELEMENT METHOD. J.ACOUSTICAL
        SOC. AMERICA 46, 693-700, 1969.

1352 MEINARDUS G., APPROXIMATION OF FUNCTIONS. SPRINGER VERLAG, BERLIN,
        1967.

1353 MEINARDUS G. AND TAYLOR G.D., LOWER ESTIMATES FOR THE ERROR OF
        BEST UNIFORM APPROXIMATION. TECH.REPORT. CS-73-376, COMPUTER
        SCIENCE DEPARTMENT, STANFORD UNIVERSITY, 1973.

1354 MEIR A., AND SHARMA A., (EDS.), SPLINE FUNCTIONS AND
        APPROXIMATION THEORY. I.S.N.M. 21, BIRKHAUSER VERLAG,
        BASEL, 1973.

1355 MEIS T., AND TORNIG W., ITERATIVE LOSUNG NICHTLINEARER
        GLEICHUNGS-SYSTEME UND DISKRETISIERUNGSVERFAHREN BEI
        ELLIPTISCHEN DIFFERENTIALGLEICHUNGEN. COMPUTING 5, 1970.

1356 MEISSNER U., A MIXED FINITE ELEMENT MODEL FOR USE IN POTENTIAL
        FLOW PROBLEMS. INT.J.NUMER.METH.ENG.6, 467-473, 1973.

1357 MELKES F., THE FINITE ELEMENT METHOD IN NON-LINEAR PROBLEMS. APLIKACE
        MATEMATIKY 15, 177-189, 1970.

1358 MELKES F., REDUCED PIECEWISE BIVARIATE HERMITE INTERPOLATIONS.
        NUMER.MATH.19, 326-340,1972.

1359 MELOSH R.J., A STIFFNESS MATRIX FOR THE ANALYSIS OF THIN PLATES
        IN BENDING. J.AEROSPACE SOC. 28, 34-42, 1961.

1360 MELOSH R.J., BASIS FOR DERIVATION OF MATRICES FOR THE DIRECT SITFFNESS
        METHOD. AIAA J.1, 1631-1637, 1963.

1361 MELOSH R.J., INHERITED ERROR IN FINITE ELEMENT ANALYSIS OF STRUCTURES.
        NATIONAL SYMP. COMPUTERISED STRUCT. ANALYSIS AND DESIGN. GEORGE
        WASHINGTON UNIVERSITY, 1972.

1362 MELOSH R.J., NUMERICAL ANALYSIS OF AUTOMOBILE STRUCTURES. PP. 745-756
        OF J.T. ODEN, R.W. CLOUGH AND Y. YAMOMOTO (EDS.), ADVANCES IN
        COMPUTATIONAL METHODS IN STRUCTURAL MECHANICS AND DESIGN.
        UNIVERSITY OF ALABAMA PRESS, HUNTSVILLE, 1972.

1363 MELOSH R.J., COMPUTATIONAL TECHNIQUES FOR FINITE ELEMENT ANALYSIS.
        PAPER M4/1, PROC. 2ND STRUCTURAL MECH. IN REACTOR TECH.
        CONF., BERLIN, 1973.

1364 MELOSH R.J., STATUS REPORT ON COMPUTATIONAL TECHNIQUES FOR FINITE
        ELEMENT ANALYSES. NUCL.ENG.DESIGN 27, 274-285, 1974.

1365 MENDELSON A., EVALUATION OF THE USE OF A SINGULARITY IN THE FINITE
        ELEMENT ANALYSIS OF A CENTRE-CRACKED PLATE. N.A.S.A. TECHNICAL
        REPORT TND-6703, 1972.

1366 MERCIER B., NUMERICAL SOLUTION OF THE BIHARMONIC PROBLEM BY MIXED
        FINITE ELEMENTS OF CLASS CO. REPORT, LABORATORIO DI ANALISI
        NUMERICA DEL C.N.R., PAVIA, 1973.

1367 MERTEN K., ZUR DISKRETISIERUNG VON VARIATIONSPROBLEM. REPORT
        156, FACHBEREICH MATHEMATIK, TECHNISCHE HOCHSCHULE,
        DARMSTADT, 1974.

1368 MEYER G.H., INITIAL VALUE METHODS FOR BOUNDARY VALUE PROBLEMS.

ACADEMIC PRESS, NEW YORK, 1973.

1369 MICULA G., APPROXIMATE SOLUTION OF THE DIFFERENTIAL EQUATION Y" = F (X,Y) WITH SPLINE FUNCTIONS. MATH.COMP.27, 807-815, 1974.

1370 MIKHLIN S.G., ZUR RITZCHEN METHODE. DOKL.AKAD.NAUK.SSSR.106, 391-394,1956.

1371 MIKHLIN S.G., THE STABILITY OF THE RITZ METHOD. SOVIET MATH. DOKL.1, 1230-1233, 1960.

1372 MIKHLIN S.G., VARIATIONAL METHODS FOR SOLVING LINEAR AND NON-LINEAR BOUNDARY VALUE PROBLEMS. IN BABUSKA (ED.), DIFFERENTIAL EQUATIONS AND THEIR APPLICATIONS, ACADEMIC PRESS, NEW YORK,1963.

1373 MIKHLIN S.G., VARIATIONAL METHODS IN MATHEMATICAL PHYSICS, PERGAMON PRESS, OXFORD, 1964.

1374 MIKHLIN S.G., THE PROBLEM OF THE MINIMUM OF A QUADRATIC FUNCTIONAL. HOLDEN-DAY, SAN FRANCISCO, 1965.

1375 MIKHLIN S.G., THE NUMERICAL PERFORMANCE OF VARIATIONAL METHODS. WOLTER-NORDHOFF, GRONINGEN, 1971.

1376 MIKHLIN S.G., AND SMOLITSKIY K.L., APPROXIMATE METHODS FOR SOLUTION OF DIFFERENTIAL AND INTEGRAL EQUATIONS. AMERICAN ELSEVIER PUBLISHING CO., NEW YORK, 1967.

1377 MIKSCH M., AND SCHMITT W., APPLICATION OF THE FINITE ELEMENT METHOD FOR THE SAFETY EVALUATION OF REACTOR COMPONENTS. PAPER G5/8, PROC.2ND STRUCTURAL MECH. IN REACTOR TECH. CONF., BERLIN, 1973.

1378 MILES G.A., AND WHITE D.J., THE FINITE ELEMENT METHOD AND ITS APPLICATIONS IN AN ENGINEERING LABORATORY. J.SCIENCE AND TECHNOLOGY 37, 127-136, 1970.

1379 MILLER J.J.H. (ED.), TOPICS IN NUMERICAL ANALYSIS. ACADEMIC PRESS, LONDON,1973.

1380 MILLER W.F., AND LEWIS E.E., QUADRATIC FINITE ELEMENT IN NEUTRON TRANSPORT. TRANS.AMER.NUCL.SOC.17,235, 1973.

1381 MILLER W.F., LEWIS E.E., AND ROSSOW E.C., APPLICATION OF PHASE SPACE FINITE ELEMENTS TO THE ONE-DIMENSIONAL NEUTRON TRANSPORT EQUATION. NUCL.SCI.ENG.51, 148-156, 1973.

1382 MILLER W.F., LEWIS E.E. AND ROSSOW E.C., APPLICATION OF PHASE SPACE FINITE ELEMENTS TO THE TWO-DIMENSIONAL NEUTRON TRANSPORT EQUATION IN X-Y GEOMETRY. NUCL.SCI.ENG.52, 12-22,1973.

1383 MIRANKER W.L., GALERKIN APPROXIMATIONS AND THE OPTIMIZATION OF DIFFERENCE SCHEMES FOR BOUNDARY VALUE PROBLEMS. SIAM J.NUMER.ANAL.8, 486-496, 1971.

1384 MITCHELL A.R., AN INTRODUCTION TO THE MATHEMATICS OF THE FINITE ELEMENT METHOD. PP. 37-58 OF J.R. WHITEMAN (ED.), THE MATHEMATICS OF FINITE ELEMENTS AND APPLICATIONS. ACADEMIC PRESS, LONDON, 1973.

1385 MITCHELL A.R., THE FINITE ELEMENT METHOD. BULLETIN INST. MATH. APPLICS. 10, 76-79, 1974.

1386 MITCHELL A.R., AND MARSHALL J.A., MATCHING OF ESSENTIAL BOUNDARY CONDITIONS IN THE FINITE ELEMENT METHOD. IN J. MILLER (ED.), PROC.CONF. ON NUMERICAL ANALYSIS, DUBLIN, 1974.

1387 MITCHELL A.R. AND MCLEOD R., CURVED ELEMENTS IN THE FINITE ELEMENT METHOD. PP.89-104 OF G.A. WATSON (ED.), PROCEEDINGS OF CONF. ON NUMERICAL SOLUTION OF DIFFERENTIAL EQUATIONS. LECTURE NOTES IN MATHEMATICS, NO.363. SPRINGER-VERLAG, BERLIN, 1974.

1388 MITCHELL A.R., PHILLIPS G., AND WACHSPRESS E., FORBIDDEN SHAPES IN THE FINITE ELEMENT METHOD. J.INST.MATH.APPLICS., 8, 260-270, 1971.

1389 MITTELMANN H.D., DIE APPROXIMATION DER LOSUNGEN GEMISCHTER RANDWERTPROBLEME QUASILINEARER ELLIPTISCHER DIFFERENTIALGLEICHUNGEN, PREPRINT 114, FACHBEREICH MATHEMATIK, TECHNISCHE HOCHSCHULE DARMSTADT, 1974.

1390 MITTELMANN H.D., FINITE ELEMENT VERFAHREN BEI QUASILINEAREN ELLIPTISCHEN RANDWERTPROBLEM. PREPRINT 133, FACHBEREICH MATHEMATIK, TECHNISCHE HOCHSCHULE DARMSTADT, 1974.

1391 MITTELMANN H.D., NICHTLINEARE DIRICHLET PROBLEME UND EINFACHE FINITE ELEMENT VERFAHREN. REPORT 157, FACHBEREICH

MATHEMATIK, TECHNISCHE HOCHSCHULE DARMSTADT, 1974.

1392 MITTELMANN H.D., STABILITAT BEI DER METHODE DER FINITEN ELEMENTE
FÜR QUASILINEARE ELLIPTISCHE RANDWERTPROBLEME. REPORT
154, FACHBEREICH MATHEMATIK, TECHNISCHE HOCHSCHULE
DARMSTADT, 1974.

1393 MIYAMOTO H., APPLICATION OF FINITE ELEMENT METHOD TO FRACTURE
MECHANICS. PAPER L6/4, PROCEEDINGS 1ST STRUCTURAL MECH.
IN REACTOR TECH. CONF., BERLIN, 1971.

1394 MIYAMOTO H., AND FUKUDA S., COMPUTER PREDICTION OF FATIGUE CRACK
PROPAGATION UNDER RANDOM LOADING. PAPER M7/9, PROCEEDINGS
2ND STRUCTURAL MECH. IN REACTOR TECH. CONF., BERLIN, 1973.

1395 MIYAMOTO H., ISHIJIMA Y., SHIRATORI M., AND MIYOSHI T., THE APPLICATION
OF THE FINITE ELEMENT METHOD TO FRACTURE MECHANICS. PP. 679-702
OF J.T. ODEN, R.W. CLOUGH AND Y. YAMOMUTO (EDS.), ADVANCES IN
COMPUTATIONAL METHODS IN STRUCTURAL MECHANICS AND DESIGN,
UNIVERSITY OF ALABAMA PRESS, HUNTSVILLE, 1972.

1396 MIYOSHI T., CONVERGENCE OF FINITE ELEMENT SOLUTIONS REPRESENTED
BY A NON-CONFORMING BASIS. KUMAMOTO J.SCI.MATH.9, 11-20,
1972.

1397 MIYOSHI T., FINITE ELEMENT METHOD OF MIXED TYPE AND ITS CONVERGENCE
IN LINEAR SHELL PROBLEMS. KUMAMOTO J.SCI. (MATH) 10,
35-58, 1973.

1398 MIZUMACHI W., NONLINEAR THERMAL STRESS ANALYSIS FOR NUCLEAR POWER
PLANT BY THE FINITE ELEMENT METHOD. PROC.3RD CONF.MATRIX
METHODS IN STRUCTURAL MECHANICS, WRIGHT-PATTERSON AFB.,
OHIO, 1971.

1399 MIZUMACHI W., APPLICATION OF FINITE ELEMENT METHOD TO BOILING WATER
REACTOR POWER PLANT (JAPANESE). TOSHIBA REBYU 27, 1018-1022, 1972.

1400 MIZUMACHI W., FINITE ELEMENT ANALYSIS IN NUCLEAR POWER PLANTS. (JAPANESE).
KARYOKU HATSUDEN 23, 505-514, 1972.

1401 MOAN T., FINITE ELEMENT STRESS FIELD SOLUTION OF THE PROBLEM OF
SAINT VENANT TORSION.
INT.J.NUMER.METH.ENG.5, 455-458,1973.

1402 MOAN T., EXPERIENCES WITH ORTHOGONAL POLYNOMIALS AND BEST
NUMERICAL INTEGRATION FORMULAS ON A TRIANGLE WITH PARTICULAR
REFERENCE TO FINITE ELEMENT APPROXIMATIONS.
INSTITUTE FOR SHIPCONSTRUCTION, TRONDHEIM,1973.

1403 MOISEIWITSCH B.L., VARIATIONAL PRINCIPLES. INTERSCIENCE, NEW YORK,
1966.

1404 MONFORTON G.R., AND SCHMIT L.A., FINITE ELEMENT ANALYSIS OF
SANDWICH PLATES AND CYLINDRICAL SHELLS WITH LAMINATED
FACES. PROC. 2ND CONF. MATRIX METHODS IN STRUCTURAL
MECHANICS, WRIGHT-PATTERSON AFB., OHIO, AFFDL-TR-68-150,
1968.

1405 MONFORTON G.R. AND SCHMIT L.A., FINITE ELEMENT ANALYSIS OF SKEW
PLATES IN BENDING. AIAA J.6, 1150-1151, 1968.

1406 MORI K., CODE FOR STRESS ANALYSIS BY THE THREE DIMENSIONAL FINITE ELEMENT
METHOD (JAPANESE). FAPIG (TOKYO) 55, 186-192, 1969.

1407 MORI K., NAKAI Y., NAGATO K., SASAKI T., TAKENAKA Y., EXAMPLES OF STRESS
ANALYSIS OF STRUCTURES BY THE FINITE ELEMENT METHOD, (JAPANESE).
FAPIG (TOKYO) 59, 182-187, 1970.

1408 MORLEY L.S.D., A TRIANGULAR EQUILIBRIUM ELEMENT WITH LINEARLY
VARYING BENDING MOMENTS FOR PLATE BENDING PROBLEMS.
J.ROYAL AERO.SOC.71, 715, 1967.

1409 MORLEY L.S.D., THE TRIANGULAR EQUILIBRIUM ELEMENT IN THE SOLUTION
OF PLATE BENDING PROBLEMS. AERO. QUART.19, 149-169, 1968.

1410 MORLEY L.S.D., A MODIFICATION OF THE RAYLEIGH-RITZ METHOD FOR
STRESS CONCENTRATION PROBLEMS IN ELASTOPLASTICS. J.MECH.
PHYS.SOLIDS 17, 73-82, 1969.

1411 MORLEY L.S.D., A FINITE ELEMENT APPLICATION OF THE MODIFIED
RAYLEIGH-RITZ METHOD. INT.J.NUMER.METH.ENG.2, 85-98, 1970.

1412 MORLEY L.S.D., EXTENDED INTERPOLATION IN FINITE ELEMENT ANALYSIS.
PROC 3RD. CONF. MATRIX METHODS IN STRUCTURAL MECHANICS.
WRIGHT-PATTERSON AFB., OHIO, 1971.

1413 MORLEY L.S.D., THE CONSTANT BENDING MOMENT PLATE BENDING ELEMENT.
J. STRAIN ANALYSIS 6, 20-24, 1971.

1414   MORLEY L.S.D., FINITE ELEMENT SOLUTION OF BOUNDARY VALUE PROBLEMS
                WITH NON-REMOVABLE SINGULARITIES. REPORT 73034, ROYAL
                AIRCRAFT ESTABLISHMENT, FARNBOROUGH, 1972.
1415   MORREY C.B., MULTIPLE INTEGRALS IN THE CALCULUS OF VARIATIONS
                SPRINGER-VERLAG, BERLIN, 1966.
1416   MORREY C.B., DIFFERENTIABILITY THEOREMS FOR WEAK SOLUTIONS OF
                NONLINEAR ELLIPTIC DIFFERENTIAL EQUATIONS.
                BULL.AMER.MATH.SOC., 4, 684-705, 1969.
1417   MORSE M., VARIATIONAL ANALYSIS. WILEY,NEW YORK,1973.
1418   MOSCO U., AN INTRODUCTION TO THE APPROXIMATE SOLUTION OF VARIATIONAL
                INEQUALITIES. CONSTRUCTIVE ASPECTS OF FUNCTIONAL ANALYSIS
                II, CENTRO INTERNAZIONALE MATEMATICO ESTIVO (C.I.M.E.) ROME,
                1973.
1419   MOSCO U., AND SCARPINI F., COMPLEMENTARY SYSTEMS AND APPROXIMATION
                OF VARIATIONAL INEQUALITIES, REPORT, INSTITUTO MATEMATICO
                G. CASTELNUOVO, UNIVERSITA DEGLI STUDI DI ROMA, 1974.
1420   MOSCO U., AND STRANG G., ONE-SIDED APPROXIMATION AND VARIATIONAL
                INEQUALITIES. BULL.AMERICAN MATH.SOC.80, 308-312, 1974.
1421   MOSCO U., AND TROIANIELLO G.M., ON THE SMOOTHNESS OF SOLUTIONS
                OF UNILATERAL DIRICHLET PROBLEMS. BOLLETTINO U.M.I.8,
                57-67, 1973.
1422   MOTE C.D., GLOBAL-LOCAL FINITE ELEMENT. INT.J.NUMER.METH.ENG.3,
                565-574,1971.
1423   MOTE C.D., NONCONSERVATIVE STABILITY BY FINITE ELEMENTS. J.ENG.MECH.DIV.
                A.S.C.E. 97, EM3, 645-656, 1971.
1424   MOWBRAY D.F. AND MCCONNELLE J.E., APPLICATION OF FINITE ELEMENT
                ELASTIC-PLASTIC STRESS ANALYSIS TO NOTCHED FATIGUE SPECIMEN
                BEHAVIOUR. PAPER M6/8, PROCEEDINGS 1ST STRUCTURAL MECH.
                IN REACTOR TECH. CONF., BERLIN, 1971.
1425   MUNTEANU M.J., AND SCHUMAKER L.L., ON A METHOD OF CARASSO AND LAURENT
                FOR CONSTRUCTING INTERPOLATING SPLINES. MATH.COMP.27, 317-
                325, 1973.
1426   MURFIN W.B., EFFECT OF GEOLOGIC IRREGULARITIES ON SEISMIC RESPONSE.
                REPORT SLA-74-24, SANDIA LABORATORIES, ALBUQUERQUE, 1974.
1427   MURRAY D.W. AND WILSON E.L., FINITE ELEMENT LARGE DEFLECTION
                ANALYSIS OF PLATES. J.ENG.MECH.DIV.A.S.C.E.95, EM1, 143-166,
                1969.
1428   MURRAY D.W., AND WILSON E.L., FINITE ELEMENT POST BUCKLING
                ANALYSIS OF THIN ELASTIC PLATES. AIAA J. 4, 1915-1920,
                1969.
1429   MURRAY K.H., COMMENTS ON THE CONVERGENCE OF FINITE ELEMENT SOLUTIONS.
                AIAA J.8, 815-816, 1970.
1430   MURTHY M.V.V., RAO K.P., AND RAO A.K., ACCURATE DETERMINATION OF STRESS
                CONCENTRATIONS AROUND ELLIPTIC HOLES IN CYLINDRICAL SHELLS.
                PAPER G2/1, PROCEEDINGS 2ND STRUCTURAL MECH. IN REACTOR TECH.
                CONF., BERLIN, 1973.
1431   MUSKHELISHVILI N.I., SOME BASIC PROBLEMS IN THE MATHEMATICAL THEORY
                OF ELASTICITY. NOORDHOFF, GRONINGEN, 1953.
1432   MYERS G.E., ANALYTICAL METHODS IN CONDUCTION HEAT TRANSFER. MCGRAW HILL,
                NEW YORK, 1971.
1433   NAEHRIG T.H. AND GASCHEN J.P., THE CALCULATION OF THREE DIMENSIONAL
                TEMPERATURE DISTRIBUTIONS AND THERMAL STRESSES USING
                FINITE ELEMENT METHODS. PAPER M5/2, PROCEEDINGS 1ST
                STRUCTURAL MECH. IN REACTOR TECH.CONF., BERLIN, 1971.
1434   NAGARAJAN S., AND POPOV E.P., NONLINEAR ELASTIC-VISCOPLASTIC
                ANALYSIS USING THE FINITE ELEMENT METHOD. PROC. CONF.
                COMP. METHODS IN NONLINEAR MECHANICS, AUSTIN, TEXAS, 1974.
1435   NAGATO K., DYNAMIC ELASTIC-PLASTIC ANALYSIS OF AXISYMMETRIC
                STRUCTURES BY FINITE ELEMENT METHOD COMPUTER CODE. PAPER
                H3/3, PROCEEDINGS 2ND STRUCTURAL MECH. IN REACTOR TECH.
                CONF., BERLIN, 1973.
1436   NAGTEGALL J.C., PARKS D.M., AND RICE J.R., ON NUMERICALLY ACCURATE
                FINITE ELEMENT SOLUTIONS IN THE FULLY PLASTIC RANGE.
                COMP.METH.APPL.MECH.ENG.4, 153-177, 1974.
1437   NAKAMURA S., ITERATIVE SOLUTIONS FOR THE FINITE ELEMENT METHOD. PP.639-
                656 OF NUMERICAL REACTOR CALCULATIONS, INTERNATIONAL ATOMIC ENERGY

AGENCY, VIENNA, 1972.

1438   NAKAO Y., AND KAWASHIMA M.,   PRACTICAL ANALYSIS OF STEEL STRUCTURES
USING HIGHER-ORDER ELEMENTS. PP. 573-590 OF J.T. ODEN, R.W.CLOUGH
AND Y. YAMOMOTO (EDS.), ADVANCES IN COMPUTATIONAL METHODS IN
STRUCTURAL MECHANICS AND DESIGN. UNIVERSITY OF ALABAMA PRESS,
HUNTSVILLE, 1972.

1439   NAPOLITANO L.G.,   FINITE ELEMENT METHODS IN FLUID DYNAMICS. AGARD-VKI
LECTURE SERIES, ADVANCES IN NUMERICAL FLUID DYNAMICS. VON KARMAN
INSTITUTE FOR FLUID DYNAMICS, BRUSSELS, 1973.

1440   NASSIF N.,   FINITE ELEMENT METHOD FOR TIME DEPENDENT PROBLEMS.
PH.D. THESIS, HARVARD UNIVERSITY, 1972.

1441   NASSIF N.,   NUMERICAL SOLUTION OF PARABOLIC PROBLEMS BY THE
GENERALIZED CRANK NICHOLSON SCHEME. (TO APPEAR).

1442   NATH B.,   FUNDAMENTALS OF FINITE ELEMENT METHODS FOR ENGINEERS. ATHLONE
PRESS, LONDON, 1974.

1443   NATO-NRC INTERNATIONAL RESEARCH SEMINAR, THE THEORY AND APPLICATION
OF FINITE ELEMENT METHODS. LECTURE NOTES ON THE PRELIMINARY
PROGRAM. UNUVERSITY OF CALGARY, 1973.

1444   NATTERER F.,   NUMERISCHE BEHANDLUNG SINGULAERER STURM-LIOUVILLE
PROBLEME. NUMER.MATH.13, 434-447, 1969.

1445   NATTERER F., AND WERNER B.,   EINE ERWEITERUNG DES MAXIMUMPRINZIPS
FUR DEN LAPLACESCHEN OPERATOR. NUMER.MATH.22, 149-156,
1974.

1446   NAVARATNA D.R.,   COMPUTATION OF STRESS RESULTANTS IN FINITE
ELEMENT ANALYSIS. AIAA J. 4, 2058-2060, 1966.

1447   NAVARATNA D.R.,   STABILITY ANALYSIS OF SHELLS OF REVOLUTION BY
THE FINITE ELEMENT METHOD. AIAA J.6, 355-360, 1968.

1448   NAYAK G.C.,   THE METHOD OF GENERALIZED MODULUS AND GENERALIZED
STRESSES   AND   SOME   EXPERIENCES   IN
NONLINEAR SOLUTION ALGORITHMS. PROC.CONF.COMP.METHODS IN
NONLINEAR MECH., AUSTIN, TEXAS, 1974.

1449   NAYAK G.C., AND ZIENKIEWICZ O.C.,   ELASTIC-PLASTIC STRESS ANALYSIS, A
GENERALISATION USING ISOPARAMETRIC ELEMENTS AND VARIOUS
CONSTITUTIVE RELATIONS. REPORT C/R/148/71, DEPARTMENT OF CIVIL
ENGINEERING, UNIVERSITY OF WALES, SWANSEA, 1971.

1450   NAYAK G.C. AND ZIENKIEWICZ O.C.,   NOTE ON THE ALPHA-CONSTANT
STIFFNESS METHOD FOR THE ANALYSIS OF NON-LINEAR PROBLEMS.
INT.J.NUMER.METH.ENG.4, 579-597, 1972.

1451   NAYLOR D.J.,   STRESSES IN NEARLY INCOMPRESSIBLE MATERIALS BY FINITE ELEMENTS
WITH APPLICATION TO THE CALCULATION OF EXCESS PORE PRESSURES.
INT.J.NUMER.METH.ENG.8, 443-460, 1974.

1452   NEALE B.K.,   A FINITE ELEMENT MESH GENERATION PROGRAM FOR ARBITRARY
TWO-AND THREE-DIMENSIONAL STRUCTURES. PAPER M4/5,
PROCEEDINGS 2ND STRUCTURAL MECH. IN REACTOR TECH. CONF.,
BERLIN, 1973.

1453   NEALE B.K.,   BERGEN AN AUTOMATIC FINITE ELEMENT MESH GENERATION PROGRAM
FOR ARBITRARY STRUCTURES. REPORT RD/B/N 2432, CENTRAL ELECTRICITY

1454   NEALE B.K.,   FINITE ELEMENT MESH GENERATION PROGRAM FOR ARBITRARY
TWO-AND THREE-DIMENSIONAL STRUCTURES. PAPER M4/5, PROC.
2ND STRUCTURAL MECH. IN REACTOR TECH.CONF., BERLIN, 1973.
GENERATING BOARD, BERKELEY NUCLEAR LABORATORIES, 1972.

1455   NEBRENSKY J., PITTMAN J.F.T., AND SMITH J.M.,   FLOW AND HEAT TRANSFER IN
AN EXTRUDER CHANNEL. A VARIATIONAL ANALYSIS APPLIED IN HELICAL
CO-ORDINATES. PROC. EUROMECH 37, NAPLES, 1972.

1456   NECAS J., LES METHODES DIRECTES EN THEORIE DES EQUATIONS ELLIPTIQUES.
ACADEMIA,PRAGUE, 1967.

1457   NEILSON G.M.,   MULTIVARIATE SMOOTHING AND INTERPOLATING SPLINES. SIAM
J. NUMER. ANAL.11, 435-446, 1974.

1458   NEMAT-NASSER S. AND LEE K.N.,   APPLICATIONS OF GENERAL VARIATIONAL
METHODS WITH DISCONTINUOUS FIELDS TO BENDING, BUCKLING, AND
VIBRATION OF BEAMS. COMP.METH.APPL.MECH.ENG.2, 33-42, 1973

1459   NEMAT-NASSER S., AND LEE K.N.,   FINITE ELEMENT FORMULATIONS FOR ELASTIC
PLATES BY GENERAL VARIATIONAL STATEMENTS WITH DISCONTINUOUS
FIELDS. REPORT 30, TECHNICAL UNIVERSITY OF DENMARK, LYNGBY,
1973.

1460   NETHERCOT D.A., AND ROCKEY K.C.,   FINITE ELEMENT SOLUTIONS FOR THE

BUCKLING OF COLUMNS AND BEAMS. INT.J.MECH.SCI.13, 945-949, 1971.

1461 NEU H., AND WERNER H., ANALYTISCHES NAHERUNGSVERFAHREN ZUR
BERECHNUNG AZIMUTAL PERIODISCHER MAGNETFELDER IN
KREISBESCHLEUNIGERN. NUCLEAR INSTRUMENTS AND METHODS 10,
329-334, 1961.

1462 NEUMAN C.P., RECENT DEVELOPMENTS IN DISCRETE WEIGHTED RESIDUAL
METHODS. PROC.CONF.COMP.METHODS IN NONLINEAR MECH., AUSTIN,
TEXAS, 1974.

1463 NEUMAN C.P., AND SCHONBACH D.I., DISCRETE LEGENDRE ORTHOGONAL
POLYNOMIALS. INT.J.NUMER.METH.ENG.8, 743-770, 1974.

1464 NEWMARK N.M., NUMERICAL METHODS OF ANALYSIS IN BARS, PLATES AND
ELASTIC BODIES. IN GRINTER (ED.), NUMERICAL METHODS OF
ANALYSIS IN ENGINEERING,MACMILLAN, LONDON, 1949.

1465 NEWMARK N.M., A METHOD OF COMPUTATION FOR STRUCTURAL DYNAMICS.
PROC. A.S.C.E. 85, EM3, 67-94, 1959.

1466 NEY R.A., AND UTKU S., AN ALTERNATIVE FOR THE FINITE ELEMENT METHOD.
PROC.SYMP. VARIATIONAL METHODS. UNIVERSITY OF SOUTHAMPTON, 1972.

1467 NGO D., AND SCORDELIS A.C., FINITE ELEMENT ANALYSIS OF REINFORCED CONCRETE
BEAMS. J.AMERICAN CONCRETE INST. 64, 152-163, 1967.

1468 NICKELL R.E., STRESS WAVE ANALYSIS IN LAYERED THERMOVISCOELASTIC
MATERIALS BY EXTENDED RITZ METHOD. TECH.REPORT S-175,
ROHM AND HAAS CO., HUNTSVILLE, ALABAMA, 1968.

1469 NICKEL R.E., AND SECOR G.A., CONVERGENCE OF CONSISTENTLY DERIVED
TIMOSHENKO BEAM FINITE ELEMENTS. INT.J.NUMER.METH.ENG.5,
243-253, 1972.

1470 NICOLAIDES R.A., ON LAGRANGIAN INTERPOLATION IN N VARIABLES.
TECH.NOTE 274, UNIV.OF LONDON INST.COMP.SCI., 1970.

1471 NICOLAIDES R.A., ON A CLASS OF FINITE ELEMENTS GENERATED BY LAGRANGE
INTERPOLATION. SIAM J. NUMER. ANAL.9,435-445,1972.

1472 NICOLAIDES R.A., ON A CLASS OF FINITE ELEMENTS GENERATED BY LAGRANGE
INTERPOLATION II. TECH.NOTE 329, UNIV. OF LONDON INST.COMP.
SCI., 1971. ALSO SIAM J. NUMER. ANAL.10, 182-189, 1973.

1473 NIELSON G.M., COMPUTATIONS OF NU-SPLINES. TECHNICAL REPORT 11,
DEPARTMENT OF MATHEMATICS, ARIZONA STATE UNIVERSITY, 1974.

1474 NIELSON G.M., SOME PIECEWISE POLYNOMIAL ALTERNATIVES TO SPLINES UNDER
TENSION. TECHNICAL REPORT 10, DEPARTMENT OF MATHEMATICS, ARIZONA
STATE UNIVERSITY, 1974.

1475 NIELSON H.B., A FINITE ELEMENT METHOD FOR CALCULATING OPEN CHANNEL FLOW.
REPORT 19, TECHNICAL UNIVERSITY OF DENMARK,1969.

1476 NILSON A.H., NONLINEAR ANALYSIS OF REINFORCED CONCRETE BY THE FINITE
ELEMENT METHOD. J. AMERICAN CONCRETE INST.65, 757-766, 1968.

1477 NIRENBERG L., REMARKS ON STRONGLY ELLIPTIC PARTIAL DIFFERENTIAL
EQUATIONS. COMM.PURE APPL.MATH.8, 649-674, 1955.

1478 NITSCHE J. EN KRITERIUM FUR DIE QUASI-OPTIMALITAT DES RITZSCHEN
VERFAHRENS. NUMER.MATH.11, 346-348, 1968.

1479 NITSCHE J., INTERPOLATION IN SOBOLEVSCHEN FUNKTIONENRAUMEN.
NUMER.MATH.13, 334-343, 1969.

1480 NITSCHE J., ORTHOGONAL RIEHENENTWICKLUNG NACH LINEAREN SPLINE
FUNKTIONEN. J.APPROX. THEORY 2, 66-78, 1969.

1481 NITSCHE J., UMKEHRSATZE FUR SPLINE APPROXIMATION. COMPOSITIO
MATHEMATICA 21, 400-416, 1969.

1482 NITSCHE J., VERGLEICH DER KONVERGENZGESCHWINDIGKEIT DES RITZCHEN
VERFAHRENS UND DER FEHLERQUADRATMETHODE. ZAMM.49,591-596,
1969.

1483 NITSCHE J., KONVERGENZ DES RITZ-GALERKINSCHEN VERFAHRENS BEI
NICHTLINEAREN OPERATORGLEICHUNGEN. ITERATIONSVERFAHREN,
NUMERISCHE MATHEMATIK, APPROXIMATIONSTHEORIE. I.S.N.M.
15, BIRKHAUSER VERLAG, BERLIN, 1970.

1484 NITSCHE J.A., A PROJECTION METHOD FOR DIRICHLET PROBLEMS USING
SUBSPACES WITH NEARLY ZERO BOUNDARY CONDITIONS. PP.603-
627 OF A.K. AZIZ (ED.), THE MATHEMATICAL FOUNDATIONS OF
THE FINITE ELEMENT METHOD WITH APPLICATIONS TO PARTIAL
DIFFERENTIAL EQUATIONS. ACADEMIC PRESS, NEW YORK, 1972.

1485 NITSCHE J.A., INTERIOR ERROR ESTIMATES OF PROJECTION METHODS.
PP. 235-239, PROC. EQUADIFF 3, J.E. PURKYNE UNIVERSITY,
BRNO, CZECHOSLOVAKIA, 1973.

1486   NITSCHE J.A.,   CONVERGENCE OF NONCONFORMING METHODS.   PROC.SYMP.
          MATHEMATICAL ASPECTS OF FINITE ELEMENTS IN PARTIAL
          DIFFERENTIAL EQUATIONS.   UNIVERSITY OF WISCONSIN, MADISON,
          1975.
1487   NITSCHE J.,   LINEARE SPLINE-FUNKTIONEN UND DIE METHODEN VON RITZ
          FUR ELLIPTISCHE RANDWERT PROBLEME.   TO APPEAR.
1488   NITSCHE J.A.,   POLLUTION EFFECTS OF THE RITZ METHOD. (TO APPEAR).
1489   NITSCHE J., AND NITSCHE J.C.C.,   ERROR ESTIMATES FOR THE NUMERICAL
          SOLUTION OF ELLIPTIC DIFFERENTIAL EQUATIONS.   ARC.RATIONAL
          MECH.ANAL.,5, 293-306, 1960.
1490   NITSCHE J.C.C., AND NITSCHE J.,   FEHLERABSCHATZUNG FUR DIE NUMERISCHE
          BERECHNUNG VON INTEGRALEN, DIE LOSUNGEN ELLIPTISCHER
          DIFFERENTIALGLEICHUNGEN ENTHALTEN. ARCH. RATIONAL MECH.
          ANAL.5, 307-314, 1960.
1491   NITSCHE J., AND SCHATZ A.,   ON LOCAL APPROXIMATION PROPERTIES OF
          L2 PROJECTION ON SPLINE SUBSPACES.   APPLICABLE ANAL.2,
          161-168, 1972.
1492   NITSCHE J., AND SCHATZ A.,   INTERIOR ESTIMATES FOR RITZ-GALERKIN
          METHODS.   SIAM J.NUMER.ANAL.1974, (TO APPEAR).
1493   NIURA I., AND ASAOKA K.,   SIMULATION OF THE TENSIBLE BEHAVIOUR OF
          TWO-PHASE ALLOY CONTAINING A FIBRE UNDER TENSIBLE LOAD.
          J.JAPANESE INST.MET.37, 1212-1216, 1973.
1494   NOBLE B.,   COMPLEMENTARY VARIATIONAL PRINCIPLES FOR BOUNDARY VALUE
          PROBLEMS I: BASIC PRINCIPLES, WITH AN APPLICATION TO ORDINARY
          DIFFERENTIAL EQUATIONS.   II NONLINEAR NETWORKS. MATHEMATICS
          RESEARCH CENTER, TECHNICAL REPORTS 473 AND 643, 1964 AND 1966.
1495   NOBLE B.,   A BIBLIOGRAPHY ON: METHODS FOR SOLVING INTEGRAL EQUATIONS -
          AUTHOR LISTING.   TECH REPT.NO.1176, MATHEMATICS RESEARCH
          CENTER, UNIVERSITY OF WISCONSIN, MADISON, 1971.
1496   NOBLE B.,   A BIBLIOGRAPHY ON: METHODS FOR SOLVING INTEGRAL EQUATIONS -
          SUBJECT LISTING.   TECH REPT.NO.1177,MATHEMATICS RESEARCH
          CENTER, UNIVERSITY OF WISCONSIN, MADISON, 1971.
1497   NOBLE B.,   VARIATIONAL FINITE ELEMENT METHODS FOR INITIAL VALUE
          PROBLEMS. PP.143-151 OF J.R. WHITEMAN (ED.), THE MATHEMATICS
          OF FINITE ELEMENTS AND APPLICATIONS.   ACADEMIC PRESS,
          LONDON, 1973.
1498   NOBLE B., AND SEWELL M.J.,   ON DUAL EXTREMUM PRINCIPLES IN APPLIED
          MATHEMATICS.   J.INST.MATH.APPLICS 9, 123-193, 1972.
1499   NOBLE B., AND WHITEMAN J.R.,   SOLUTION OF DUAL TRIGONOMETRICAL
          SERIES USING ORTHOGONALITY RELATIONS.   SIAM J.APPL.MATH.
          18, 372-379, 1970.
1500   NOBLE B., AND WHITEMAN J.R.,   THE SOLUTION OF DUAL COSINE SERIES BY
          THE USE OF ORTHOGONALITY RELATIONS.   PROC.EDINBURGH MATH.
          SOC. 17, 47-51, 1970.
1501   NOOR A.K. AND MATHERS M.D.,   NONLINEAR FINITE ELEMENT ANALYSIS OF
          LAMINATED COMPOSITE SHELLS.   PROC.CONF.COMP.METHODS IN
          NONLINEAR MECH., AUSTIN, TEXAS, 1974.
1502   NOOR M.A.,   NONLINEAR VARIATIONAL INEQUALITIES, BULL. UNIONE MATH.
          ITALIAN 9, 736-741, 1974.
1503   NOPPEN R.,   FINEL: UNIVERSELLES PROGRAMSYSTEM ZUR BERECHNUNG DER
          ELASTISCHEN EIGENSCHAFTEN VON MASCHINENBAUTEILEN NACH DER
          METHODE FINITER ELEMENTER.   INDUSTRIE-ANZEIGER 24, 1972.
1504   NOPPEN R.,   BERECHNUNG DER ELASTIZITATSEIGENSCHAFTEN VON
          MASCHINENBAUTEILEN NACH DER METHODE FINITER ELEMENTE.
          PH.D THESIS, TECHNISCHE HOCHSCHULE AACHEN, 1973.
1505   NORRIE D.H., AND DE VREIS G.,   THE FINITE ELEMENT METHOD.   ACADEMIC
          PRESS, NEW YORK, 1973.
1506   NORRIE D.H. AND DE VRIES G.,   A LAGRANGIAN FINITE ELEMENT SOLUTION TO
          UNSTEADY FLOW.   MECH.ENG.REPORT 17, DEPARTMENT OF MECHANICAL
          ENGINEERING, UNIVERSITY OF CALGARY,1974.
1507   NORRIE D.H. AND DE VRIES G.,   APPLICATION OF THE PSEUDO-FUNCTIONAL
          FINITE ELEMENT METHOD TO NON-LINEAR PROBLEMS.   (MANUSCRIPT).
1508   ORAID P.,   THREE DIMENSIONAL FINITE ELEMENT ANALYSIS.   PH.D. THESIS,
          CASE WESTERN RESERVE UNIVERSITY, 1971.
1509   ODEN J.T.,   CALCULATION OF GEOMETRIC STIFFNESS MATRICES FOR
          COMPLEX STRUCTURES. AIAA J.4, 1480-1482, 1966.

1510   ODEN J.T.,   NUMERICAL FORMULATION OF NONLINEAR ELASTICITY PROBLEMS.
                J.STRUCT.DIV.ASCE, 93, 235-255, 1967.
1511   ODEN J.T.,   CALCULATION OF STIFFNESS MATRICES FOR FINITE ELEMENTS
                OF THIN SHELLS OF ARBITRARY SHAPE. AIAA J.6, 969-971,1968.
1512   ODEN J.T.,   FINITE PLANE STRAIN OF INCOMPRESSIBLE ELASTIC SOLIDS BY
                THE FINITE ELEMENT METHOD. THE AERONAUTICAL QUARTERLY 19,
                254-264, 1968.
1513   ODEN J.T.,   A GENERAL THEORY OF FINITE ELEMENTS   I TOPOLOGICAL
                CONSIDERATIONS II APPLICATIONS. INT.J.NUMER.METHODS
                ENG.1, 205-221 AND 247-260, 1969.
1514   ODEN J.T.,   FINITE ELEMENT ANALYSIS OF NONLINEAR PROBLEMS IN THE
                DYNAMICAL THEORY OF COUPLE THERMOELASTICITY. NUCLEAR ENG.
                AND DESIGN 10, 465-475, 1969.
1515   ODEN J.T.,   FINITE ELEMENT LARGE DELECTION ANALYSIS OF PLATES.
                J.ENG.MECH.DIV.PROC.A.S.C.E.,95, 143, 1969.
1516   ODEN J.T.,   A GENERALIZATION OF THE FINITE ELEMENT CONCEPT AND ITS
                APPLICATION TO A CLASS OF PROBLEMS IN NONLINEAR VISCOELASTICITY.
                IN D.FREDERICK (ED.), DEVELOPMENT IN THEORETICAL AND APPLIED
                MECHANICS VOL.8, PERGAMON PRESS, OXFORD 1970.
1517   ODEN J.T.,   NOTE ON AN APPROXIMATE METHOD FOR COMPUTING
                NONCONSERVATIVE GENERALIZED FORCES ON FINITELY DEFORMED
                FINITE ELEMENTS. AIAA J.8, 2088-2090, 1970.
1518   ODEN J.T.,   FINITE ELEMENT FORMULATION OF PROBLEMS OF FINITE
                DEFORMATION AND IRREVERSIBLE THERMODYNAMICS OF NONLINEAR
                CONTINUA. FROM RECENT ADVANCES IN MATRIX METHODS OF
                STRUCTURAL ANALYSIS AND DESIGN, UNIVERSITY OF ALABAMA
                PRESS, 1971.
1519   ODEN J.T.,   FINITE ELEMENT MODELS OF NONLINEAR OPERATOR
                EQUATIONS. PROC.3RD CONF. MATRIX METHODS IN STRUCTURAL
                MECHANICS. WRIGHT-PATTERSON AFB, OHIO, 1971.
1520   ODEN J.T.,   LECTURES ON FINITE ELEMENT METHODS IN CONTINUUM
                MECHANICS. NATO ADVANCED STUDY INSTITUTE ON FINITE ELEMENT
                METHODS IN CONTINUUM MECHANICS, LISBON, 1971.
1521   ODEN J.T.,   FINITE ELEMENTS OF NONLINEAR CONTINUA. MCGRAW-HILL,
                NEW YORK. 1972.
1522   ODEN J.T.,   GENERALIZED CONJUGATE FUNCTIONS FOR MIXED FINITE
                ELEMENT APPROXIMATIONS OF BOUNDARY VALUE PROBLEMS. PP.629-669
                OF A.K. AZIZ (ED.), THE MATHEMATICAL FOUNDATIONS OF THE
                FINITE ELEMENT METHOD WITH APPLICATIONS TO PARTIAL
                DIFFERENTIAL EQUATIONS. ACADEMIC PRESS, NEW YORK, 1972.
1523   ODEN J.T.,   SOME ASPECTS OF THE MATHEMATICAL THEORY OF FINITE ELEMENTS.
                PP.3-38 OF J.T. ODEN, R.W. CLOUGH AND Y. YAMAMOTO (EDS.),
                ADVANCES IN COMPUTATIONAL METHODS IN STRUCTURAL MECHANICS AND
                DESIGN. UNIVERSITY OF ALABAMA PRESS, HUNTSVILLE, 1972.
1524   ODEN J.T.,   VARIATIONAL PRINCIPLES IN NONLINEAR CONTINUUM
                MECHANICS. PROC.CONF.VARIATIONAL METHODS IN ENGINEERING,
                UNIVERSITY OF SOUTHAMPTON, 1972.
1525   ODEN J.T.,   APPROXIMATIONS AND NUMERICAL ANALYSIS OF FINITE
                DEFORMATIONS OF ELASTIC SOLIDS. PP.175-228 OF L. BALL (ED.),
                NONLINEAR ELASTICITY, ACADEMIC PRESS, NEW YORK, 1973.
1526   ODEN J.T.,   FINITE ELEMENT APPLICATIONS IN MATHEMATICAL PHYSICS.
                PP.239-282 OF J.R. WHITEMAN (ED.), THE MATHEMATICS OF
                FINITE ELEMENTS AND APPLICATIONS. ACADEMIC PRESS,
                LONDON, 1973.
1527   ODEN J.T.,   FINITE ELEMENT APPROXIMATIONS IN NONLINEAR THERMOVISCOELASTICITY.
                PP. 77-120 OF J.T. ODEN AND E.R. DE A. OLIVEIRA (EDS.), LECTURES
                ON FINITE ELEMENT METHODS IN CONTINUUM MECHANICS. UNIVERSITY
                OF ALABAMA PRESS, HUNTSVILLE, 1973.
1528   ODEN J.T.,   THE FINITE ELEMENT METHOD IN FLUID MECHANICS. PP. 151-186 OF
                J.T. ODEN AND E.R. DE A. OLIVEIRA (EDS), LECTURES ON FINITE
                ELEMENT METHODS IN CONTINUUM MECHANICS. UNIVERSITY OF ALABAMA
                PRESS, HUNTSVILLE, 1973.
1529   ODEN J.T.,   SOME CONTRIBUTIONS TO THE MATHEMATICAL THEORY OF
                MIXED FINITE ELEMENT APPROXIMATIONS. PROC.TOKYO SEMINAR
                ON FINITE ELEMENTS, UNIVERSITY OF TOKYO PRESS, 1973.
1530   ODEN J.T.,   THEORY OF CONJUGATE PROJECTIONS IN FINITE ELEMENT ANALYSIS.

PP.41-76 OF J.T. ODEN AND E.R. DE A. OLIVEIRA (EDS.), LECTURES ON
FINITE ELEMENT METHODS IN CONTINUUM MECHANICS. FINITE ELEMENT
METHODS IN CONTINUUM MECHANICS. UNIVERSITY OF ALABAMA PRESS,
HUNTSVILLE, 1973.

1531 ODEN J.T., FINITE ELEMENT FORMULATION OF NONLINEAR BOUNDARY VALUE PROBLEMS,
      PP. 187-208 OF J.T. ODEN AND E.R. DE A. OLIVEIRA (EDS.), LECTURES
      ON FINITE ELEMENT METHODS IN CONTINUUM MECHANICS. UNIVERSITY OF
      ALABAMA PRESS, HUNTSVILLE, 1973.

1532 ODEN J.T., AN IMBEDDING TECHNIQUE FOR THE GENERATION OF WEAK-WEAK
      FINITE APPROXIMATIONS OF LINEAR AND NONLINEAR OPERATORS.
      (TO APPEAR.)

1533 ODEN J.T., AND AGUIRRE-RAMIREZ G., FORMULATION OF GENERAL DISCRETE
      MODELS OF THERMOMECHANICAL BEHAVIOUR OF MATERIALS WITH MEMORY.
      INT.J.SOLIDS STRUCTURES 5, 1077-1093, 1969.

1534 ODEN J.T., AND ARMSTRONG W.H., ANALYSIS OF NONLINEAR DYNAMIC
      COUPLED THERMOVISCOELASTICITY PROBLEMS BY THE FINITE
      ELEMENT METHOD. COMPUTERS AND STRUCTURES 1, 603-622, 1971.

1535 ODEN J.T., AND BRAUCHLI H.J., A NOTE ON ACCURACY AND CONVERGENCE
      OF FINITE ELEMENT APPROXIMATIONS. INT.J.NUMER.METH.ENG.3,
      291-292,1971.

1536 ODEN J.T., AND BRAUCHLI H.J., ON THE CALCULATION OF CONSISTENT
      STRESS DISTRIBUTION IN FINITE ELEMENT APPLICATIONS.
      INT.J.NUMER.METH.IN ENG.3, 317-326, 1971.

1537 ODEN J.T. CHUNG T.J. AND KEY J.E., ANALYSIS OF NONLINEAR
      THERMOELASTIC AND THERMOPLASTIC BEHAVIOUR OF SOLIDS OF
      REVOLUTION BY THE FINITE ELEMENT METHOD. PAPER M5/6,
      PROCEEDINGS 1ST STRUCTURAL MECH. IN REACTOR TECH. CONF.,
      BERLIN, 1971.

1538 ODEN J.T., CLOUGH R.W., AND YAMAMOTO Y. (EDS.), ADVANCES IN
      COMPUTATIONAL METHODS IN STRUCTURAL MECHANICS AND DESIGN.
      UNIVERSITY OF ALABAMA PRESS, HUNTSVILLE, 1973.

1539 ODEN J.T., AND FOST R.B., CONVERGENCE, ACCURACY AND STABILITY OF
      FINITE ELEMENT APPROXIMATIONS OF A CLASS OF NON-LINEAR
      HYPERBOLIC EQUATIONS. INT.J.NUMER.METH.ENG.6, 357-365, 1973.

1540 ODEN J.T., AND KELLEY B.E., FINITE ELEMENT FORMULATION OF GENERAL
      ELECTROTHERMOELASTICITY PROBLEMS. INT.J.NUMER.METH.ENG.2,
      161-180, 1971.

1541 ODEN J.T., AND KEY J.E., NUMERICAL ANALYSIS OF FINITE AXISYMMETRIC
      DEFORMATIONS OF INCOMPRESSIBLE ELASTIC SOLIDS OF REVOLUTION.
      INT.J.SOLIDS AND STRUCTURES 6, 497-518, 1970.

1542 ODEN J.T. AND KEY J.E., ANALYSIS OF FINITE DEFORMATIONS OF ELASTIC
      SOLIDS BY THE FINITE ELEMENT METHOD. UNIV.OF ALABAMA RESEARCH
      INSTITUTE REPT.NO.9, 1971.

1543 ODEN J.T. AND KEY J.N., ON SOME GENERALIZATIONS OF THE
      INCREMENTAL STIFFNESS RELATIONS FOR FINITE DEFORMATIONS
      OF COMPRESSIBLE AND INCOMPRESSIBLE FINITE ELEMENTS.
      NUCLEAR ENG. AND DESIGN 15, 121-134, 1971.

1544 ODEN J.T. AND KEY J.E., A NOTE ON THE FINITE DEFORMATION OF A THICK
      ELASTIC SHELL OF REVOLUTION. PAPER J2/5, PROCEEDINGS 1ST
      STRUCTURAL MECH. IN REACTOR TECH. CONF., BERLIN, 1971.

1545 ODEN J.T., KEY J.E., AND FOST R.B., A NOTE ON THE ANALYSIS OF
      NONLINEAR DYNAMICS OF ELASTIC MEMBRANES BY THE FINITE
      ELEMENT METHOD. COMPUTERS AND STRUCTURES 4, 445-452, 1974.

1546 ODEN J.T. AND KROSS D.A., ANALYSIS OF GENERAL COUPLED
      THERMOELASTICITY PROBLEMS BY THE FINITE ELEMENT METHOD.
      PROC.2ND CONF. MATRIX METHODS IN STRUCTURAL MECHANICS.
      WRIGHT-PATTERSON AFB., OHIO, AFFDL-TR-68-150,1968.

1547 ODEN J.T. AND KUBITZA W.K., NUMERICAL ANALYSIS OF NONLINEAR PNEUMATIC
      STRUCTURES, PROC.1ST INTERNATIONAL COLLOQUIUM ON PNEUMATIC
      STRUCTURES, STUTTGART, MAY 1967.

1548 ODEN J.T. AND OLIVEIRA E.R. DE A., LECTURES ON FINITE ELEMENT
      METHODS IN CONTINUUM MECHANICS. UNIVERSITY OF ALABAMA
      PRESS, HUNTSVILLE, 1973.

1549 ODEN J.T., AND REDDY J.N., NOTE ON AN APPROXIMATE METHOD FOR
      COMPUTING CONSISTENT STRESSES IN ELASTIC FINITE ELEMENTS.
      INT.J.NUMER.METH.ENG.6, 55-61, 1973.

1550   ODEN J.T., RIGSBY D.M., AND CORNETT D., ON THE NUMERICAL
             SOLUTION OF A CLASS OF PROBLEMS IN A LINEAR FIRST STRAIN
             GRADIENT THEORY OF ELASTICITY. INT.J.NUMER.METH.ENG,2,
             159-174, 1970.
1551   ODEN J.T. AND SATO T., FINITE STRAINS AND DISPLACEMENTS OF
             ELASTIC MEMBRANES BY THE FINITE ELEMENT METHOD. INT.J.
             SOLIDS STRUCTURES 3, 471-488, 1967.
1552   ODEN J.T., AND SOMOGYI D., FINITE ELEMENT APPLICATIONS IN FLUID
             MECHANICS. J.ENG.MECH DIV. ASCE. 95, 821-826, 1969.
1553   ODEN J.T., AND WELLFORD L.C., THE ANALYSIS OF FLOW OF VISCOUS FLUIDS
             BY THE FINITE ELEMENT METHOD. TO APPEAR.
1554   OGANESJAN L.A., CONVERGENCE OF DIFFERENCE SCHEMES IN CASE OF
             IMPROVED APPROXIMATION OF THE BOUNDARY. (RUSSIAN)
             Z. VYCISL.MAT.I MAT.FYZ.6,1029-1042, 1966.
1555   OGANESJAN L.A., AND RUHOVEC L.A., VARIATIONAL DIFFERENCE SCHEMES
             FOR SECOND ORDER LINEAR ELLIPTIC EQUATIONS IN A TWO-
             DIMENSIONAL REGION WITH A PIECEWISE SMOOTH BOUNDARY. Z.
             VYCISL.MAT.I MAT.FIZ.8, 97-114, 1968.
1556   OGANESJAN L.A., AND RUCHOVEC P.A., INVESTIGATION OF THE CONVERGENCE
             RATE OF VARIATIONAL - DIFFERENCE SCHEMES FOR ELLIPTIC SECOND
             ORDER EQUATIONS IN A TWO DIMENSIONAL DOMAIN WITH A SMOOTH
             BOUNDARY. (RUSSIAN). Z.VYCISL.MAT.I.FYZ.9,1102-1120,1969.
1557   OHYA H.G., YAGAWA G., AND ANDO Y., SNAP-THROUGH BUCKLING ANALYSIS
             OF SHELLS OF REVOLUTION USING INCREMENTAL STIFFNESS
             MATRICES. PROC.CONF.COMP.METHODS IN NONLINEAR MECH.,
             AUSTIN, TEXAS, 1974.
1558   OHNISHI T., APPLICATION OF FINITE ELEMENT SOLUTION TECHNIQUE TO NEUTRON
             DIFFUSION AND TRANSPORT EQUATIONS. PROC.CONF. ON NEW DEVELOPMENTS
             IN REACTOR MATHEMATICS AND APPLICATIONS, CONF 710302, IDAHO
             FALLS, 1971.
1559   OHNISHI T., FINITE ELEMENT METHOD APPLIED TO REACTOR PHYSICS PROBLEMS.
             J. NUCL. SCI. TECH. (TOKYO) 8, 712-720,1971.
1560   OHNISHI T., FINITE ELEMENT SOLUTION TECHNIQUE FOR NEUTRON TRANSPORT
             EQUATION. PP. 629-638 OF NUMERICAL REACTOR CALCULATIONS,
             INTERNATIONAL ATOMIC ENERGY AGENCY, VIENNA, 1972.
1561   OLDHAM K.B., AND SPANIER J., THE FRACTIONAL CALCULUS. ACADEMIC
             PRESS, NEW YORK, 1974.
1562   OLIVEIRA E.R. DE A., COMPLETENESS AND CONVERGENCE IN THE FINITE
             ELEMENT METHOD. PROC. 2ND. CONF.MATRIX METHODS IN STRUCTURAL
             MECHANICS, WRIGHT-PATTERSON AFB., OHIO, AFFDL-TR-68-150, 1968.
1563   OLIVEIRA E.R. DE A., THEORETICAL FOUNDATIONS OF THE FINITE ELEMENT
             METHOD. INT.J. SOLIDS STRUCTURES 4, 929-952,1968.
1564   OLIVEIRA E.R.DE A. THEORETICAL FOUNDATIONS OF THE FINITE ELEMENT
             METHOD. TECNICA, NOS.399-400, VOL.32,1970.
1565   OLIVEIRA E.R.DE A., THE CONVERGENCE THEOREMS AND THEIR ROLE IN
             THE THEORY OF STRUCTURES. PROC.IUTAM SYMPOSIUM ON HIGH
             SPEED COMPUTING OF ELASTIC STRUCTURES. LIEGE,BELGIUM,1970.
1566   OLIVEIRA E.R. DE A., A METHOD OF FICTITIOUS FORCES FOR THE
             GEOMETRICALLY NONLINEAR ANALYSIS OF STRUCTURES. PROC.CONF.
             COMP. METHODS IN NONLINEAR MECH., AUSTIN, TEXAS, 1974.
1567   OLIVEIRA E.R. DE A., CONVERGENCE OF FINITE ELEMENT SOLUTIONS IN
             VISCOUS FLOW PROBLEMS.
             PROCEEDINGS OF SYMPOSIUM ON FINITE ELEMENT ELEMENT METHODS
             IN FLOW PROBLEMS,SWANSEA,1974.
1568   OLIVEIRA A. DE E., THE ROLE OF CONVERGENCE IN THE THEORY OF
             SHELLS. J.SOLIDS STRUCTURES 10, 531-553, 1974.
1569   OLSON M.D. FINITE ELEMENTS APPLIED TO PANEL FLUTTER. AIAA J,5,
             2267-2269, 1967.
1570   OLSON M.D., SOME FLUTTER SOLUTIONS USING FINITE ELEMENTS. AIAA,J.8,
             747-752,1970.
1571   OLSON M.D., A VARIATIONAL FINITE ELEMENT METHOD FOR TWO-DIMENSIONAL
             STEADY VISCOUS FLOWS. REPORT NO.5, STRUCTURAL RESEARCH SERVICES,
             DEPARTMENT OF CIVIL ENGINEERING, UNIVERSITY OF BRITISH
             COLUMBIA, 1972.
1572   OLSON M.D. AND LINDBERG G.M., VIBRATION ANALYSIS OF CANTILEVERED
             CURVED PLATES USING A NEW CYLINDRICAL SHELL FINITE ELEMENT.

PROC 2ND CONF.MATRIX METHODS IN STRUCTURAL MECHANICS,
WRIGHT-PATTERSON AFB., OHIO, AFFDL-TR-65-150, 1968.

1573   OLSON M.D., AND LINDBERG G.M.,   ANNULAR AND CIRCULAR SECTOR FINITE
       ELEMENTS FOR PLATE BENDING. INT.J.MECH.SCI.12, 17-34,
       1970.

1574   OLSON M.D., AND LINDBERG G.M.,   DYNAMIC ANALYSIS OF SHALLOW SHELLS WITH
       A DOUBLY CURVED TRIANGULAR FINITE ELEMENT. J. SOUND VIB,19,
       299-318, 1971.

1575   ORRIS R.M., AND PETYT M.,   A FINITE ELEMENT STUDY OF HARMONIC
       WAVE PROPAGATION IN PERIODIC STRUCTURES. J.SOUND VIB.
       33, 223-236, 1974.

1576   ORSZAG S.A.,   GALERKIN APPROXIMATIONS TO FLOWS WITHIN SLABS,
       SPHERES AND CYLINDERS. PHYSICS REV. LETTERS 26, 1100-1103,
       1971.

1577   ORSZAG S.A.,   NUMERICAL SIMULATION OF INCOMPRESSIBLE FLOWS WITHIN
       SIMPLE BOUNDARIES. 1. GALERKIN (SPECTRAL) REPRESENTATIONS.
       STUDIES IN APPLIED MATHEMATICS 50, 293-327, 1971.

1578   ORTEGA J., AND RHEINBOLDT W.,   ITERATIVE SOLUTION OF NONLINEAR
       EQUATIONS IN SEVERAL VARIABLES. ACADEMIC PRESS, NEW YORK,
       1970.

1579   OWEN D.R.J., NAYAK G.C., KFOURI A.P., AND GRIFFITHS J.R.,   STRESSES
       IN A PARTLY YIELDED NOTCHED BAR, AN ASSESSMENT OF THREE
       ALTERNATIVE PROGRAMS. INT.J.NUMER.METH.ENG.6, 63-73, 1973.

1580   OWEN D.R.J., AND PRAKASH A.,   THE ANALYSIS OF DISLOCATION SYSTEMS
       BY THE FINITE ELEMENT METHOD. INT.J.NUMER.METH.ENG.6,
       117-128, 1973.

1581   OWEN D.R.J., AND PRAKASH A.,   THE FINITE ELEMENT ANALYSIS OF
       ELASTO-PLASTIC MATERIALS BY THE USE OF DISLOCATION DIPOLE
       SYSTEMS. INT.J.NUMER.METH.ENG.8, 277-288, 1974.

1582   PADOVAN J.,   SEMI-ANALYTICAL FINITE ELEMENT PROCEDURE FOR
       CONDUCTION IN ANISOTROPIC AXISYMMETRIC SOLIDS. INT.J.
       NUMER.METH.ENG.8, 295-310, 1974.

1583   PALACOL E.L., AND STANTON E.L.,   ANISOTROPIC PARAMETRIC PLATE
       DISCRETE ELEMENTS. INT.J.NUMER.METH.ENG.6, 413-425, 1973.

1584   PALIT K., AND FENNER R.T.,   FINITE ELEMENT ANALYSIS OF SLOW NON-NEWTONIAN
       CHANNEL FLOW. A.I.CH.E. J. 18, 628-633, 1972.

1585   PALMERTON J.B., AND BANKS D.S.,   APPLICATION OF FINITE ELEMENT METHOD IN
       DETERMINING STABILITY OF CRATER SLOPES. REPORT AD-73718, ARMY
       ENGINEER WATERWAYS EXPERIMENT STATION, VICKSBURG, 1972.

1586   PAPAMICHAEL N., AND WHITEMAN J.R.,   A CUBIC SPLINE TECHNIQUE FOR
       THE ONE DIMENSIONAL HEAT CONDUCTION EQUATION. J.INST.
       MATH.APPLICS.11, 111-113, 1973.

1587   PAPAMICHAEL N., AND WHITEMAN J.R.,   A NUMERICAL CONFORMAL
       TRANSFORMATION METHOD FOR HARMONIC MIXED BOUNDARY VALUE
       PROBLEMS IN POLYGONAL DOMAINS. Z.A.M.P. 24, 304-316, 1973.

1588   PAPAMICHAEL N., AND WHITEMAN J.R.,   CUBIC SPLINE INTERPOLATION TO
       HARMONIC FUNCTIONS. B.I.T. 14, XXX-XXX, 1974.

1589   PAREKH C.J.,   A FINITE ELEMENT SOLUTION OF TIME DEPENDENT FIELD PROBLEMS.
       M.SC THESIS, UNIV. OF WALES,SWANSEA,1967.

1590   PARIKH S.K.,   ANALYSIS OF EARTHEN DAMS BY THE METHOD OF FINITE ELEMENTS.
       J.SOIL.MECH.AND FOUNDATION ENG. DIV. A.S.C.E. 96, SM2, 155-169,
       1970.

1591   PARKER J.V.,   PRACTICAL APPLICATIONS OF FINITE ELEMENT METHODS IN
       THE STRESS ANALYSIS OF THIN SHELL PRESSURE VESSELS.
       PAPER J1/3, PROCEEDINGS 1ST STRUCTURAL MECH. IN REACTOR
       TECH. CONF., BERLIN, 1971.

1592   PARKER R. AND MA Y.C.,   NORMAL GRADIENT BOUNDARY CONDITIONS IN FINITE
       DIFFERENCE CALCULATIONS. INT.J.NUMER.METH.ENG.7.,395-411,1973.

1593   PATIL S.H.,   NON-POLYNOMIAL LAGRANGIANS. PP.431-445, PROC.
       1ST SYMP.HIGH ENERGY PHYSICS, TATA INST.FUNDAMENTAL
       RESEARCH, BOMBAY,1972.

1594   PATIL S.S., RAO A.K. AND RAJU I.S.,   SPECIAL FINITE ELEMENTS FOR
       FLAW AND DISCONTINUITY STRESSES IN COMPOSITES. PAPER M3/6,
       PROCEEDINGS 2ND STRUCTURAL MECH. IN REACTOR TECH.CONF.,
       BERLIN, 1973.

1595   PAULLING J.R., AND PAYER H.G.,   HULL-DECKHOUSE INTERACTION BY

FINITE-ELEMENT CALCULATIONS. TRANS SOC.NAVAL ARCHITECTS
MARINE ENGRS. 76, 281-308,1968.

1596 PAWSEY S.F., AND CLOUGH R.W., IMPROVED NUMERICAL INTEGRATION OF
THICK SHELL FINITE ELEMENTS. INT.J.NUMER.METH.ENG.3,
575-586, 1971.

1597 PECKNOLD D.A. AND SCHNOBRICH W.C., FINITE ELEMENT ANALYSIS OF
SKEWED SHALLOW SHELLS. J. STRUCT. DIV. A.S.C.E. 95, ST4,
715-745, 1969.

1598 PEDERSEN P., SOME PROPERTIES OF LINEAR STRAIN TRIANGLES AND OPTIMAL
FINITE ELEMENT MODELS. INT.J.NUMER.METH.ENG.7, 415-430,1973.

1599 PEETRE J., ESPACES D'INTERPOLATION ET THEOREME DE SOBOLEV, ANN.
INST.FOURIER (GRENOBLE) 16, 279-317, 1966.

1600 PENDERZINI A., RUSSO S., AND VIOTTI V., CONCOR A NUMERICAL CODE
TO PREDICT THE CONFIGURATION OF A LMFBR CORE TAKING INTO
ACCOUNT SUBASSEMBLIES INTERACTION, SWELLING, THERMAL
BOWING AND CREEP. PROC. 2ND STRUCTURAL MECH. IN REACTOR
TECH. CONF., BERLIN, 1973.

1601 PERCY J.H., QUADRILATERAL FINITE ELEMENTS IN ELASTIC-PLASTIC
PLANE STRESS ANALYSIS. AIAA J.5, 367 , 1967

1602 PERCY J.H., PIAN T.H.H., KLEIN S., AND NAVARATHA D.R., APPLICATION
OF MATRIX DISPLACEMENT METHOD TO LINEAR ELASTIC ANALYSIS
OF SHELLS OF REVOLUTION. AIAA J.3, 2183-2145, 1965.

1603 PERIAUX J., THREE DIMENSIONAL ANALYSIS OF COMPRESSIBLE POTENTIAL FLOWS
WITH THE FINITE ELEMENT METHOD. REPORT AVIONS MARCEL DASSAULT-
BREGUET AVIATION, 1973.

1604 PERRIN F.M., PRICE H.S., AND VARGA R.S., ON HIGHER-ORDER NUMERICAL
METHODS FOR NON-LINEAR TWO-POINT BOUNDARY VALUE PROBLEMS,
NUMER.MATH.13,180-198.1969.

1605 PETRYSHIN W.V., DIRECT AND ITERATIVE METHODS FOR THE SOLUTION OF
LINEAR OPERATOR EQUATIONS IN HILBERT SPACE. TRANS.AMER.
MATH.SOC.105,136-175, 1962.

1606 PETYT M. AND FLEISCHER C.C., VIBRATION OF CURVED STRUCTURES USING
QUADRILATERAL FINITE ELEMENTS. PP.367-378 OF J.R.WHITEMAN
(ED.), THE MATHEMATICS OF FINITE ELEMENTS AND APPLICATIONS,
ACADEMIC PRESS, LONDON, 1973.

1607 PETYT M., MIRZA W.H., BREBBIA C.A., AND TOTTENHAM H., FINITE
ELEMENT MODELLING TECHNIQUES FOR THE ANALYSIS OF BUILDING
VIBRATIONS. PROCEEDINGS VARIATIONAL METHODS IN ENGINEERING
CONF., UNIVERSITY OF SOUTHAMPTON, 1972.

1608 PHILLIPS D.V., TECHNIQUES IN TRIANGULAR MESH GENERATION FOR FLAT AND
CURVED SURFACES. M.SC.THESIS, UNIVERSITY OF WALES, SWANSEA,
1969.

1609 PHILLIPS Z., AND PHILLIPS D.V., AN AUTOMATIC GENERATION SCHEME
FOR PLANE AND CURVED SURFACES BY ISOPARAMETRIC COORDINATES.
INT.J.NUMER.METH.ENG.3, 519-528, 1971.

1610 PIAN T.H.H., DERIVATION OF ELEMENT STIFFNESS MATRICES.
AIAA. J.2, 576-577, 1964.

1611 PIAN T.H.H., DERIVATION OF ELEMENT STIFFNESS MATRICES BY ASSUMED
STRESS DISTRIBUTIONS. AIAA.J.2, 1333-1336, 1964.

1612 PIAN T.H.H., ELEMENT STIFFNESS MATRICES FOR BOUNDARY COMPATIBILITY
AND FOR PRESCRIBED BOUNDARY STRESSES. PROC.1ST CONF.MATRIX
IN STRUCTURAL MECHANICS. WRIGHT-PATTERSON AFB, OHIO,
AFFDL TR 66-80, 1965.

1613 PIAN T.H.H., FINITE ELEMENT STIFFNESS METHODS BY DIFFERENT
VARIATIONAL PRINCIPLES IN ELASTICITY. IN BIRKHOFF AND
VARGA (EDS.), NUMERICAL SOLUTION OF FIELD PROBLEMS IN
CONTINUUM MECHANICS. SIAM-AMS PROCEEDINGS, VOL. 11,
PROVIDENCE, R.I., 1970.

1614 PIAN T.H.H., FORMULATIONS OF FINITE ELEMENT METHODS FOR SOLID CONTINUA.
IN R.H. GALLAGHER, Y. YAMADA AND J.T. ODEN (EDS.), RECENT ADVANCES
IN MATRIX METHODS OF STRUCTURAL ANALYSIS OF DESIGN. UNIVERSITY
OF ALABAMA PRESS, HUNTSVILLE, 1971.

1615 PIAN T.H.H., VARIATIONAL FORMULATIONS OF NUMERICAL METHODS IN SOLID
CONTINUA. PP.421-448 OF G.M.L. GLADWELL (ED.), COMPUTER AIDED
ENGINEERING, STUDY NO.5, SOLID MECHANICS DIVISION, UNIVERSITY OF
WATERLOO,1971.

1616   PIAN T.H.H.,   FINITE ELEMENT FORMULATION BY VARIATIONAL PRINCIPLES
                WITH RELAXED CONTINUITY REQUIREMENTS.   PP.671-687 OF A.K.
                AZIZ (ED.), THE MATHEMATICAL FOUNDATIONS OF THE FINITE
                ELEMENT METHOD WITH APPLICATIONS TO PARTIAL DIFFERENTIAL
                EQUATIONS.   ACADEMIC PRESS, NEW YORK, 1972.
1617   PIAN T.H.H.,   FINITE ELEMENT METHODS BY VARIATIONAL PRINCIPLES WITH
                RELAXED CONTINUITY REQUIREMENT.   IN C.A. BREBBIA AND H.
                TOTTENHAM (EDS.), VARIATIONAL METHODS IN ENGINEERING,
                UNIVERSITY OF SOUTHAMPTON PRESS, 1972.
1618   PIAN T.H.H., AND MAU S.T.,   SOME RECENT STUDIES IN ASSUMED STRESS HYBRID
                MODELS.   PP. 87-106 OF J.T. ODEN, R.W. CLOUGH, AND Y. YAMAMOTO
                (EDS.), ADVANCES IN COMPUTATIONAL METHODS IN STRUCTURAL
                MECHANICS AND DESIGN. UNIVERSITY OF ALABAMA PRESS,
                HUNTSVILLE, 1973.
1619   PIAN T.H.H., AND TONG P.,   THE CONVERGENCE OF FINITE ELEMENT METHOD
                IN SOLVING LINEAR ELASTIC PROBLEMS. INT.J.SOLIDS STRUCTURES
                3, 865-880, 1967.
1620   PIAN T.H.H. AND TONG P.,   RATIONALIZATION IN DRIVING ELEMENT
                STIFFNESS MATRIX BY ASSUMED STRESS APPROACH. PROC. 2ND
                CONF. MATRIX METHODS IN STRUCTURAL MECHANICS. WRIGHT-
                PATTERSON AFB.,   OHIO, AFFDL-TR-68-150, 1968.
1621   PIAN T.H.H., AND TONG P.,   BASIS OF FINITE ELEMENT METHODS FOR SOLID
                CONTINUA. INT.J.NUMER.ENG.1, 3-28, 1969.
1622   PIAN T.H.H., AND TONG P.,   VARIATIONAL FORMULATION OF FINITE
                DISPLACEMENT ANALYSIS. PP. 43-63 OF B. FRAEIJS DE VEUBEKE
                (ED.), HIGH SPEED COMPUTING OF ELASTIC STRUCTURES.
                UNIVERSITY OF LIEGE, 1971.
1623   PIAN T.H.H., AND TONG P.,   FINITE ELEMENT METHODS IN CONTINUUM
                MECHANICS.   ADVANCES IN APPLIED MECHANICS 12, 1-58, 1972.
1624   PIAN T.H.H. TONG P. AND LUK C.H.,   ELASTIC CRACK ANALYSIS BY A
                FINITE ELEMENT HYBRID METHOD. PROC. 3RD CONF. MATRIX
                METHODS IN STRUCTURAL MECHANICS. WRIGHT-PATTERSON AFB.,
                OHIO, 1971.
1625   PIERCE J.G., AND VARGA R.S.,   HIGHER ORDER CONVERGENCE RESULTS
                FOR THE RAYLEIGH-RITZ METHOD APPLIED TO EIGENVALUE
                PROBLEMS, I, ESTIMATES RELATING RAYLEIGH-RITZ AND GALERKIN
                APPROXIMATIONS TO EIGENFUNCTIONS. SIAM J. NUMER.ANAL.9,
                137-151, 1972.
1626   PIERCE J.G. AND VARGA R.S.,   HIGHER ORDER CONVERGENCE RESULTS FOR THE
                RAYLEIGH-RITZ METHOD APPLIED TO EIGENVALUE PROBLEMS:2 IMPROVED
                ERROR BOUNDS FOR EIGENFUNCTIONS. NUMER.MATH.12, 155-169, 1972.
1627   PIFKO A.B. AND ISAKSON G.,   A FINITE ELEMENT METHOD FOR THE PLASTIC
                BUCKLING ANALYSIS OF PLATES. AIAA J.7, 1950-1957, 1969.
1628   PINSON L.D.,   EVALUATION OF A FINITE ELEMENT ANALYSIS FOR THE
                LONGITUDINAL VIBRATIONS OF LIQUID-PROPELLANT LAUNCH
                VEHICLES. N.A.S.A. TECHNICAL NOTE TN D-5803, 1970.
1629   PIROTIN S.D.,   INCREMENTAL LARGE DEFLECTION ANALYSIS OF ELASTIC
                STRUCTURES.   PH.D. THESIS, DEPARTMENT OF AERONAUTICS AND
                ASTRONAUTICS, MASSACHUSETTS INSTITUTE OF TECHNOLOGY, 1971.
1630   PITKAERANTA J., AND SILVENNOINEN P.,   COMPUTATIONAL EXPERIMENTATION ON THE
                FINITE ELEMENT METHOD IN BARE SLAB CRITICALITY CALCULATIONS.
                NUCL.SCI.ENG.50, 297-300, 1973.
1631   PITKARANTA J., AND SILVENNOINEN P.,   VACUUM DESCRIPTION IN
                SYMMETRIZED TRANSPORT CALCULATIONS. TRANS.AMER.NUCL.SOC,
                17, 235-236, 1973.
1632   PITKARANTA J. AND SILVENNOINEN P.,   FINITE ELEMENT ANALYSIS OF SOME
                CRITICAL FAST ASSEMBLIES. NUCLEAR SCIENCE AND ENG, 52,
                447-453, 1974.
1633   PLANK R.J., AND WITTRICK W.H.,   BUCKLING UNDER COMBINED LOADING
                OF THIN FLAT-WALLED STRUCTURES BY A COMPLEX FINITE STRIP
                METHOD.   INT.J.NUMER.METH.ENG.8, 323-339, 1974.
1634   PLATTEN J.K., FLANDROY P., AND VANDERBORCK P.,   VARIATIONAL AND
                ACCURATE SOLUTION OF THE ORR-SOMMERFELD EQUATION.
                INT. J. ENG. SCI. 12, 995-1006, 1974.
1635   POLYA G.,   SUR UNE INTERPOLATION DE LA METHODE DES DIFFERENCES
                FINIES QUI PEUT FOURNIR DES BORNES SUPERIEURES OU INFERIEURES.

COMPTES RENDUS 235, 995-997, 1952.

1636 POLYA G., AND SZEGO G., ISOPERIMETRIC INEQUALITIES IN MATHEMATICAL PHYSICS. ANNALS OF MATHEMATICS STUDIES 27, PRINCETON UNIVERSITY PRESS, 1951.

1637 PONTER A.R.S. THE APPLICATION OF DUAL MINIMUM THEOREMS TO THE FINITE ELEMENT SOLUTION OF POTENTIAL PROBLEMS WITH SPECIAL REFERENCE TO SEEPAGE. INT.J.NUMER.METH.ENG.4, 85-93, 1972.

1638 PONTRYAGIN L.S., ORDINARY DIFFERENTIAL EQUATIONS. PRENTICE HALL, NEW JERSEY, 1966.

1639 POPE G.G., THE APPLICATION OF THE MATRIX DISPLACEMENT METHOD IN PLANE ELASTOPLASTIC STRESS PROBLEMS. PROC.1ST CONF.MATRIX METHODS IN STRUCTURAL MECHANICS, WRIGHT-PATTERSON AFB, OHIO, AFFDL TR66-80, 1965.

1640 POPE G., A DISCRETE ELEMENT METHOD FOR ANALYSIS OF PLANE ELASTIC-PLASTIC STRESS PROBLEMS. ROYAL AERONAUTICAL ESTABLISHMENT TR65028, 1965.

1641 POPOV E.P., KHOJASTCH-BAKHT M., AND YAGHMAI S., BENDING OF CIRCULAR PLATES OF HARDENING MATERIAL. INT.J.SOLIDS STRUCT.3, 975-988, 1967.

1642 POPOV E.P., AND SHARIFT P., A REFINED CURVED ELEMENT FOR THIN SHELLS OF REVOLUTION. INT.J.NUMER.METH.ENG.3,495-508, 1971.

1643 POPPLEWELL N., AND MCDONALD D., CONFORMING RECTANGULAR AND TRIANGULAR PLATE BENDING ELEMENTS. J.SOUND VIB.19, 333-347, 1971.

1644 POWELL G.H., AND CHI H.M., COMPUTATIONAL PROCEDURES FOR INELASTIC STRESS ANALYSIS. PAPER L4/4, PROC.2ND STRUCTURALMECH. IN REACTOR TECH. CONF., BERLIN, 1973.

1645 PRAGER W. AND SYNGE J.L., APPROXIMATIONS IN ELASTICITY BASED ON THE CONCEPT OF A FUNCTION SPACE. QUART.APPL.MATH.5, 241-269, 1947.

1646 PRASAD K.S.R.K., KRISHNA MURTY A.V. AND RAO A.K., A FINITE ELEMENT ANALOGUE OF THE MODIFIED RAYLEIGH-RITZ METHOD FOR VIBRATION PROBLEMS. INT.J.NUMER.METH.ENG.5, 163-169,1972.

1647 PRATO C.A., SHELL FINITE ELEMENT METHOD VIA REISSNER'S PRINCIPLE. INT.J.SOLIDS STRUCTURES 5, 1119-1133, 1969.

1648 PRENTER P.M., LAGRANGE AND HERMITE INTERPOLATION IN BANACH SPACES. J. APPROX. THEORY 4, 419-432, 1971.

1649 PRENTER P.M., AND RUSSELL R.D., COLLOCATION FOR ELLIPTIC PARTIAL DIFFERENTIAL EQUATIONS. (TO APPEAR).

1650 PRICE H.S., CAVENDISH J.C., AND VARGA R.S., NUMERICAL METHODS OF HIGHER-ORDER ACCURACY FOR DIFFUSION-CONVECTION EQUATIONS. SOC.PETROLEUM ENGRG. 8, 293-303, 1968.

1651 PRICE H.S., AND VARGA R.S., ERROR BOUNDS FOR SEMIDISCRETE GALERKIN APPROXIMATIONS FOR PARABOLIC PROBLEMS WITH APPLICATIONS TO PETROLEUM RESERVOIR MECHANICS. IN BIRKHOFF AND VARGA (EDS.), NUMERICAL SOLUTION OF FIELD PROBLEMS IN CONTINUUM MECHANICS. SIAM-AMS PROCEEDINGS, VOL.11, PROVIDENCE, R.I. 1970.

1652 PROFANT M. AND WALTHER H., UBER DIE STATIONARE LOSUNG DER DIFFUSIONGLEICHUNG MIT DRIFT-TERM. TECHNICAL REPORT JUL -1027-MA, ZENTRALINSTITUT FUR ANGEWANDTE MATHEMATIK, KERNFORSCHUNGSANLAGE, JULICH, 1973.

1653 PRYCE J.D., BASIC METHODS OF LINEAR FUNCTIONAL ANALYSIS. HUTCHINSON, LONDON, 1973.

1654 PRYOR C.W. AND BARKER R.M., A FINITE ELEMENT ANALYSIS INCLUDING TRANSVERSE SHEAR EFFECTS FOR APPLICATIONS TO LAMINATED PLATES. AIAA J.9, 912-916, 1971.

1655 PRYOR C.W., BARKER R.M., AND FREDRICK D., FINITE ELEMENT BENDING ANALYSIS OF REISSNER PLATES. J.ENG.MECH.DIV.A.S.C.E. 96, EM6, 967-983, 1970.

1656 PRZEMIENIECKI J.S., TRIANGULAR PLATE ELEMENTS IN THE MATRIX FORCE METHOD OF STRUCTURAL ANALYSIS. AIAA J.1, 1895-1897,1963.

1657 PRZEMIENIECKI J.S., TETRAHEDRON ELEMENTS IN THE MATRIX FORCE METHOD OF STRUCTURAL ANALYSIS. AIAA J.2, 1152-1154, 1964.

1658 PRZEMIENIECKI J.S., EQUIVALENT MASS MATRICES FOR RECTANGULAR PLATES IN BENDING. AIAA J.4, 949-950, 1966.

1659   PRZEMIENIECKI J.S.,   THEORY OF MATRIX STRUCTURAL ANALYSIS,
           MCGRAW HILL, NEW YORK, 1967.
1660   PRZEMIENIECKI J.S.,   MATRIX ANALYSIS OF LOCAL INSTABILITY IN
           PLATES STIFFENED PANELS AND COLUMNS.   INT.J.NUMER.METH.
           ENG.5, 209-216, 1972.
1661   RACHFORD H.H.,   TWO-LEVEL DISCRETE-TIME GALERKIN APPROXIMATIONS FOR
           SECOND ORDER NON-LINEAR PARABOLIC PARTIAL DIFFERENTIAL EQUATIONS,
           SIAM.J.NUMER.ANAL 10, 1010-1026,1973.
1662   RACHFORD H.H. AND WHEELER MARY F.,   A 1/H GALERKIN PROCEDURE FOR
           TWO POINT BOUNDARY VALUE PROBLEMS. PROC SYMPOSIUM ON
           MATHEMATICAL ASPECTS OF THE FINITE ELEMENT METHOD.
           UNIVERSITY OF WISCONSIN, MADISON, 1974.
1663   RADAJ D.,   ACCURACY OF THE FINITE ELEMENT ANALYSIS FOR THE
           ELASTIC PLATE WITH A CIRCULAR HOLE.   INT.J.NUMER.METH.ENG.
           6, 443-446, 1973.
1664   RAFALSKI P.,   THE ORTHOGONAL PROJECTION METHOD FOR THE NUMERICAL
           SOLUTION OF A FIELD PROBLEM,
           INT.J.NUMER.METH.ENG.6, 287-296,1973.
1665   RAINER K.,   ZUR NUMERISCHEN INTEGRATION PERIODISCHER FUNCTIONEN
           NACH DER RECHTECKREGEL. NUMER.MATH.20, 87-92, 1972
1666   RAINVILLE E.D.,   SPECIAL FUNCTIONS.   MACMILLAN, NEW YORK, 1960.
1667   RAJU I.S.,   FINITE ELEMENT ANALYSIS OF STRESS CONCENTRATIONS AND
           SINGULARITIES.   PH.D. THESIS, INDIAN INSTITUTE OF SCIENCE,
           BANGALORE, 1972.
1668   RAJU I.S., KRISHNA MURTY A.V., AND RAO A.K.,   SECTOR ELEMENTS FOR
           MATRIX DISPLACEMENT ANALYSIS.   INT.J.NUMER.METH.ENG.6,
           553-563, 1973.
1669   RAJU I.S., PRAKASO B., AND VENKATESWARA G.,   AXISYMMETRIC
           VIBRATIONS OF LINEARLY TAPERED ANNULAR PLATES.   J.SOUND
           VIB.32, 507-512, 1974.
1670   RAJU I.S., AND RAO A.K.,   STIFFNESS MATRICES FOR SECTOR ELEMENTS.
           AIAA J.7, 156-157, 1969.
1671   RAJU P.P., AND PALUSAMY S.,   APPROXIMATE METHODS FOR ACCIDENT
           CONDITION ANALYSIS OF PWR PRIMARY COOLANT LOOP COMPONENTS.
           PROC. 2ND STRUCTURAL MECH. IN REACTOR TECH. CONF., BERLIN,
           1973.
1672   RALL L.B.,   ON COMPLEMENTARY VARIATIONAL PRINCIPLES.   J.MATH.ANAL.APPL.
           14, 174-184, 1966.
1673   RALL L.B.,   COMPUTATIONAL SOLUTION OF NONLINEAR OPERATOR EQUATIONS,
           WILEY, NEW YORK, 1969.
1674   RALL L.B. (ED.),   NONLINEAR FUNCTIONAL ANALYSIS AND APPLICATIONS,
           ACADEMIC PRESS, NEW YORK, 1971.
1675   RAMANI D.T.,   TRIANGULAR THICK SHELL FINITE ELEMENT FOR THE THREE-
           DIMENSIONAL STRESS ANALYSIS OF PRESSURE VESSELS.   PAPER
           M3/7, PROC.2ND STRUCTURAL MECH IN REACTOR TECH.CONF.
           BERLIN, 1973.
1676   RAO A.K.,   STRESS CONCENTRATIONS AND SINGULARITIES AT INTERFACE
           CORNERS.   Z.A.M.M. 51, 395-406, 1971.
1677   RAO A.K.,   REVIEW OF CONTINUUM, FINITE ELEMENT AND HYBRID
           TECHNIQUES IN THE ANALYSIS OF STRESS CONCENTRATIONS IN
           STRUCTURES.   PAPER M5/1, PROCEEDINGS 2ND STRUCTURAL MECH.
           IN REACTOR TECH. CONF., BERLIN, 1973.
1678   RAO A.K., DATTAGURU B., RAJAIAH K. AND VENKATARAMAN N.S.,
           DETERMINATION OF STRESSES DUE TO DISCONTINUITIES IN
           FINITE PLATES OF ISOTROPIC AND ORTHOTROPIC MATERIALS.
           PAPER M5/2, PROCEEDING 2ND STRUCTURAL MECH. IN REACTOR
           TECH. CONF., BERLIN, 1973.
1679   RAO A.K., KRISHNA MURTY A.V. AND RAJU I.S.,   SPECIAL FINITE
           ELEMENTS FOR THE ANALYSIS OF STRESS CONCENTRATIONS AND
           SINGULARITIES.   PAPER M6/6, PROCEEDINGS 1ST STRUCTURAL
           MECH. IN REACTOR TECH. CONF., BERLIN, 1972.
1680   RAO A.K., RAJU I.S. AND KRISHNA MURTY A.V.,   A POWERFUL HYBRID METHOD
           IN FINITE ELEMENT ANALYSIS.
           INT.J.NUMER.METH.ENG.3, 389-403,1971.
1681   RASHID Y.R.,   ANALYSIS OF AXISYMMETRIC COMPOSITE STRUCTURES BY THE
           FINITE ELEMENT METHOD.   NUCLEAR ENG. AND DESIGN 3, 163-

                182, 1966.
1682   RASHID Y.R.,   THREE DIMENSIONAL ANALYSIS OF ELASTIC SOLIDS,
                I - ANALYSIS PROCEDURE, II - THE COMPUTATIONAL PROBLEM.
                INT.J.SOLIDS STRUCTURES 5, 1311-1333, 1969 AND
                6,195-207, 1970.
1683   RASHID Y.R., AND GILMAN J.D.,   THREE-DIMENSIONAL ANALYSIS OF REACTOR
                PRESSURE VESSELS. PP.193-213, VOL.4, PROCEEDINGS 1ST STRUCTURAL
                MECH. IN REACTOR TECH. CONF., BERLIN, 1971.
1684   RASHID Y.R., AND ROCKENHAUSER W.,   PRESSURE VESSEL ANALYSIS BY FINITE
                ELEMENT METHOD.   PP.375-383 OF M.S. UDALL (ED.), PRESTRESSED
                CONCRETE PRESSURE VESSELS, INST. OF CIVIL ENGINEERS, LONDON, 1968.
1685   RAUTMANN R.,   ON THE CONVERGENCE OF A GALERKIN METHOD TO SOLVE THE
                INITIAL PROBLEM OF A STABILIZED NAVIER-STOKES EQUATION.   (TO
                APPEAR).
1686   RAVIART P.A.,   METHODE DES ELEMENTS FINIS.   REPORT 73005,
                LABORATOIRE ANALYSE NUMERIQUE, UNIVERSITE DE PARIS, 1973.
1687   RAVIART P.A.,   THE USE OF NUMERICAL INTEGRATION IN FINITE ELEMENT METHODS
                FOR SOLVING PARABOLIC EQUATIONS. PP.233-264 OF J.J.H. MILLER (ED.),
                TOPICS IN NUMERICAL ANALYSIS.   ACADEMIC PRESS, LONDON,1973.
1688   RAVIART P.A.,   HYBRID FINITE ELEMENT METHODS FOR SOLVING 2ND ORDER
                ELLIPTIC EQUATIONS.   REPORT 74015, LABORATOIRE ANALYSE
                NUMERIQUE, UNIVERSITY OF PARIS, 1974.
1689   RAZZAQUE A.,   PROGRAMME FOR TRIANGULAR BENDING ELEMENT WITH DERIVATIVE
                SMOOTHING.   INT.J.NUMER.METH.ENG.6, 333-345, 1973.
1690   REDDY J.,   SOME MATHEMATICAL PROPERTIES OF CERTAIN MIXED GALERKIN
                APPROXIMATIONS IN NONLINEAR ELASTICITY.   PROC.CONF.COMP.
                METHODS IN NONLINEAR MECHANICS, AUSTIN, TEXAS, 1974.
1691   REED M., AND SIMON B.,   METHODS IN MODERN MATHEMATICAL PHYSICS :1 FUNCTIONAL
                ANALYSIS.   ACADEMIC PRESS, NEW YORK, 1972.
1692   REED W.H., AND HILL T.R.,   TRIANGULAR MESH METHODS FOR THE NEUTRON
                TRANSPORT EQUATION.   PROC.AMER.NUCL.SOC. (TO APPEAR).
1693   REED W.H., HILL T.R., BRINKLEY F.W., AND LATHROP K.D.,   TRIPLET A
                TWO-DIMENSIONAL, MULTIGROUP, TRIANGULAR MESH, PLANAR
                GEOMETRY, EXPLICIT TRANSPORT CODE.   REPORT LA-5428-MS,
                LOS ALAMOS SCIENTIFIC LABORATORY, NEW MEXICO, 1973.
1694   REEVES M., AND DUGUID J.O.,   WATER FLOW IN SATURATED-UNSATURATED
                SOILS BY FINITE ELEMENT METHODS.   PROC. CONF.COMP. METHODS
                IN NONLINEAR MECHANICS AUSTIN, TEXAS, 1974.
1695   REID J.K.,   ON THE CONSTRUCTION AND CONVERGENCE OF A FINITE ELEMENT
                SOLUTION OF LAPLACE'S EQUATION.   J.INST.MATH.APPLICS 9,
                1-13, 1972.
1696   REID J.K., AND TURNER A.B.,   FORTRAN SUBROUTINES FOR THE SOLUTION
                OF LAPLACE'S EQUATION OVER A GENERAL REGION IN TWO
                DIMENSIONS. A.E.R.E. HARWELL, REPORT T.P.422, 1970.
1697   REINSCH C.,   TWO EXTENSIONS OF THE SARD-SCHOENBERG THEORY OF BEST
                APPROXIMATION. SIAM J. NUMER.ANAL.11, 45-51, 1974.
1698   REISSNER E.,   ON THE THEORY OF BENDING OF ELASTIC PLATES.   J.MATH.
                PHYS 23, 184, 1944.
1699   REISSNER E.,   NOTE ON THE METHOD OF COMPLEMENTARY ENERGY.   J. MATH.
                PHYS. 27, 159-160, 1948.
1700   REISSNER E.,   ON A VARIATIONAL THEOREM IN ELASTICITY.   J. MATH.
                PHYS.29, 1950.
1701   REISSNER E.,   ON SOME VARIATIONAL THEOREMS IN ELASTICITY. PROBLEMS
                OF CONTINUUM MECHANICS 370-381,   SIAM, PHILADELPHIA,
                PENNSYLVANIA, 1961.
1702   REKTORYS K.,   ON APPLICATION OF DIRECT VARIATIONAL METHODS TO THE
                SOLUTION OF PARABOLIC BOUNDARY VALUE PROBLEMS OF ARBITRARY ORDER
                IN THE SPACE VARIABLES.   CZECHOSLOVAK MATH.J.21, 318-339, 1971.
1703   REKTORYS K., AND ZAHRADNIK V.,   SOLUTION OF THE FIRST BIHARMONIC
                PROBLEM BY THE METHOD OF LEAST SQUARES ON THE BOUNDARY.
                APLIKACE MATEMATIKY 19, 101-130, 1974.
1704   REMEDIOS F.E., BELL R. AND LOVATT J.D.,   AN ENGINEERING APPROACH
                TO FINITE ELEMENT ANALYSIS OF NUCLEAR COMPONENTS.
                PAPER M2/2, PROCEEDINGS 2ND STRUCTURAL MECH IN REACTOR
                TECH. CONF., BERLIN, 1973.
1705   REYES S.F., AND DEERE D.U.,   ELASTO-PLASTIC ANALYSIS OF UNDERGROUND

OPENINGS BY THE FINITE ELEMENT METHOD. PROC.1ST INT.
CONF. ROCK MECHANICS II, 477-486, 1966.

1706 REYNOLDS A.C., CONVERGENT FINITE DIFFERENCE SCHEMES FOR NONLINEAR
PARABOLIC EQUATIONS. SIAM J. NUMER.ANAL.9, 523-533, 1972.

1707 RHEINBOLDT W.C., ON NUMERICAL METHODS FOR SOLVING LARGE SPARSE
SETS OF NONLINEAR EQUATIONS. PROC.CONF.COMP. METHODS IN
NONLINEAR MECHANICS, AUSTIN, TEXAS, 1974.

1708 RICE J.R., THE APPROXIMATION OF FUNCTIONS, VOLS I AND II.
ADDISON-WESLEY, READING, MASSACHUSETTS, 1964,1968.

1709 RICHARD R.M. AND BLACKLACK J.R., FINITE ELEMENT ANALYSIS OF
INELASTIC STRUCTURES. AIAA J.7, 432-438, 1969.

1710 RICHTMEYER R.D., AND MORTON K.W., DIFFERENCE METHODS FOR INITIAL VALUE
PROBLEMS. INTERSCIENCE, NEW YORK   , 1967.

1711 RIESZ F., AND NAGY B.S., FUNCTIONAL ANALYSIS. UNGAR, NEW YORK, 1955.

1712 RIIKONEN I., ANALYSIS OF THICK-WALLED PRESSURE VESSELS USING THE CONSTANT
STRAIN AXISYMMETRIC FINITE ELEMENT. PROC. 4TH NORDIC SYMP.
STRENGTH OF MATERIALS, OTANIEMI, FINLAND, 1971.

1713 RIGBY G.L., AND MCNEICE G.M., FURTHER LIMITS ON GENERAL
HEXAHEDRON FINITE ELEMENTS WITH DERIVATIVE DEGREES OF
FREEDOM. INT.J.NUMER.METH.ENG.5, 137-139, 1972.

1714 ROBINSON J., BASIS FOR ISOPARAMETRIC STRESS ELEMENTS. COMP.METH.APPL.
MECH.ENG.2, 43-63,1973

1715 ROBINSON J., INTEGRATED THEORY OF FINITE ELEMENT METHODS. WILEY,
LONDON, 1973.

1716 ROBINSON J., AND HAGGENMACHER G.W., BASIS FOR ELEMENT
INTERCHANGEABILITY IN FINITE ELEMENT PROGRAMS. PROC.3RD
CONF. MATRIX METHODS IN STRUCTURAL MECHANICS, WRIGHT-
PATTERSON AFB., OHIO, 1971.

1717 ROBINSON P.D., COMPLEMENTARY VARIATIONAL PRINCIPLES. PP. 507-576 OF
L.B. RALL (ED.), NONLINEAR FUNCTIONAL ANALYSIS AND
APPLICATIONS. ACADEMIC PRESS, NEW YORK, 1971.

1718 ROSE D.J., AND WILLOUGHBY R.A., (EDS.), SPARSE MATRICES AND THEIR
APPLICATIONS. PLENUM PRESS, NEW YORK, 1972.

1719 ROULIER J., LINEAR OPERATORS INVARIANT ON NONNEGATIVE MONOTONE
FUNCTIONS. SIAM J.NUMER.ANAL.8, 30-35, 1971.

1720 ROWE G.H., MATRIX DISPLACEMENT METHODS IN FRACTURE MECHANICS ANALYSIS
OF REACTOR VESSELS. NUCL.ENG.DESIGN 20, 251-263, 1972.

1721 RUDIN W., FUNCTIONAL ANALYSIS. MCGRAW HILL, NEW YORK, 1973.

1722 RUSSELL R.D. AND VARAH J.M., A COMPARISON OF GLOBAL METHODS FOR LINEAR
TWO-POINT BOUNDARY VALUE PROBLEMS. TECHNICAL REPORT 74-01,
DEPARTMENT OF COMPUTER SCIENCE, UNIVERSITY OF BRITISH COLUMBIA,
VANCOUVER, 1974.

1723 SAATY T.L., AND BRAM J., NONLINEAR MATHEMATICS. MCGRAW HILL, NEW YORK,
1974.

1724 SABIR A., AN EXTENSION OF THE SHALLOW TO THE NON-SHALLOW STIFFNESS
MATRIX FOR A CYLINDRICAL SHELL FINITE ELEMENT.
INT.J.MECH.SCI.12, 287-292,1970.

1725 SABIR A.B., AND ASHWELL D.G., A STIFFNESS MATRIX FOR SHALLOW SHELL
FINITE ELEMENTS. INT.J.MECH.SCI.11, 269-279, 1969.

1726 SABIR A.B. AND ASHWELL D.G., A COMPARISON OF CURVED BEAM FINITE
ELEMENTS WHEN USED IN VIBRATION PROBLEMS. J.SOUND VIB.18,
555-563,1971.

1727 SABIR A.B. AND ASHWELL D.G., FINITE ELEMENT ANALYSIS OF A BROAD ARCH
BRIDGE WITH A SLAB DECK. PROCEEDINGS CONFERENCE ON
DEVELOPMENTS IN BRIDGE DESIGN, UNIVERSITY COLLEGE, CARDIFF,
1971.

1728 SABIR A.B., AND LOCK A.C., A CURVED CYLINDRICAL SHELL FINITE
ELEMENT. INT.J.MECH.SCI.14, 125-135, 1972.

1729 SABIR A.B., AND LOCK A.C., LARGE DEFLECTION GEOMETRICALLY NON-
LINEAR FINITE ELEMENT ANALYSIS OF CIRCULAR ARCHES. INT.
J.MECH.SCI.15, 37-47, 1973.

1730 SAKAI M., ON RAYLEIGH-RITZ-GALERKIN PROCEDURES FOR NONLINEAR
TWO POINT BOUNDARY VALUE PROBLEMS.
(MANUSCRIPT).

1731 SAKAI M. PIECEWISE CUBIC INTERPOLATION AND TWO-POINT BOUNDARY
VALUE PROBLEMS. PUBL.R.I.M.S. KYOTO UNIVERSITY, 7, 345-

362, 1971.

1732 SAKAI M., PIECEWISE CUBIC INTERPOLATION AND DEFERRED CORRECTION. MEMOIRS FAC.SCI, KYUSHU UNIVERSITY, A 26, 339-350,1972.

1733 SAKAI M., RITZ METHOD FOR TWO-POINT BOUNDARY VALUE PROBLEM MEMOIRS FAC.SCI, KYUSHU UNIVERSITY, A 27, 83-97,1973

1734 SAKAI M., CUBIC SPLINE FUNCTION AND DIFFERENCE METHOD. MEMOIRS FAC. SCI, KYUSHU A 28, 43-58, 1974.

1735 SAKANO K., AND TORII H., AN EXAMPLE OF THE EVALUATION OF FATIGUE STRENGTH OF A HIGH TEMPERATURE STRUCTURE BY ASME CODE. FAPIG (TOKYO) 69, 23-29, 1973.

1736 SALINAS E., NGUYEN D.H., AND SOUTHWORTH T.W., FINITE ELEMENT SOLUTIONS OF A NONLINEAR REACTOR DYNAMICS PROBLEM. PROC. CONF.COMP.METHODS IN NONLINEAR MECHANICS, AUSTIN, TEXAS, 1974.

1737 SALONEN E.-M., A RECTANGULAR PLATE BENDING ELEMENT THE USE OF WHICH IS EQUIVALENT TO THE USE OF THE FINITE DIFFERENCE METHOD. INT.J.NUMER.METH.ENG.1, 261-274, 1969.

1738 SANDER C., BORNES SUPERIEURES ET INFERIEURES DANS L'ANALYSE MATRICIELLE DES PLAQUES EN FLEXION-TORSION. BULL.SOC. ROY.SCI.LIEGE 33, 456-494, 1964.

1739 SANDER G., AND BECKERS P., IMPROVEMENTS OF FINITE ELEMENT SOLUTIONS FOR STRUCTURAL AND NONSTRUCTURAL APPLICATIONS. PROC. 3RD CONF. MATRIX METHODS IN STRUCTURAL MECHANICS. WRIGHT-PATTERSON AFB, OHIO, 1971.

1740 SANDER G., BON C. AND GERADIN M., FINITE ELEMENT ANALYSIS OF SUPERSONIC PANEL FLUTTER. INT.J.NUMER.METH.ENG.7, 379-394, 1973.

1741 SANDER G., AND GERADIN M., FINITE ELEMENTS IN ELASTODYNAMICS. OXFORD UNIVERSITY PRESS, TO APPEAR.

1742 SANDHU R.S. AND PISTER K.S., A VARIATIONAL PRINCIPLE FOR LINEAR COUPLED FIELD PROBLEMS IN CONTINUUM MECHANICS. INT.J.ENG.SCI, 8, 989-999, 1970.

1743 SANDHU R.S., AND PISTER K.S., VARIATIONAL PRINCIPLES FOR BOUNDARY VALUE AND INITIAL VALUE PROBLEMS IN CONTINUUM MECHANICS. INT.J.SOLIDS STRUCTURES 7, 639-654, 1971.

1744 SANDHU R.S., AND WILSON E.L., FINITE ELEMENT ANALYSIS OF SEEPAGE IN ELASTIC MEDIA. J.ENG.MECH.DIV.,PROC.ASCE 95, 641-651, 1969.

1745 SARD A., LINEAR APPROXIMATION. MATHEMATICAL SURVEY 9, AMERICAN MATHEMATICAL SOCIETY, PROVIDENCE, RHODE ISLAND, 1963.

1746 SARD A., AND WEINTRAUB S., A BOOK OF SPLINES. WILEY, NEW YORK, 1971.

1747 SAROJ SINGH AND RAMASWAMY G.S., A SECTOR ELEMENT FOR THIN PLATE FLEXURE. INT.J.NUMER.METH.ENG.4,133-142,1972.

1748 SAUER R., AND SZABO I., MATHEMATISCHE HILFSMITTEL DES INGENIEURS. VOLS 1-4, SPRINGER-VERLAG, BERLIN, 1967.

1749 SAUGY E., AND ZIMMERMANN T., NONLINEAR ANALYSIS OF SOLID STRUCTURES. ANN.INST.TECH.BATIMENT TRAV. PUBLICS 292, 125-143, 1972.

1750 SAWKO F., AND MERRIMAN P.A., AN ANNULAR SEGMENT ELEMENT FOR PLATE BENDING. INT.J.NUMER.METH.ENG.3, 119-129, 1971.

1751 SAWKO F., AND COPE R.J., THE ANALYSIS OF SKEW BRIDGE DECKS; A NEW FINITE ELEMENT APPROACH. STRUCTURAL ENGINEER 47, 215-224, 1969.

1752 SCAIFFE B.K.P. (ED.), STUDIES IN NUMERICAL ANALYSIS. ACADEMIC PRESS, NEW YORK, 1974.

1753 SCHADE D., ON THE ELASTICITY THEORY FOR PLANE FLEXURALLY STIFF SYSTEMS OF PARTICLES WITH RECTANGULAR NETWORK. ING. ARCH.42, 296-308, 1973.

1754 SCHAEFFER H., LATTENINTERPOLATION BEI EINER FUNKTION VON ZWEI VERANDERLICHEN. Z.A.M.P.14, 90-96, 1963.

1755 SCHAUER D.A., ELFED A COMPUTER CODE TO GENERATE FINITE ELEMENT MESH FOR PROBLEMS OF COMPLEX AXISYMMETRIC GEOMETRY. PRESSURE VESSELS AND PIPING MATERIALS NUCLEAR CONF., MIAMI, 1973.

1756 SCHAUER D.A., ELFED A COMPUTER CODE TO GENERATE A FINITE ELEMENT MESH FOR PROBLEMS OF COMPLEX AXISYMMETRIC GEOMETRY. REPORT UCRL-74879, UNIVERSITY OF CALIFORNIA, LAWRENCE

                    LIVERMORE LABORATORY, 1974.
1757   SCHECHTER M., ON LP ESTIMATES AND REGULARITY, II. MATH. SCAND. 13,
                    47-69, 1963.
1758   SCHECHTER R.S., THE VARIATIONAL METHOD IN ENGINEERING.
                    MCGRAW-HILL, NEW YORK, 1967.
1759   SCHERER K., CHARACTERIZATION OF GENERALIZED LIPSCHITZ CLASSES BY BEST
                    APPROXIMATION WITH SPLINES. SIAM J.NUMER.ANAL.11,283-304,
                    1974.
1760   SCHIOP A.I., STABILITY OF RITZ PROCEDURE FOR NONLINEAR TWO POINT
                    BOUNDARY VALUE PROBLEM. NUMER.MATH.20,208-212,1973.
1761   SCHMELTER J., THE ENERGY METHOD OF NETWORK OF ARBITRARY SHAPE
                    IN PROBLEMS OF THEORY OF ELASTICITY. IN OLSZAK (ED),
                    PROC IUTAM SYMPOSIUM ON NON-HOMOGENEITY IN ELASTICITY
                    AND PLASTICITY. PERGAMON PRESS, OXFORD, 1959.
1762   SCHMIDT F.A.R., AND FRANKE H.P., FINITE ELEMENTS VERSUS FINITE DIFFERENCES,
                    A COMPARISON OF THE TWO METHODS FOR THE SOLUTION OF THE DIFFUSION
                    EQUATION. VP. LEOPOLDSHAFEN GER., ZENTRALSTELLE FUER ATOMKERNENERGIE
                    -DOKUMENTATION, 1973.
1763   SCHMIDT G.H., APPLICATION OF THE FINITE ELEMENT METHOD TO THE
                    EXTENSIONAL VIBRATIONS OF PIEZOELECTRIC PLATES.
                    PP.351-366 OF J.R.WHITEMAN (ED.), THE MATHEMATICS OF
                    FINITE ELEMENTS AND APPLICATIONS. ACADEMIC PRESS,LONDON,1973.
1764   SCHMIDT W.F., PROJECTIVE METHOD APPLIED TO THREE-DIMENSIONAL
                    ELASTICITY EQUATIONS. INT.J.NUMER.METH.ENG.8, 697-711,
                    1974.
1765   SCHMIT L.A., FINITE DEFLECTION STRUCTURAL ANALYSIS USING PLATE AND
                    SHELL DISCRETE ELEMENTS. AIAA J.6, 781-791, 1968.
1766   SCHMIT L.A., BOYNER F.K., AND FOX R.L., FINITE DEFLECTION
                    STRUCTURAL ANALYSIS USING PLATE AND CYLINDRICAL SHELL
                    DISCRETE ELEMENTS. AIAA J.5, 1525-1527, 1968.
1767   SCHOENBERG I.J., CONTRIBUTIONS TO THE PROBLEM OF APPROXIMATION
                    OF EQUIDISTANT DATA BY ANALYTIC FUNCTIONS.
                    QUART.APPL.MATH.4, 45-99 AND 122-141, 1946.
1768   SCHOENBERG I.J., ON HERMITE-BIRKHOFF INTERPOLATION. J. MATH.
                    ANAL. APPL. 16, 538-543, 1968.
1769   SCHOENBERG I.J., APPROXIMATIONS WITH SPECIAL EMPHASIS ON SPLINE
                    FUNCTIONS. ACADEMIC PRESS, NEW YORK, 1969.
1770   SCHOLZ R., ABSCHATZUNGEN LINEARER DURCHMESSER IN SOBOLEV UND
                    BESOV RAUMEN. MANUSCRIPTA MATH. 11, 1-14, 1974.
1771   SCHOLZ R., DURCHMESSERABSCHATZUNGEN FUR DIE EINHEITSKUGEL DES
                    SOBOLEV RAUMES WQR IN LP. APPLICABLE ANALYSIS (TO APPEAR).
1772   SCHOMBURG U., THE FINITE ELEMENT METHOD AND LOCAL BOUNDS FOR
                    BOUNDARY VALUE PROBLEMS OF ELASTIC STRUCTURES. PAPER M2/9,
                    PROCEEDINGS 2ND STRUCTURAL MECH. IN REACTOR TECH.CONF.,
                    BERLIN, 1973.
1773   SCHONHAGE A., APPROXIMATIONSTHEORIE. W.DE GRUYTER AND CO.,
                    BERLIN, 1971.
1774   SCHULTZ M.H., ERROR-BOUNDS FOR THE RAYLEIGH-RITZ-GALERKIN METHOD.
                    J.MATH.ANAL.APPL.,27,524-533,1969.
1775   SCHULTZ M.H., LINFINITY MULTIVARIATE APPROXIMATION THEORY.
                    SIAM J. NUMER.ANAL.6, 161-183, 1969.
1776   SCHULTZ M.H., MULTIVARIATE SPLINE FUNCTIONS AND ELLIPTIC PROBLEMS,
                    SIAM J. NUMER.ANAL.6, 523-538, 1969.
1777   SCHULTZ M.H., MULTIVARIATE SPLINE FUNCTIONS AND ELLIPTIC PROBLEMS,
                    PP 279-347 IN SCHOENBERG I.J.(ED.), APPROXIMATIONS WITH
                    SPECIAL EMPHASIS ON SPLINE FUNCTIONS, ACADEMIC PRESS,
                    NEW YORK, 1969.
1778   SCHULTZ M.H., RAYLEIGH-RITZ-GALERKIN METHODS FOR MULTIDIMENSIONAL
                    PROBLEMS. SIAM J. NUMER.ANAL.6, 570-582, 1969.
1779   SCHULTZ M.H., ELLIPTIC SPLINE FUNCTIONS AND THE RAYLEIGH-RITZ-GALERKIN
                    METHOD. MATH.COMP. 24, 65-80, 1970.
1780   SCHULTZ M.H., ERROR BOUNDS FOR POLYNOMIAL SPLINE INTERPOLATION,
                    MATH.COMP.24,507-515,1970.
1781   SCHULTZ M.H. L2 ERROR BOUNDS FOR THE RAYLEIGH-RITZ-GALERKIN METHOD.
                    SIAM J. NUMER.ANAL.8, 737-748, 1971.
1782   SCHULTZ M.H., ERROR BOUNDS FOR A BIVARIATE INTERPOLATION SCHEME,

                    J.APPROX THEORY 8, 189-194, 1973.
1783 SCHULTZ M.H.,   APPROXIMATION THEORY OF MULTIVARIATE SPLINE FUNCTIONS
                    IN SOBOLEV SPACES. SIAM J.NUMER.ANAL. TO APPEAR.
1784 SCHULTZ M.H.,   ERROR BOUNDS FOR THE GALERKIN METHOD FOR LINEAR
                    PARABOLIC EQUATIONS. TO APPEAR
1785 SCHULTZ M.H.,   ERROR BOUNDS FOR THE GALERKIN METHOD FOR NONLINEAR
                    PARABOLIC EQUATIONS. TO APPEAR
1786 SCHULTZ M.H.,   THE GALERKIN METHOD FOR NONSELFADJOINT DIFFERENTIAL
                    EQUATIONS. J.MATH.ANAL.APPL. TO APPEAR.
1787 SCHULTZ M.H., AND VARGA R.S.,   L-SPLINES. NUMERISCHE MATH., 10,
                    345-369, 1967.
1788 SCHUMAKER L.L.,   CONSTRUCTIVE ASPECTS OF DISCRETE POLYNOMIAL SPLINE
                    FUNCTIONS. PP.469-476 OF G.G.LORENTZ (ED.), APPROXIMATION
                    THEORY. ACADEMIC PRESS, NEW YORK, 1973.
1789 SCHWARTZ L.,   THEORIE DES DISTRIBUTIONS, VOLS.I AND II. HERMANN,
                    PARIS, 1957.
1790 SCHWARTZ L.,   MATHEMATICS FOR THE PHYSICAL SCIENCES. ADDISON-
                    WESLEY, MASSACHUSETTS, 1967.
1791 SCHWARZ H.R.,   THE EIGENVALUE PROBLEM (A-LAMBDAB)X=0 FOR SYMMETRIC
                    MATRICES OF HIGH ORDER. COMP.METH.APPL.MECH.ENG.3, 11-28, 1974.
1792 SCHWIEKERT D.G.,   AN INTERPOLATION CURVE USING A SPLINE IN TENSION.
                    J.MATH.PHYS.45, 312-317, 1966.
1793 SCOTT R.,   FINITE ELEMENT CONVERGENCE FOR SINGULAR DATA.
                    NUMER.MATH.21, 317-327,1973.
1794 SEGUI W.T.,   COMPUTER PROGRAMS FOR THE SOLUTION OF SYSTEMS OF LINEAR
                    ALGEBRAIC EQUATIONS. INT.J.NUMER.METH.ENG.7, 479-490,1973.
1795 SEITELMAN L.H.,   SOME PRACTICAL SOLUTION TECHNIQUES FOR FINITE
                    ELEMENT ANALYSIS. PP. 471-481 OF J.R. WHITEMAN (ED.), THE
                    MATHEMATICS OF FINITE ELEMENTS AND APPLICATIONS. ACADEMIC
                    PRESS, LONDON, 1973.
1796 SEMENZA L.A.,   APPLICATION OF THE FINITE ELEMENT METHOD TO TWO-DIMENSIONAL
                    MULTIGROUP NEUTRON DIFFUSION. REPORT 72-32, 573, NORTHWESTERN
                    UNIVERSITY, EVANSTON, 1972.
1797 SEMENZA L.A., LEWIS E.E., AND ROSSOW F.C.,   APPLICATION OF THE FINITE
                    ELEMENT METHOD TO THE MULTIGROUP NEUTRON DIFFUSION EQUATION.
                    NUCL.SCI.ENG.47, 302-310, 1972.
1798 SEN S.K. AND GOULD P.L.,   CRITERIA FOR FINITE ELEMENT DISCRETIZATION
                    OF SHELLS OF REVOLUTION.
                    INT.J.NUMER.METH.ENG.6, 265-274,1973.
1799 SERBIN L.,   A COMPUTATIONAL INVESTIGATION OF LEAST SQUARES AND OTHER
                    PROJECTION METHODS FOR THE APPROXIMATE SOLUTION OF BOUNDARY
                    VALUE PROBLEMS. PH D. THESIS,CORNELL UNIVERSITY, 1971.
1800 SEVERN R.T., AND TAYLOR D.R.,   THE FINITE ELEMENT METHOD FOR FLEXURE
                    OF SLABS WHEN STRESS DISTRIBUTIONS ARE ASSUMED.
                    PROC.INST.CIV.ENG.34, 153-170, 1966.
1801 SEVERN R.T., AND TAYLOR P.R.,   PLATE VIBRATION PROBLEMS USING THE
                    FINITE ELEMENT METHOD WITH ASSUMED STRESS DISTRIBUTIONS AND
                    TRIANGULAR ELEMENTS. NUMERICAL METHODS FOR VIBRATION PROBLEMS,
                    I.S.V.R.,2, 38-54, 1966.
1802 SEWELL M.J.,   THE GOVERNING EQUATIONS AND EXTREMUM PRINCIPLES OF
                    ELASTICITY AND PLASTICITY GENERATED FROM A SINGLE FUNCTIONAL.
                    TECH.RPT.1227, MATHEMATICS RESEARCH CENTER, UNIVERSITY OF
                    WISCONSIN,MADISON,WISCONSIN,1972.
1803 SHAMPINE L.F.,   ERROR BOUNDS AND VARIATIONAL METHODS FOR NONLINEAR
                    BOUNDARY VALUE PROBLEMS. NUMER.MATH.12, 410-415, 1968.
1804 SHATOFF H.D.,   THREE-DIMENSIONAL ELASTIC FINITE ELEMENT ANALYSIS
                    USING GRADIENT DEGREES OF FREEDOM. PAPER M2/10,
                    PROCEEDINGS 2ND STRUCTURAL MECH. IN REACTOR TECH. CONF.,
                    BERLIN, 1973.
1805 SHIEH W.J.,   ANALYSIS OF PLATE BENDING BY TRIANGULAR ELEMENTS. J.ENG.
                    MECH.DIV.A.S.C.E. 94, EM5, 1089-1108,1968.
1806 SHIMAZAKI K.,   PRE-SEISMIC CRUSTAL DEFORMATION CAUSED BY AN
                    UNDERTHRUSTING OCEANIC PLATE IN EASTERN HOKKAIDO. PHYS.
                    EARTH AND PLANET INT.8, 148-157, 1974.
1807 SHIMIZU S.,   FREE VIBRATION ANALYSIS OF STIFFENED PLATES. PP. 219-236
                    OF J.T. ODEN, R.W. CLOUGH AND Y. YAMOMOTO (EDS.), ADVANCES IN

COMPUTATIONAL METHOD IN STRUCTURAL MECHANICS AND DESIGN,
UNIVERSITY OF ALABAMA PRESS, HUNTSVILLE, 1972.

1808 SIBONY M.,   CONTROLE DES SYSTEMES GOUVERNES PAR DES EQUATIONS AUX
      DERIVEES PARTIELLES. REND. SEMINARIO MATEMATICO, UNIVERSITA
      PADOVA, VOL 43, 277-338, 1970.

1809 SIBONY M.,   METHODES ITERATIVES POUR LES EQUATIONS ET INEQUATIONS
      AUX DERIVEES PARTIELLES NON LINEAIRES DE TYPE MONOTONE.
      ESTRATTO DA CALCOLO (GENNAIO-GIUGNO) 65-183, 1970.

1810 SIBONY M.,   MINIMISATION DE FONCTIONNELLES NON DIFFERENTIABLES.   ISRAEL
      J.MATH.8, 105-126, 1970.

1811 SIBONY M.,   SUR L'APPROXIMATION D'EQUATIONS ET INEQUATIONS AUX
      DERIVEES PARTIELLES NON LINEAIRES DE TYPE MONOTONE. J.MATH.
      ANAL. APPL.34, 502-564,1971.

1812 SIH G.C. (ED.), MECHANICS OF FRACTURE METHODS OF ANALYSIS AND SOLUTION
      OF CRACK PROBLEMS. NOORDHOFF, LEYDEN, 1972.

1813 SILVESTER P.,   FINITE ELEMENT SOLUTION OF HOMOGENEOUS WAVEGUIDE PROBLEMS.
      ALTA FREQUENZA 38, 313-317,1969.

1814 SILVESTER P., A GENERAL HIGH-ORDER FINITE-ELEMENT WAVEGUIDE ANALYSIS
      PROGRAM. I.E.E.E. TRANS. MTT-A, 204-210, 1969.

1815 SILVESTER P., HIGH-ORDER POLYNOMIAL TRIANGULAR FINITE ELEMENTS FOR
      POTENTIAL PROBLEMS. INT.J.ENG.SCI., 7,849-861, 1969.

1816 SILVESTER P., AND HSEIH M-S.,   FINITE ELEMENT SOLUTION OF
      TWO-DIMENSIONAL EXTERIOR FIELD PROBLEMS. PROC.I.E.E. 118,
      1743-1747, 1971.

1817 SILVESTER P., AND KONRAD A.,   AXISYMMETRIC TRIANGULAR FINITE
      ELEMENTS FOR THE SCALAR HELMHOLTZ EQUATION. INT J. NUMER.
      METH.ENG.5, 481-497, 1973.

1818 SINGA-RAO K., AND AMBA-RAO C.L.,   VIBRATION OF BEAMS WITH OVERHANDS.
      AIAA J. 11, 1445-1446, 1973.

1819 SLAGTER W., FAST REACTOR PROGRAMME, FINITE ELEMENT TEMPERATURE
      ANALYSIS IN FUEL PINS. REPORT EURFNR-1158, SLICHTING
      REACTOR CENTRUM, NEDERLAND, 1973.

1820 SLETTEN R., AND PEDERSEN B.,   APPLICATION OF THE FINITE ELEMENT
      METHOD TO OFF-SHORE STRUCTURES. COMPUTERS AND STRUCTURES
      4, 131-148, 1974.

1821 SLOAN D.M., EXTREMUM PRINCIPLES FOR MAGNETOHYDRODYNAMIC CHANNEL FLOW.
      Z.A.M.P. 24, 689-698,1973.

1822 SLOBODECKII M.I., GENERALIZED SOBOLEV SPACE AND THEIR APPLICATIONS
      TO BOUNDARY PROBLEMS FOR PARTIAL DIFFERENTIAL EQUATIONS,
      LENINGR.GOS.UNIV.197, 54-112, 1958, AND AMER.MATH.SOC.
      TRANSL.2, 57, 207-275, 1966.

1823 SLYPER H.A., DEVELOPMENT OF EXPLICIT STIFFNESS AND MASS MATRICES FOR A
      TRIANGULAR PLATE ELEMENT. INT.J.SOLIDS STRUCTURES 5, 241-249,
      1969.

1824 SMIRNOV M.M., MATHEMATICAL ANALYSIS. PERGAMON PRESS, OXFORD,
      1965.

1825 SMIRNOV V.I., A COURSE OF HIGHER MATHEMATICS, VOL.V.PERGAMON
      PRESS, OXFORD, 1964.

1826 SMITH I.M., A FINITE ELEMENT ANALYSIS FOR MODERATELY RECTANGULAR
      PLATES IN BENDING. INT.J.MECH.SCI.10,563-570, 1968.

1827 SMITH I.M., AND DUNCAN W., THE EFFECTIVENESS OF EXCESSIVE NODAL
      CONTINUITIES IN THE FINITE ELEMENT ANALYSIS OF THIN
      RECTANGULAR AND SKEW PLATES IN BENDING. INT.J.NUMER.
      METH.ENG.2, 253-258, 1970.

1828 SMITH J.H., NONLINEAR BEAM AND PLATE ELEMENTS. J.STRUCT.DIV.A.S.C.E.
      98, ST3, 553-568, 1972.

1829 SMITH P.D., SEMIGROUP SCHEMES FOR DYNAMIC PROBLEMS. INT.J.NUMER.METH.
      ENG.7, 556-560,1973.

1830 SMITH R.G., AND WEBSTER J.A., MATRIX ANALYSIS OF BEAM COLUMNS BY THE
      METHOD OF FINITE ELEMENTS. PROC.ROY.SOC.EDINBURGH A 67, 156-173,
      1964.

1831 SNEDDON I.N., MIXED BOUNDARY VALUE PROBLEMS IN POTENTIAL THEORY,
      WILEY, NEW YORK, 1966.

1832 SNEDDON I.N., AND LOWENGRUB M., CRACK PROBLEMS IN THE CLASSICAL THEORY
      OF ELASTICITY. WILEY, NEW YORK, 1969.

1833 SOBOLEV S.L., UBER EINE ABSCHATZUNG GEWISSER SUMMEN VON

GITTERFUNCTIONEN. ISWESTJA AKAD.NAUK.SSR 4, 1940.

1834 SOBOLEV S.L., FUNCTIONAL ANALYSIS IN MATHEMATICAL PHYSICS.
AMERICAN MATH.SOC. MONOGRAPH TRANSLATION 7,
PROVIDENCE, RHODE ISLAND, 1963.

1835 SOBOLEV S.L., THE DENSITY OF FUNCTIONS WITH COMPACT SUPPORT IN THE
SPACE LP(M)(EN). SIBIRSK. MAT.ZH.4, 673-682, 1963.

1836 SOBOLEV S.L., PARTIAL DIFFERENTIAL EQUATIONS OF
MATHEMATICAL PHYSICS. PERGAMON PRESS, OXFORD, 1964.

1837 SOBOLEVSKII P.E., APPROXIMATE METHODS OF SOLVING DIFFERENTIAL
EQUATIONS IN BANACH SPACES. (RUSSIAN). DOKL.AKAD.NAUK.SSR
115, 240-243, 1957.

1838 SOKOLNIKOFF I.S., MATHEMATICAL THEORY OF ELASTICITY. MCGRAW HILL, NEW
YORK, 1956.

1839 SOLOMON A.D., ON VARIATIONAL APPROACHES TO STEADY STATE HEAT
CONDUCTION. J.ENG.MATHEMATICS 5, 33-37, 1971.

1840 SOMERVAILLE I.J., A TECHNIQUE FOR MESH GRADING APPLIED TO
CONFORMING PLATE BENDING FINITE ELEMENTS. INT.J.NUMER.
METH.ENG.6, 310-311, 1973.

1841 SONTVEDT T., PROPELLER BLADE STRESSES, APPLICATION OF FINITE
ELEMENT METHODS. COMPUTERS AND STRUCTURES 4, 193-204,
1974.

1842 SOUTHWELL R.V., RELAXATION METHODS IN THEORECTICAL PHYSICS.
CLARENDON PRESS, OXFORD, 1946.

1843 SPEARS R.K., FINITE ELEMENT VISCOELASTIC STRESS ANALYSIS STUDY USING
THE GENERALIZED KELVIN SOLID MODEL. REPORT GEPP-111, GENERAL
ELECTRIC CO., ST. PETERSBURG, FLORIDA, 1973.

1844 STACEY W.M., VARIATIONAL METHODS IN NUCLEAR REACTOR PHYSICS.
ACADEMIC PRESS, NEW YORK, 1974.

1845 STADINKOVA N.A., ON THE CHARACTER OF CONVERGENCE OF THE RITZ METHOD.
VISCISL, MAT. 7, 187-190, 1961.

1846 STAGG K.G., AND ZIENKIEWICZ O.C.,(EDS.), CONTINUUM MECHANICS
AS AN APPROACH TO ROCK MASS PROBLEMS. WILEY, LONDON,1968.

1847 STAKGOLD I., BOUNDARY VALUE PROBLEMS OF MATHEMATICAL PHYSICS.
MACMILLAN, NEW YORK, 1968.

1848 STAMPACCHIA G., EQUATIONS ELLIPTIQUES DU SECOND ORDRE A COEFFICIENTS
DISCONTINUS. SEMINAIRE DE MATHEMATIQUES SUPERIEURES, MONTREAL
UNIVERSITY PRESS, 1965.

1849 STANCU D.D., THE REMAINDER OF CERTAIN LINEAR APPROXIMATION
FORMULAS IN TWO VARIABLES. SIAM J.NUMER.ANAL.B,1,
137-163, 1964.

1850 STANTON E.L., AND SCHMIT L.A., A DISCRETE ELEMENT STRESS AND
DISPLACEMENT ANALYSIS OF ELASTOPLASTIC PLATES. AIAA J.8, 1245-
1251, 1970.

1851 STEFANOU G.D., YU C.W., AND ENGLAND G.L., TWO-DIMENSIONAL TIME-DEPENDENT
ANALYSIS OF PERFORATED END CAPS FOR NUCLEAR REACTOR PRESSURE VESSELS
BY THE FINITE ELEMENT METHOD. PP. 167-186, VOL.4, PROCEEDINGS
1ST STRUCTURAL MECH. IN REACTOR TECH.CONF.,BERLIN, 1971.

1852 STEINMUELLER G., RESTRICTIONS IN THE APPLICATION OF AUTOMATIC
MESH GENERATION SCHEMES BY ISOPARAMETRIC CO-ORDINATES.
INT. J.NUMER.METH.ENG.8, 289-294, 1974.

1853 STEPLEMAN R., FINITE DIMENSIONAL ANALOGUES OF VARIATIONAL AND
QUASI-LINEAR ELLIPTIC DIRICHLET PROBLEMS. TECH.REPORT
69-88, COMPUTER SCIENCE CENTER, UNIVERSITY OF MARYLAND,
1969.

1854 STEPLEMAN R.S., FINITE-DIMENSIONAL ANALOGUES OF VARIATIONAL
PROBLEMS IN THE PLANE. SIAM J.NUMER.ANAL.8, 11-24, 1971.

1855 STETTER H.J., ANALYSIS OF DISCRETIZATION METHODS FOR ORDINARY
DIFFERENTIAL EQUATIONS. SPRINGER-VERLAG, BERLIN, 1973.

1856 STEWART G.W., INTRODUCTION TO MATRIX COMPUTATIONS. ACADEMIC PRESS,
NEW YORK, 1973.

1857 STODDART W.C.T., TRANSIENT RESPONSE OF LINEAR ELASTIC STRUCTURES
DETERMINED BY THE MATRIX EXPONENTIAL METHOD. PAPER E1/6,
PROCEEDINGS 1ST STRUCTURAL MECH. IN REACTOR TECH. CONF.,
BERLIN, 1971.

1858 STONE M.H., LINEAR TRANSFORMATIONS IN HILBERT SPACE. AMERICAN MATH. SOC.
COLLOQUIUM PUBLICATIONS 15, 1932.

1859   STORDAHL H., AND CHRISTENSEN H.,   FINITE ELEMENT ANALYSIS OF
          AXISYMMETRIC ROTORS.   J.STRAIN ANALYSIS 4, 163-168, 1969,
1860   STRANG G.,   FOURIER ANALYSIS OF THE FINITE ELEMENT METHOD IN RITZ-
          GALERKIN THEORY.   STUDIES IN APPLIED MATH.48, 265-273, 1969,
1861   STRANG G.,   VARIATIONAL CRIMES IN THE FINITE ELEMENT METHOD.   PP.
          689-710 OF A.K. AZIZ (ED.),   THE MATHEMATICAL FOUNDATIONS
          OF THE FINITE ELEMENT METHOD WITH APPLICATIONS TO PARTIAL
          DIFFERENTIAL EQUATIONS.   ACADEMIC PRESS, NEW YORK, 1972,
1862   STRANG G.,   APPROXIMATION IN THE FINITE ELEMENT METHOD.
          NUMER.MATH.19,81-98, 1972,
1863   STRANG G.,   THE DIMENSION OF PIECEWISE POLYNOMIAL SPACES AND ONE-SIDED
          APPROXIMATION.   PP.144-152 OF G.A. WATSON (ED.), PROCEEDINGS OF
          CONF. ON NUMERICAL SOLUTION OF DIFFERENTIAL EQUATIONS. LECTURE
          NOTES IN MATHEMATICS, NO.363.   SPRINGER-VERLAG,BERLIN,1974.
1864   STRANG G.,   THE FINITE ELEMENT METHOD, LINEAR AND NONLINEAR
          APPLICATIONS.   PROCEEDINGS, INTERNATIONAL CONGRESS OF
          MATHEMATICIANS, VANCOUVER, 1974.
1865   STRANG G., AND BERGER A.E.,   THE CHANGE IN SOLUTION DUE TO CHANGE
          IN DOMAIN. PROC. A.M.S. SYMPOSIUM ON PARTIAL DIFFERENTIAL
          EQUATIONS. UNIVERSITY OF CALIFORNIA, BERKELEY, 1971,
1866   STRANG G., AND FIX G.,   AN ANALYSIS OF THE FINITE ELEMENT METHOD,
          PRENTICE HALL, NEW JERSEY, 1973.
1867   STRICKLAND G., AND LODEN W.,   A DOUBLY-CURVED TRIANGULAR SHELL ELEMENT,
          PROC.2ND CONF.MATRIX METHODS IN STRUCTURAL MECHANICS,
          WRIGHT-PATTERSON AFB, OHIO, 1968.
1868   STRICKLIN J.A.,   COMPUTATION OF STRESS RESULTANTS FROM ELEMENT STIFFNESS
          MATRICES.   AIAA J. 4, 1095-1096, 1966.
1869   STRICKLIN J.A.,   INTEGRATION OF AREA COORDINATES IN MATRIX
          STRUCTURAL ANALYSIS.   AIAA J.6, 2023, 1968.
1870   STRICKLIN J.A.,   NONLINEAR ANALYSIS OF SHELLS OF REVOLUTION BY THE
          MATRIX DISPLACEMENT METHOD.   AIAA J.6, 2306-2311, 1968.
1871   STRICKLIN J.A., HAISLER W.E., TISDALE P.R., AND GUNDERSON R.,   A
          RAPIDLY CONVERGING TRIANGULAR PLATE ELEMENT.   AIAA J.7,
          180-181, 1969.
1872   STRICKLIN J.A., NAVARATNA D.R., AND PIAN T.H.H.,   IMPROVEMENTS IN
          THE ANALYSIS OF SHELLS OF REVOLUTION BY MATRIX DISPLACEMENT
          METHOD (CURVED ELEMENTS).   AIAA J.4, 2069-2072, 1966.
1873   STROUD A.H.,   APPROXIMATE CALCULATION OF MULTIPLE INTEGRALS.
          PRENTICE HALL, NEW JERSEY, 1971.
1874   STROUD A.H.,   TWO FIFTH DEGREE HARMONIC INTERPOLATION FORMULAS
          FOR THE N-BOX.   SIAM J.NUMER.ANAL.9, 518-521, 1972,
1875   STROUD A.H.,   NUMERICAL QUADRATURE AND SOLUTION OF ORDINARY
          DIFFERENTIAL EQUATIONS.   SPRINGER VERLAG, BERLIN, 1974.
1876   STROUD A.H., CHEN K-W., WANG P-L, AND MAO Z.,   SOME SECOND AND
          THIRD DEGREE HARMONIC INTERPOLATION FORMULAS.   SIAM J.
          NUMER.ANAL.8, 681-692, 1971.
1877   STROUD A.H., AND SECREST D.,   GAUSSIAN QUADRATURE FORMULAS,
          PRENTICE HALL, NEW JERSEY, 1966.
1878   SUHARA J., AND FUKUDA J.,   AUTOMATIC MESH GENERATION FOR FINITE ELEMENT
          ANALYSIS.   PP. 607-624 OF J.T. ODEN, R.W. CLOUGH AND Y. YAMOMOTO
          (EDS.), ADVANCES IN COMPUTATIONAL METHODS IN STRUCTURAL MECHANICS
          AND DESIGN.   UNIVERSITY OF ALABAMA PRESS, HUNTSVILLE, 1972,
1879   SUN C.T., AND YANG T.Y.,   A CONTINUUM APPROACH TOWARD DYNAMICS
          OF GRIDWORKS.   TRANS ASME SER.E, 40, 186-192, 1973.
1880   SVEC O.J. AND MCNEICE G.M.,   FINITE ELEMENT ANALYSIS OF FINITE SIZED
          PLATES BOUNDED TO AN ELASTIC HALF SPACE. COMP.METH. APPL.
          MECH.ENG. 1, 265-277, 1972.
1881   SWANSON J.A., AND PATTERSON J.F.,   APPLICATION OF FINITE ELEMENT METHODS
          FOR THE ANALYSIS OF THERMAL CREEP, IRRADIATION INDUCED CREEP
          AND SWELLING FOR LMF REACTOR DESIGN.   PAPER L4/3 PROCEEDINGS 1ST
          STRUCTURAL MECH. IN REACTOR TECH.CONF., BERLIN, 1971,
1882   SWANSON J.A., AND PATTERSON J.F.,   APPLICATION OF FINITE ELEMENT METHODS
          FOR THE ANALYSIS OF THERMAL CREEP, IRRADIATIATION INDUCED CREEP,
          ABD SWELLING FOR LMFBR DESIGN.   PP. 293-310, VOL. 6 OF PROCEEDINGS
          1ST STRUCTURAL MECH. IN REACTOR TECH. CONF., BERLIN, 1971,
1883   SWARTZ B.,   O(H2N+2-L) BOUNDS ON SOME SPLINE INTERPOLATION ERRORS,

BULL.AMER.MATH.SOC.,74, 1072-1078, 1968.

1884 SWARTZ B.K., CREATION AND COMPARISON OF FINITE DIFFERENCE
ANALOGS OF SOME FINITE ELEMENT SCHEMES. PROC.CONF.
MATHEMATICAL ASPECTS OF FINITE ELEMENTS IN PARTIAL
DIFFERENTIAL EQUATIONS. MADISON, 1974.

1885 SWARTZ B., AND VARGA R.S., ERROR BOUNDS FOR SPLINE AND L-SPLINE
INTERPOLATION. J.APPROX.THEORY 6, 6-49, 1972.

1886 SWARTZ B., AND WENDROFF B., COMPARATIVE EFFICIENCY OF CERTAIN
FINITE ELEMENT AND FINITE DIFFERENCE METHODS FOR A
HYPERBOLIC PROBLEM. REPORT LA-UR-73-918, LOS ALAMOS, 1973.
SCIENTIFIC LABORATORY, NEW MEXICO, 1973.

1887 SWARTZ B. AND WENDROFF B. THE RELATIVE EFFIENCY OF FINITE-
DIFFERENCE AND FINITE ELEMENT METHODS. I-HYPERBOLIC
PROBLEMS AND SPLINES. REPORT LA-UR-73-837, LOS ALAMOS.
1973.

1888 SWARTZ B. AND WENDROFF B., THE COMPARATIVE EFFICIENCY OF CERTAIN FINITE
ELEMENT AND FINITE DIFFERENCE METHODS FOR A HYPERBOLIC PROBLEM.
PP.153-163 OF G.A. WATSON (ED.), PROCEEDINGS OF CONF. ON NUMERICAL
SOLUTION OF DIFFERENTIAL EQUATIONS. LECTURE NOTES IN MATHEMATICS,
NO.363. SPRINGER-VERLAG,BERLIN,1974.

1889 SYNGE J.L., TRIANGULATION IN THE HYPERCIRCLE METHOD FOR PLANE
PROBLEMS. PROC. ROYAL IRISH ACAD. 54A, 341-367, 1952.

1890 SYNGE J.L., THE HYPERCIRCLE IN MATHEMATICAL PHYSICS, CAMBRIDGE
UNIVERSITY PRESS, LONDON, 1957.

1891 SYNGE J.L., THE HYPERCIRCLE METHOD. IN B.K.P. SCAIFF (ED.), STUDIES
IN NUMERICAL ANALYSIS. ACADEMIC PRESS, NEW YORK, 1974.

1892 SZABO B.A., RECENT DEVELOPMENTS IN FINITE ELEMENT ANALYSIS.
COMPUTERS AND MATHEMATICS, (TO APPEAR).

1893 SZABO B.A., AND LEE G.C., DERIVATION OF STIFFNESS MATRICES FOR
PROBLEMS OF PLANE ELASTICITY BY THE GALERKIN METHOD.
INT.J.NUMER.METH.ENG.1, 301-310, 1969.

1894 SZABO B.A., AND LEE G.C., STIFFNESS METHODS FOR PLATES BY GALERKIN'S
METHOD. J.ENG.MECH.DIV, A.S.C.E. 95, EM3, 571-586, 1969.

1895 SZABO B.A., AND TSAI C.T., THE QUADRATIC PROGRAMMING APPROACH TO THE
FINITE ELEMENT METHOD. INT.J.NUMER. METH.ENG. 5, 375-381, 1973.

1896 SZEGO G., ORTHOGONAL POLYNOMIALS. COLLOQUIUM PUBLICATIONS,
AMERICAN MATH.SOC., PROVIDENCE, RHODE ISLAND, 1939.

1897 SZEGO G.P. (ED.), MINIMIZATION ALGORITHMS MATHEMATICAL THEORY AND
COMPUTER RESULTS. ACADEMIC PRESS, NEW YORK, 1972.

1898 SZMELTER J., ENERGY METHOD OF NETWORKS OF ARBITRARY SHAPE IN
PROBLEMS OF THE THEORY OF ELASTICITY. IN OLSZAK(ED.),
PROC. I.U.T.A.M., SYMPOSIUM ON NON-HOMOGENEITY IN
ELASTICITY AND PLASTICITY, PERGAMON, OXFORD, 1959.

1899 TABARROK B., AND SODHI D.S., ON THE GENERALIZATION OF STRESS
FUNCTION PROCEDURE FOR DYNAMIC ANALYSIS OF PLATES. INT.
J.NUMER.METH.ENG.5, 523-542, 1973.

1900 TADA Y., AND LEE G.A., FINITE ELEMENT SOLUTION TO AN ELASTICA.
PROBLEMS OF BEAMS. INT.J.NUMER.METH.ENG.2, 229-242,1970.

1901 TALBOT A.(ED.), APPROXIMATION THEORY. ACADEMIC PRESS, LONDON, 1970.

1902 TAY A.O., AND DAVIS G. DE V., APPLICATION OF THE FINITE ELEMENT
METHOD TO CONVECTION HEAT TRANSFER BETWEEN PARALLEL PLANES.
INT.J.HEAT MASS TRANSFER 14, 1057-1070, 1971.

1903 TAYLOR C., AND AL-MASHIDANI G., AN ANALYSIS OF TWO-DIMENSIONAL
SURFACE RUN-OFF BY FINITE ELEMENTS. PROC.CONF.COMP.
METHODS IN NONLINEAR MECH., AUSTIN, TEXAS, 1974.

1904 TAYLOR C., AL-MASHIDANI G., AND DAVIS J.M., A FINITE ELEMENT APPROACH
TO WATERSHED RUNOFF. J.HYDROLOGY 21, 231-246, 1974.

1905 TAYLOR C., AND DAVIS J.M., TIDAL AND LONG WAVE PROPAGATION - A FINITE
ELEMENT APPROACH. REPORT C/R/189, DEPARTMENT OF CIVIL
ENGINEERING, UNIV. OF WALES, SWANSEA, 1973.

1906 TAYLOR C., FRANCE P.W., AND ZIENKIEWICZ O.C., SOME FREE SURFACE
TRANSIENT FLOW PROBLEMS OF SEEPAGE AND IRROTATIONAL FLOW.
PP.313-326 OF WHITEMAN (ED.), THE MATHEMATICS OF FINITE ELEMENTS AND
APPLICATIONS, ACADEMIC PRESS, LONDON, 1973.

1907 TAYLOR C. AND HOOD P., A NUMERICAL SOLUTION OF THE NAVIER
STOKES EQUATIONS USING THE FINITE ELEMENT TECHNIQUE.

COMPUTERS AND FLUIDS 1, 73-100,1973.

1908 TAYLOR C., PATIL B.S. AND ZIENKIEWICZ O.C., HARBOUR OSCILLATION IN
      A NUMERICAL TREATMENT FOR UNDAMPED MODES. PROC.INST.CIV.
      ENG.43, 141-155, 1969.

1909 TAYLOR R.L., ON THE COMPLETENESS OF SHAPE FUNCTIONS FOR FINITE
      ELEMENT ANALYSIS. INT.J.NUMER.METH.ENG.4, 17-22, 1972.

1910 TAYLOR R.L., PISTER K.S., AND GOODREAU G.I., THERMOMECHANICAL
      ANALYSIS OF VISCOELASTIC SOLIDS. INT.J.NUMER.METH.ENG.2,
      45-60, 1970.

1911 TELEGA J.J., FINITE ELEMENT METHOD IN SOIL AND ROCK MECHANICS.
      (POLISH). MECH.TEOR. AND STOSOW 11, 195-210, 1973.

1912 TEMAM R., NUMERICAL ANALYSIS.
      D.REIDEL PUBLISHING CO., DORDRECHT,1973.

1913 TEZCAN S.S., AGRAWAL K.M., AND KOSTRO G., FINITE ELEMENT ANALYSIS OF
      HYPERBOLIC PARABOLOID SHELLS. J.STRUCT.DIV. A.S.C.E. 94, ST1,
      407-424, 1971.

1914 TEZCAN S.S. AGRAWAL K.M., AND KOSTRO G., FINITE ELEMENT ANALYSIS OF
      HYPERBOLIC PARABOLOID SHELLS. J. STRUCT. DIV. A.S.C.E. 98,
      ST3, 671-690, 1972.

1915 THATCHER P.W., THE THEORY AND APPLICATION OF THE FINITE ELEMENT
      METHOD. PH.D. THESIS, UNIVERSITY OF LONDON, 1971.

1916 THIERAUF G., ELASTIC-PLASTIC DEFORMATIONS OF FLEXURALLY STIFF
      FRAMEWORKS FROM SECOND ORDER STRESS THEORY. ING.ARCH.
      42, 285-295, 1973.

1917 THOMAS D.H., AND WIXOM J.A., DOMAIN MAPPING VIA SPLINE-BLENDING AND
      INTERACTIVE GRAPHICS. REPORT, GENERAL MOTORS RESEARCH
      LABORATORIES, WARREN, MICHIGAN, 1974.

1918 THOMEE V., DISCRETE INTERIOR SCHAUDER ESTIMATES FOR ELLIPTIC
      DIFFERENCE OPERATORS. SIAM J.NUMER.ANAL.5, 626-645, 1968.

1919 THOMEE V., SPLINE APPROXIMATIONS AND DIFFERENCE SCHEMES FOR
      THE HEAT EQUATIONS. PP. 711-746 OF A.K. AZIZ (ED.), THE
      MATHEMATICAL FOUNDATIONS OF THE FINITE ELEMENT METHOD
      WITH APPLICATIONS TO PARTIAL DIFFERENTIAL EQUATIONS.
      ACADEMIC PRESS, NEW YORK, 1972.

1920 THOMEE V., APPROXIMATE SOLUTION OF DIRICHLET'S PROBLEM USING
      APPROXIMATING POLYGONAL DOMAINS. PP.311-328 OF J.J.H.
      MILLER (ED.), TOPICS IN NUMERICAL ANALYSIS. ACADEMIC PRESS,
      LONDON, 1973.

1921 THOMEE V., POLYGONAL DOMAIN APPROXIMATION IN DIRICHLET'S PROBLEM.
      J.INST.MATH.APPLICS.11, 33-44,1973.

1922 THOMEE V., CONVERGENCE ESTIMATES FOR SEMI-DISCRETE GALERKIN
      METHODS FOR INITIAL VALUE PROBLEMS. PP.243-262 OF R.
      ANSORGE AND W. TORNIG (EDS.), NUMERISCHE, INSBESONDERE
      APPROXIMATIONSTHEORETISCHE BEHANDLUNG VON
      FUNCTIONALGLEICHUNGEN. LECTURE NOTES IN MATHEMATICS, N. 333,
      SPRINGER-VERLAG, 1973.

1923 THOMEE V., SPLINE-GALERKIN METHODS FOR INITIAL-VALUE PROBLEMS WITH
      CONSTANT COEFFICIENTS. PP.164-175 OF G.A. WATSON (ED.),
      PROCEEDINGS OF CONF. ON NUMERICAL SOLUTION OF DIFFERENTIAL
      EQUATIONS. LECTURE NOTES IN MATHEMATICS, NO. 363.
      SPRINGER-VERLAG, BERLIN, 1974.

1924 THOMEE V., AND WESTERGREN B., ELLIPTIC DIFFERENCE EQUATIONS AND
      INTERIOR REGULARITY. NUMER.MATH.11, 196-210, 1968.

1925 THOMPSON E.G., AND HAQUE M.I., A HIGHER ORDER FINITE ELEMENT FOR
      COMPLETELY INCOMPRESSIBLE CREEPING FLOW. INT.J.NUMER.METH.
      ENG.6, 315-321, 1973.

1926 THOMPSON E.G., MACK L.R. AND LIN F.S., FINITE ELEMENT METHOD FOR
      INCOMPRESSIBLE SLOW VISCOUS FLOW WITH A FREE SURFACE. PP.93-111 OF
      DEVELOPMENTS IN MECHANICS. PROC.11TH MIDWESTERN MECHANICS CONF.5,
      IOWA STATE UNIVERSITY PRESS, 1969.

1927 THOMPSON J.J., AND CHEN P.Y.P., DISCONTINUOUS FINITE ELEMENTS IN THERMAL
      ANALYSIS. NUCL.ENG.DESIGN 14, 211-222, 1970.

1928 TIKHONOV A.N. AND SAMARSKY A.A., EQUATIONS OF MATHEMATICAL
      PHYSICS. PERGAMON PRESS, OXFORD, 1963.

1929 TILLERSON J.R., STRICKLIN J.A., AND HAISLER W.E., NUMERICAL
      METHODS FOR THE SOLUTION OF NONLINEAR PROBLEMS IN STRUCTURAL

ANALYSIS. IN R.F. HARTUNG (ED.), NUMERICAL SOLUTION OF
NONLINEAR STRUCTURAL PROBLEMS. AMERICAN SOC.MECH.ENG.,AMD
6, 1973.

1930 TIMOSHENKO S. AND GOODIER J.N., THEORY OF ELASTICITY. MCGRAW HILL,
NEW YORK, 1951.

1931 TIMOSHENKO S., AND WOINOWSKY-KRIEGER S., THEORY OF PLATES AND SHELLS,
MCGRAW HILL, NEW YORK, 1959.

1932 TING T.W., ELASTIC-PLASTIC TORSION OF A SQUARE BAR. TRANS.AMER.
MATH.SOC.123, 369-401,1966.

1933 TING T.W., ELASTIC-PLASTIC TORSION PROBLEM II. ARCH.RAT.MECH.ANAL,25,
343-366,1967.

1934 TINGLEFF O., A METHOD FOR SIMULATING CONCENTRATED FORCES AND
LOCAL REINFORCEMENTS IN STRESS COMPUTATION.
PP. 463-470 OF J.R.WHITEMAN (ED.) THE MATHEMATICS OF
FINITE ELEMENTS AND APPLICATIONS. ACADEMIC PRESS,LONDON,1973.

1935 TIROSH J., THE EFFECT OF PLASTICITY AND CRACK BUNTING ON THE
STRESS DISTRIBUTION IN ORTHOTROPIC COMPOSITE MATERIALS.
TRANS.ASME SER.E.40, 785-790, 1973.

1936 TOCHER J.L., ANALYSIS OF PLATE BENDING USING TRIANGULAR ELEMENTS.
PH.D. THESIS, DEPARTMENT OF CIVIL ENGINEERING, UNIVERSITY
OF CALIFORNIA, BERKELEY, 1962.

1937 TOCHER J.L., AND FELIPPA C.A., COMPUTER GRAPHICS APPLIED TO
PRODUCTION STRUCTURAL ANALYSIS. PAPER AT SYMP.HIGH SPEED
COMPUTING OF ELASTIC STRUCTURES, IUTAM,LEIGE, BELGIUM, 1970.

1938 TOCHER J.L., AND HARTZ B.J., HIGHER ORDER FINITE ELEMENT FOR PLANE
STRESS. J.ENG.MECH.DIVISION, ASCE 93, EM4, 149-172, 1967.

1939 TOMLIN G.R., AN OPTIMAL SUCCESSIVE OVERRELAXATION TECHNIQUE FOR
SOLVING SECOND ORDER FINITE DIFFERENCE EQUATIONS FOR
TRIANGULAR MESHES. INT.J.NUMER.METH.ENG.5, 25-39, 1972.

1940 TONG P., AN ASSUMED STRESS HYBRID FINITE ELEMENT METHOD FOR AN
INCOMPRESSIBLE AND NEAR INCOMPRESSIBLE MATERIAL. INT J. SOLIDS
STRUCTURES 5., 455-462, 1969.

1941 TONG P., EXACT SOLUTION OF CERTAIN PROBLEMS BY FINITE ELEMENT
METHOD. AIAA J. 7, 178-180, 1969.

1942 TONG P., NEW DISPLACEMENT HYBRID FINITE ELEMENT MODELS FOR
SOLID CONTINUA. INT.J.NUMER.METH.ENG.2,73-84, 1970.

1943 TONG P., THE FINITE ELEMENT METHOD FOR FLUID FLOW. PAPER US5-4 OF R.H.
GALLAGHER (ED.), PROCEEDINGS OF JAPAN-U.S. SEMINAR ON MATRIX
METHODS IN STRUCTURAL ANALYSIS AND DESIGN. UNIVERSITY OF ALABAMA
PRESS, HUNTSVILLE, 1970.

1944 TONG P., ON THE NUMERICAL PROBLEMS OF THE FINITE ELEMENT METHODS.
IN G.M.L. GLADWELL (ED.), SYMPOSIUM ON COMPUTER-AIDED
ENGINEERING. UNIVERSITY OF WATERLOO, 1971.

1945 TONG P., THE FINITE ELEMENT METHOD IN FLUID FLOW ANALYSIS. PP. 787-808
OF R.H. GALLAGHER, Y. YAMADA AND J.T. ODEN (EDS.), RECENT ADVANCES
IN MATRIX METHODS OF STRUCTURAL ANALYSIS AND DESIGN. UNIVERSITY OF
ALABAMA PRESS, TUSCALOOSA, 1971.

1946 TONG P., MAU S.T., AND PIAN T.H.H., DERIVATION OF GEOMETRIC
STIFFNESS AND MASS MATRICES FOR FINITE ELEMENT HYBRID
MODELS. INT.J.SOLIDS STRUCTURES 10, 919-932, 1974.

1947 TONG P., AND PIAN T.H.H., THE CONVERGENCE OF THE FINITE ELEMENT METHOD
IN SOLVING LINEAR ELASTIC PROBLEMS. INT.J.SOLIDS STRUCTURES 3,
865-880, 1967.

1948 TONG P., AND PIAN T.H.H., A VARIATIONAL PRINCIPLE AND THE CONVERGENCE OF
A FINITE-ELEMENT METHOD BASED ON ASSUMED STRESS DISTRIBUTION.
INT.J.SOLIDS STRUCTURES 5, 463-472, 1969.

1949 TONG P., PIAN T.H.H., AND BUCCIARELLI L.L., MODE SHAPES AND
FREQUENCIES BY FINITE ELEMENT METHOD USING CONSISTENT AND
LUMPED MASSES. COMPUTERS AND STRUCTURES 1, 623-638, 1971.

1950 TONG P., PIAN T.H.H. AND LASRY S.J., A HYBRID-ELEMENT APPROACH TO
CRACK PROBLEMS IN PLANE ELASTICITY. INT.J.NUMER. METH.ENG.
7, 297-308, 1973.

1951 TONTI E., VARIATIONAL FORMULATION OF NONLINEAR DIFFERENTIAL
EQUATIONS I AND II. BUL.DE L'ACADEMIE ROYALE DE BELIQUE,
137-277,1969.

1952 TORIDIS T.G., AND KHOZEIMEH K., COMPUTER ANALYSIS OF RIGID FRAMES,

COMPUTERS AND STRUCTURES, 1, 193-221,1971.

1953 TOTTENHAM H. AND BREBBIA C.,   FINITE ELEMENT TECHNIQUES IN STRUCTURAL
      MECHANICS. SOUTHAMPTON UNIVERSITY PRESS,1970.

1954 TRACEY D.M., FINITE ELEMENTS FOR DETERMINATION OF CRACK TIP ELASTIC
      STRESS INTENSITY FACTORS. ENGINEERING FRACTURE MECHANICS,
      3, 255-265, 1971.

1955 TRACEY D.M., FINITE ELEMENTS FOR THREE-DIMENSIONAL ELASTIC
      CRACK ANALYSIS. NUCL.ENG.DESIGN 26, 282-290, 1974.

1956 TRLIFAJ L., A VARIATIONAL METHOD FOR HOMOGENIZATION OF A HETEROGENEOUS
      MEDIUM. J.NUCLEAR ENERGY 6, 142-154, 1957.

1957 TSUSHIMA Y., HAYAMIZU Y., AND NISHIYAMA K., ASEISMIC DESIGN OF NUCLEAR
      REACTOR BUILDING, STRESS ANALYSIS AND STIFFNESS EVALUATION OF THE
      ENTIRE BUILDING BY THE FINITE ELEMENT METHOD. PP.123-139,
      VOL.5 OF PROC. 1ST STRUCTURAL MECH. IN REACTOR TECH. CONF.,
      BERLIN, 1971.

1958 TURCKE D.J., MCNEICE G.M., BREBBIA C.A. AND TOTTENHAM H., A
      VARIATIONAL APPROACH TO GRID OPTIMIZATION IN THE FINITE
      ELEMENT METHOD. PROCEEDINGS VARIATIONAL METHODS IN
      ENGINEERING CONF., UNIVERSITY OF SOUTHAMPTON, 1972.

1959 TURNER, H.J., CLOUGH R.W., MARTIN H.C., AND TODD L.J.,
      STIFFNESS AND DEFLECTION ANALYSIS OF COMPLEX STRUCTURES.
      J.AERO.SCI.23, 805-823, 1956.

1960 TURNER M.J., DILL E.H., MARTIN H.C., AND MELOSH R.J.,
      LARGE DEFLECTION OF STRUCTURES SUBJECT TO HEATING AND
      EXTERNAL LOADS. J.AERO.SCI.27, 97-106, 1960.

1961 TWZCAN S.S., AGRAWAL K.M., AND KOSTRO G., FINITE ELEMENT ANALYSIS OF
      HYPERBOLIC PARABOLOID SHELLS. J.STRUCT.DIV. A.S.C.E. 97, 407-424,
      1971.

1962 TYCHONOV A.N., AND SAMARSKII A.A., PARTIAL DIFFERENTIAL EQUATIONS
      OF MATHEMATICAL PHYSICS. VOLS.I AND II. HOLDEN-DAY,
      SAN FRANCISCO, 1964.

1963 UDOGUCHI T., ASADA Y., AND NOZUE Y., AN APPROACH TO INVESTIGATE
      NOTCH EFFECT ON LOW-CYCLE FATIGUE WITH FINITE ELEMENTS.
      PP.785-800, PROC.2ND INT.CONF.PRESSURE VESSEL TECHNOLOGY,
      SAN ANTONIO, 1973.

1964 UKAI S., SOLUTION OF MULTI-DIMENSIONAL NEUTRON TRANSPORT EQUATION BY
      FINITE ELEMENT METHOD. J.NUCL.SCI.TECH.(TOKYO) 9, 366-373, 1972.

1965 USUKI S., APPLICATIONS OF FINITE ELEMENT METHODS TO UNSTEADY
      VISCOUS FLOW AROUND A BOX-GIRDER BRIDGE OSCILLATING IN
      UNIFORM FLOW. PROC.CONF.COMP.METHODS IN NONLINEAR
      MECHANICS. AUSTIN, TEXAS, 1974.

1966 UTKU S., A FINITE ELEMENT METHOD FOR BLOCK ADJUSTMENT PROBLEMS
      OF PHOTOGRAMMETRY. INT.J.NUMER.METH.ENG.6, 33-38, 1965.

1967 UTKU S. STIFFNESS MATRICES FOR THIN TRIANGULAR ELEMENTS OF NON-ZERO
      GAUSSIAN CURVATURE. AIAA J.5, 1659-1667, 1967.

1968 UTKU S., EXPLICIT EXPRESSIONS FOR TRIANGULAR TORUS ELEMENT
      STIFFNESS MATRIX. AIAA J.6, 1174-1176, 1968.

1969 UTKU S., ON DERIVATION OF STIFFNESS MATRICES WITH ROTATION FIELDS FOR
      PLATES AND SHELLS. PROC. 3RD CONF. MATRIX METHODS IN STRUCTURAL
      MECH., WRIGHT-PATTERSON AFB., OHIO, 1971.

1970 UTKU S., AND MELOSH R.J., BEHAVIOUR OF TRIANGULAR SHELL-ELEMENT
      STIFFNESS MATRICES ASSOCIATED WITH POLYHEDRAL DEFLECTION
      DISTRIBUTIONS. AIAA J.6, 374-375, 1968.

1971 VAINBERG M.M., VARIATIONAL METHODS FOR THE STUDY OF NONLINEAR
      OPERATORS. (TRANSLATED FROM RUSSIAN BY A.FEINSTEIN)
      HOLDEN DAY, SAN FRANCISCO, 1964.

1972 VAINIKKO G., CERTAIN ESTIMATES FOR THE ERROR IN THE BUBNOV-
      GALERKIN METHOD. I ASYMPTOTIC ESTIMATES; II ESTIMATES OF
      THE NTH APPROXIMATION. TARTU RIIKL. UL.TOIMETISED 150,
      188-201 AND 202-215, 1964.

1973 VARGA R.S., MATRIX ITERATIVE ANALYSIS. PRENTICE HALL, NEW JERSEY, 1962.

1974 VARGA R.S., HERMITE INTERPOLATION-TYPE RITZ METHODS IN TWO-POINT
      BOUNDARY VALUE PROBLEMS.
      P.P.365-373 OF J.H. BRAMBLE (ED.), NUMERICAL SOLUTION OF PARTIAL
      DIFFERENTIAL EQUATIONS. ACADEMIC PRESS, NEW YORK, 1966.

1975 VARGA R.S., ERROR BOUNDS FOR SPLINE INTERPOLATION. PP. 367-388 OF I.J.

SCHOENBERG (ED.), APPROXIMATIONS WITH SPECIAL EMPHASIS ON SPLINE
FUNCTIONS. ACADEMIC PRESS, NEW YORK, 1969.

1976 VARGA R.S., FUNCTIONAL ANALYSIS AND APPROXIMATION THEORY IN NUMERICAL
ANALYSIS. REGIONAL CONFERENCE SERIES IN APPLIED MATHEMATICS,
NO.3, SIAM, PHILADELPHIA, 1971.

1977 VARGA R.S., SOME RESULTS IN APPROXIMATION THEORY WITH APPLICATIONS
TO NUMERICAL ANALYSIS. IN HUBBARD(ED.), NUMERICAL SOLUTION
OF PARTIAL DIFFERENTIAL EQUATIONS-II, SYNSPADE 1970,
ACADEMIC PRESS, NEW YORK, 1971.

1978 VARGA R.S., THE ROLE OF INTERPOLATION AND APPROXIMATION THEORY IN
VARIATIONAL AND PROJECTIONAL METHODS FOR SOLVING PARTIAL
DIFFERENTIAL EQUATIONS. PROCEEDINGS IFIP 1971.

1979 VARGA R.S., CHEBYSHEV SEMI-DISCRETE APPROXIMATIONS FOR LINEAR
PARABOLIC PROBLEMS.
PP.452-460 OF P.L. BUTZER, J.P. KAHANE AND B.SZ.NAVY (EDS.),
LINEAR OPERATORS AND APPROXIMATION, ISNM.20, BIRKHAUSER VERLAG,
BASEL,1972.

1980 VARGA R.S., EXTENSIONS OF THE SUCCESSIVE OVERRELAXATION THEORY
WITH APPLICATIONS TO FINITE ELEMENT APPROXIMATIONS. PP.
329-343 OF J.J.H. MILLER (ED.), TOPICS IN NUMERICAL ANALYSIS.
ACADEMIC PRESS, LONDON, 1973.

1981 VEKUA I.N., GENERALIZED ANALYTIC FUNCTIONS. PERGAMON PRESS,
OXFORD, 1962.

1982 VENDHAN C.P., KAPOOR M.P., AND DAS Y.C., AN INTEGRATED
SEQUENTIAL SOLVER FOR LARGE MATRIX EQUATIONS. INT.J.
NUMER. METH.ENG.8, 227-248, 1974.

1983 VENKATESWARA RAO G., AND KRISHNA MURTY A.V., AN ALTERNATE FORM
OF THE ROMBERG-OSGOOD FORMULA FOR MATRIX DISPLACEMENT
ANALYSIS. NUCLEAR ENG. AND DESIGN 17,297-308, 1971.

1984 VENKATESWARA RAO G., AND KRISHNA MURTY A.V., A COMPARATIVE STUDY
OF THE CONSISTENT AND SIMPLIFIED FINITE
ELEMENT ANALYSES OF EIGENVALUE PROBLEMS.
J.AERO.SOC.INDIA 22, 183-188,1970.

1985 VENKATESWARA RAO G., AND RAJU I.S., A COMPARATIVE STUDY OF
VARIABLE AND CONSTANT THICKNESS HIGH PRECISION TRIANGULAR
PLATE BENDING ELEMENTS IN THE ANALYSIS OF VARIABLE
THICKNESS PLATES. NUCL.ENG.DESIGN 26, 299-304, 1974.

1986 VERNER E.A. AND BECKER E.B., FINITE ELEMENT STRESS FORMULATION FOR WAVE
PROPAGATION. INT.J.NUMER.METH.ENG.7, 441-460,1973.

1987 VILLAGGI P., PROPERTIES OF STABILITY AND MONOTOMY WITH FINITE ELEMENT
METHOD. AEROTECHNICA 49, ISSUE 3-6, 94,1969.

1988 VISSER W., A FINITE ELEMENT METHOD FOR THE DETERMINATION OF
NON-STATIONARY TEMPERATURE DISTRIBUTION AND THERMAL
DEFORMATIONS. PROC.1ST CONF.MATRIX METHODS IN STRUCTURAL
MECHANICS, WRIGHT-PATTERSON AFB, OHIO, AFFDL TR66-80,1965.

1989 VISSER W., THE FINITE ELEMENT METHOD IN DEFORMATION AND HEAT
CONDUCTION PROBLEMS. DELFT, 1968.

1990 VISSER W., A REFINED MIXED TYPE PLATE BENDING ELEMENT.
AIAA J.7, 1801-1803, 1969.

1991 VISSER W., THE APPLICATION OF A CURVED MIXED TYPE SHELL ELEMENT.
I.U.T.A.M. SYMPOSIUM, HIGH SPEED COMPUTING OF ELASTIC STRUCTURES,
LIEGE, 1970.

1992 VLADIMIROV V.S., EQUATIONS OF MATHEMATICAL PHYSICS. DEKKER,
NEW YORK, 1971.

1993 VOLKER R.E., NONLINEAR FLOW IN POROUS MEDIA BY FINITE ELEMENTS. J.
HYDRAULICS DIV. A.S.C.E. 95, HY6, 2093-2114, 1969.

1994 VON FUCHS G., ROY J.R. AND SCHREM E., HYPERMATRIX SOLUTION OF
LARGE SETS OF SYMMETRIC POSITIVE-DEFINITE LINEAR EQUATIONS.
COMP.METH.APPL.MECH.ENG.1, 197-216, 1972.

1995 VOS R.G., GENERALIZATION OF PLATE FINITE ELEMENTS TO SHELLS. J.ENG.
MECH. DIV. A.S.C.E. 98, EM2, 385-400, 1972.

1996 VOS R.G., FINITE ELEMENT SOLUTION OF NONLINEAR STRUCTURES BY
PERTURBATION TECHNIQUE. PROC.CONF.COMP.METHODS IN
NONLINEAR MECHANICS, AUSTIN, TEXAS, 1974.

1997 VOS R.G., AND VANN W.P., A FINITE ELEMENT TENSOR APPROACH TO PLATE
BUCKLING AND POSTBUCKLING INT.J.NUMER.METH.ENG.5, 351-365,

           1973.

1998   WACHSPRESS E., ITERATIVE SOLUTIONS OF ELLIPTIC SYSTEMS.
           PRENTICE HALL, NEW JERSEY, 1966.

1999   WACHSPRESS E.L., A RATIONAL BASIS FOR FUNCTION APPROXIMATION.
           J.INST.MATH.APPLICS.8,57-68, 1971.

2000   WACHSPRESS E.L., A RATIONAL BASIS FOR FUNCTION APPROXIMATION;
           CURVED SIDES. TO APPEAR.

2001   WACHSPRESS E.L., ALGEBRAIC-GEOMETRY FOUNDATIONS FOR FINITE ELEMENT
           COMPUTATION. PP.177-188 OF G.A. WATSON (ED.), PROCEEDINGS OF
           CONF. ON NUMERICAL SOLUTION OF DIFFERENTIAL EQUATIONS. LECTURE
           NOTES IN MATHEMATICS, NO.363. SPRINGER-VERLAG,BERLIN,1974.

2002   WAKE J.C. AND RAYNER M.E., VARIATIONAL METHODS FOR NONLINEAR
           EIGENVALUE PROBLEMS ASSOCIATED WITH THERMAL IGNITION,
           J. DIFF. EQUATIONS 13, 247-256, 1973.

2003   WAIT R., THE FINITE ELEMENT METHOD IN PARTIAL DIFFERENTIAL
           EQUATIONS. PH.D. THESIS, UNIVERSITY OF DUNDEE, 1970.

2004   WAIT R., AND MITCHELL A.R., THE SOLUTION OF TIME DEPENDENT PROBLEMS
           BY GALERKIN METHODS. J.INST.MATH.APPLICS.7,241-250,1971.

2005   WAIT R., AND MITCHELL A.R., CORNER SINGULARITIES IN ELLIPTIC
           PROBLEMS BY FINITE ELEMENT METHODS. J.COMP.PHYS.8, 45-52,1971.

2006   WALKER A.C., A NONLINEAR FINITE ELEMENT ANALYSIS OF SHALLOW CIRCULAR
           ARCHES. INT.J.SOLIDS STRUCTURES 5, 97-108, 1969.

2007   WALKER A.C.,AND HALL D.G., AN ANALYSIS OF THE LARGE DEFLECTIONS OF
           BEAMS USING THE RAYLEIGH RITZ FINITE ELEMENT METHOD. AERO.QUART.
           19, 357-367, 1968.

2008   WALKER T.G., QUANTITATIVE STRAIN AND STRESS STATE CRITERION FOR FAILURE
           IN THE VICINITY OF SHARP CRACKS. TRANS.AMERICAN.NUCL.SOC.16,
           100-XXX, 1973.

2009   WALLERSTEIN D.V., A GENERAL LINEAR GEOMETRIC MATRIX FOR A FULLY
           COMPATIBLE FINITE ELEMENT. AIAA J.10, 545-546, 1972.

2010   WALPOLE J.J., NEW EXTREMUM PRINCIPLES FOR LINEAR PROBLEMS, J.INST.MATH.
           APPLICS. 14, 113-118, 1974.

2011   WALTER W., DIFFERENTIAL AND INTEGRAL INEQUALITIES. SPRINGER-
           VERLAG, BERLIN, 1970.

2012   WALTON D., WOODMAN N.J., AND ELLISON F.G., A FINITE ELEMENT METHOD
           APPLIED TO PREDICTING FATIGUE CRACK GROWTH. J. STRAIN
           ANAL 8, 294-304, 1973.

2013   WALZ J.E., FULTON R.E., AND CYRUS N.J., ACCURACY AND CONVERGENCE
           OF FINITE ELEMENT APPROXIMATION. PROC. 2ND CONF. MATRIX
           METHODS IN STRUCTURAL MECHANICS, WRIGHT-PATTERSON AFB,
           OHIO, 1968.

2014   WASHIZU K., VARIATIONAL METHODS IN ELASTICITY AND PLASTICITY.
           PERGAMON PRESS, OXFORD,1968

2015   WASHIZU K., SOME CONSIDERATIONS ON BASIC THEORY FOR THE FINITE
           ELEMENT METHOD. PP. 39-54 OF J.T. ODEN, R.W. CLOUGH AND Y.
           YAMOMOTO (EDS.), ADVANCES IN COMPUTATIONAL METHODS IN STRUCTURAL
           MECHANICS AND DESIGN. UNIVERSITY OF ALABAMA PRESS, HUNTSVILLE,
           1972.

2016   WASHIZU K., AND IKEGAWA M., LIFTING SURFACE PROBLEMS ANALYSIS. PP.
           573-582 OF THEORY AND PRACTICE IN FINITE ELEMENT STRUCTURAL
           ANALYSIS. UNIVERSITY OF TOKYO PRESS, 1973.

2017   WASHIZU K., AND IKEGAWA M., SOME APPLICATIONS OF THE FINITE ELEMENT
           METHOD TO FLUID MECHANICS. THEORETICAL AND APPLIED MECHANICS
           22, 143-154, UNIVERSITY OF TOKYO, 1974.

2018   WATKINS D.S., BLENDING FUNCTIONS AND FINITE ELEMENTS. PH.D.
           THESIS, DEPARTMENT OF MATHEMATICS, STATISTICS AND COMPUTING
           SCIENCE, UNIVERSITY OF CALGARY, 1974.

2019   WATSON G.A. (ED.) CONFERENCE ON THE NUMERICAL SOLUTION OF DIFFERENTIAL
           EQUATIONS. LECTURE NOTES IN MATHEMATICS, NO 363, SPRINGER-
           VERLAG,BERLIN,1974.

2020   WATWOOD V.B., AND HARTZ B.J., AN EQUILIBRIUM STRESS FIELD MODEL FOR
           FINITE ELEMENT SOLUTIONS OF TWO-DIMENSIONAL ELASTIC PROBLEMS.
           INT. J. SOLIDS STRUCTURES 4, 854-874, 1968.

2021   WEBBER J.P.H., THERMOELASTIC ANALYSIS OF RECTANGULAR PLATES IN PLANE
           STRESS BY THE FINITE ELEMENT DISPLACEMENT METHOD. J. STRAIN
           ANALYSIS 2, 43-51, 1967.

2022 WEBBER J.P.H., STRESS ANALYSIS IN VISCOELASTIC BODIES USING FINITE
        ELEMENTS AND A CORRESPONDENCE RULE WITH ELASTICITY. J. STRAIN
        ANALYSIS 4, 236-243, 1969.

2023 WEEKS G.E., A FINITE ELEMENT MODEL FOR SHELLS BASED ON THE
        DISCRETE KIRCHHOFF HYPOTHESIS. INT.J.NUMER.METH.ENG.5,
        3-16, 1972.

2024 WEHLE L.B., AND LANSING W., A METHOD OF REDUCING THE ANALYSIS OF
        COMPLEX REDUNDANT STRUCTURES TO A ROUTINE PROCEDURE.
        J.AERO.SCI.19, 677-684, 1952.

2025 WEILER F.C., ANISOTROPIC ELASTIC-PLASTIC THERMAL STRESS ANALYSES OF
        SOLID STRUCTURES. PP.159-181, VOL.2 PROCEEDINGS 1ST
        STRUCTURAL MECH. IN REACTOR TECH.CONF., BERLIN, 1971.

2026 WEINBERGER H.F., A VARIATIONAL COMPUTATATION METHOD FOR FORCED
        VIBRATION. PROC.SYMP.APPL.MATH.,AMER.MATH.SOC.VIII, 1958.

2027 WEINBERGER H.F., VARIATIONAL METHODS FOR EIGENVALUE
        APPROXIMATION. REGIONAL CONFERENCE SERIES IN APPLIED
        MATHEMATICS, NO. 15, SIAM, PHILADELPHIA, 1974.

2028 WEINSTEIN A., LES VIBRATIONS ET LE CALCUL DES VARIATIONS. PORTUGUESE
        MATH.2, 36-55, 1941.

2029 WEISSHAAR T.A., AN APPLICATION OF FINITE ELEMENT METHODS TO PANEL FLUTTER
        OPTIMIZATION. PROC. 3RD CONF. MATRIX METHODS IN STRUCTURAL MECHANICS,
        WRIGHT-PATTERSON AFB, OHIO, 1971.

2030 WELLFORD L.C., VARIATIONAL METHODS FOR WAVE AND SHOCK PROPAGATION
        IN NONLINEAR HYPERELASTIC MATERIALS. PROC.CONF. COMP.
        METHODS IN NONLINEAR MECHANICS, AUSTIN, TEXAS, 1974.

2031 WELLFORD L.C., AND ODEN J.T., ACCURACY AND CONVERGENCE OF FINITE
        ELEMENT GALERKIN APPROXIMATIONS OF TIME DEPENDENT PROBLEMS
        WITH EMPHASIS ON DIFFUSION AND CONVECTION. IN R.H.
        GALLAGHER, J.T. ODEN, C. TAYLOR AND O.C. ZIENKIEWICZ (EDS.),
        FINITE ELEMENT METHODS IN FLOW PROBLEMS. WILEY, LONDON,
        1974.

2032 WEMPNER G.A., FINITE ELEMENTS, FINITE ROTATIONS AND SMALL STRAINS
        OF FLEXIBLE SHELLS. INT.J.SOLIDS STRUCTURES 5, 117-153,
        1969.

2033 WEMPNER, G., DISCRETE APPROXIMATIONS OF ELASTIC-PLASTIC BODIES BY
        VARIATIONAL METHODS. PROC.CONF. VARIATIONAL METHODS IN
        ENGINEERING, UNIVERSITY OF SOUTHAMPTON, 1972.

2034 WEMPNER G., ODEN J.T., AND KROSS D., FINITE ELEMENT ANALYSIS OF
        THIN SHELLS. J.ENG.MECH.DIV.PROC. A.S.C.E. 96,EM6,967-983,
        1970.

2035 WENDROFF B., BOUNDS FOR EIGENVALUES OF SOME DIFFERENTIAL
        OPERATORS BY THE RAYLEIGH-RITZ METHOD. MATH.COMP.,19,
        218-224, 1965.

2036 WENDROFF B., SPLINE-GALERKIN METHODS FOR INITIAL-VALUE PROBLEMS WITH
        VARIABLE COEFFICIENTS. PP.189-195 OF G.A. WATSON (ED.),
        PROCEEDINGS OF CONF. ON NUMERICAL SOLUTION OF DIFFERENTIAL
        EQUATIONS. LECTURE NOTES IN MATHEMATICS, NO.363, SPRINGER-VERLAG,
        BERLIN,1974.

2037 WERNER H., ANWENDUNGEN UND FEHLERABSCHATZUNGEN FUR DAS
        ALTERNIERENDE VERFAHREN VON H.A. SCHWARZ. Z.A.M.M. 43,
        55-61, 1963.

2038 WERNER H., VORLESUNG UBER APPROXIMATIONSTHEORIE. SPRINGER
        VERLAG, BERLIN, 1966.

2039 WERNER H., INTERPOLATION AND INTEGRATION OF INITIAL VALUE PROBLEMS OF
        ORDINARY DIFFERENTIAL EQUATIONS BY REGULAR SPLINES. SIAM J.
        NUMER.ANAL. (TO APPEAR).

2040 WESTBROOK D.R., A VARIATIONAL PRINCIPLE WITH APPLICATIONS IN FINITE
        ELEMENTS. J.INST.MATH.APPLICS.14, 79-82, 1974.

2041 WEYL H., THE METHOD OF ORTHOGONAL PROJECTION IN POTENTIAL THEORY. DUKE
        MATH. J. 7, 411-444,1940.

2042 WHEELER MARY F., A PRIORI L2 ERROR ESTIMATES FOR GALERKIN
        APPROXIMATIONS TO PARABOLIC PARTIAL DIFFERENTIAL
        EQUATIONS. SIAM J. NUMER.ANAL.10,723-759, 1973.

2043 WHEELER MARY F., LINFINITY ESTIMATES OF OPTIMAL ORDERS FOR GALERKIN
        METHODS FOR ONE-DIMENSIONAL SECOND ORDER PARABOLIC AND
        HYPERBOLIC EQUATIONS. SIAM J.NUMER.ANAL.10,908-913, 1973.

2044   WHEELER MARY F., AN OPTIMAL LINFINITY ERROR ESTIMATE FOT GALERKIN
            APPROXIMATIONS TO SOLUTIONS OF TWO POINT BOUNDARY VALUE PROBLEMS.
            SIAM J.NUMER.ANAL.10, 914-917,1973.
2045   WHITE D.J., AND ENDERBY L.R., FINITE ELEMENT ANALYSIS OF A MULTIPLE
            PISTON. J.STRAIN ANALYSIS 4, 33-39, 1969.
2046   WHITE J.L., FINITE ELEMENTS IN LINEAR VISCOELASTICITY. PROC. 2ND CONF.
            MATRIX METHODS IN STRUCTURAL MECHANICS, WRIGHT-PATTERSON AFB, OHIO,
            AFFDL-TR-68-150, 1968.
2047   WHITEMAN J.R., TREATMENT OF SINGULARITIES IN HARMONIC MIXED BOUNDARY
            VALUE PROBLEM BY DUAL SERIES METHODS. QUART.J.MECH.APPL.MATH.21,
            41-51, 1968.
2048   WHITEMAN J.R., NUMERICAL SOLUTION OF A HARMONIC MIXED BOUNDARY
            VALUE PROBLEM BY THE EXTENSION OF A DUAL SERIES METHOD.
            QUART.J.MECH.APPL.MATH.23, 449-455, 1970.
2049   WHITEMAN J.R., FINITE-DIFFERENCE TECHNIQUES FOR A HARMONIC MIXED
            BOUNDARY PROBLEM HAVING A RE-ENTRANT BOUNDARY.
            PROC.ROY.SOC.LOND.A. 323, 271-276,1971.
2050   WHITEMAN J.R. (ED.), THE MATHEMATICS OF FINITE ELEMENTS AND
            APPLICATIONS. ACADEMIC PRESS, LONDON, 1973.
2051   WHITEMAN J.R., NUMERICAL SOLUTION OF STEADY STATE DIFFUSION PROBLEMS
            CONTAINING SINGULARITIES. IN VOL.2 OF R.H.GALLAGHER,
            J.T.ODEN, C.TAYLOR AND O.C.ZIENKIEWICZ (EDS.),
            FINITE ELEMENT METHODS IN FLOW PROBLEMS, WILEY, LONDON,1974.
2052   WHITEMAN J.R., LAGRANGIAN FINITE ELEMENT AND FINITE DIFFERENCE
            METHODS FOR POISSON PROBLEMS. IN L. COLLATZ (ED.),
            NUMERISCHE BEHANDLUNG VON DIFFERENTIALGLEICHUNGEN, I.S.N.
            M., BIRKHAUSER VERLAG, BASEL, 1974.
2053   WHITEMAN J.R., CONFORMING FINITE ELEMENT METHODS FOR THE
            CLAMPED PLATE PROBLEM. IN L. COLLATZ (ED.), FINITE
            ELEMENTE UND DIFFERENZENVERFAHREN, I.S.N.M., BIRKHAUSER
            VERLAG, BASEL, 1975.
2054   WHITEMAN J.R. (ED.), THE MATHEMATICS OF FINITE ELEMENTS AND
            APPLICATIONS II, MAFELAP 75. ACADEMIC PRESS, LONDON,
            (TO APPEAR).
2055   WHITEMAN J.R., AND BARNHILL R.E., FINITE ELEMENT METHODS FOR
            ELLIPTIC MIXED BOUNDARY VALUE PROBLEMS CONTAINING
            SINGULARITIES. PROC. EQUADIFF 3, CZECHOSLOVAK CONFERENCE
            ON DIFFERENTIAL EQUATIONS AND THEIR APPLICATIONS, BRNO,
            1972.
2056   WHITEMAN J.R., AND PAPAMICHAEL N., CONFORMAL TRANSFORMATION
            METHODS FOR THE NUMERICAL SOLUTION OF HARMONIC MIXED BOUNDARY
            VALUE PROBLEMS. PROC.CONF.APPLICATIONS OF NUMERICAL ANALYSIS,
            DUNDEE, LECTURE NOTES IN MATHEMATICS NO.228, SPRINGER-VERLAG,
            BERLIN, 1971.
2057   WHITEMAN J.R., AND PAPAMICHAEL N., TREATMENT OF HARMONIC MIXED
            BOUNDARY PROBLEMS BY CONFORMAL TRANSFORMATION METHODS.
            Z.A.M.P.23, 655-664, 1972.
2058   WHITEMAN J.R., AND WEBB J.C., CONVERGENCE OF FINITE-DIFFERENCE
            TECHNIQUES FOR A HARMONIC MIXED BOUNDARY VALUE PROBLEM,
            B.I.T.10, 366-374, 1970.
2059   WICKS T.M., BECKER E.B., YEW C.H., AND DUNHAM R.S., ON
            APPLICATION OF THE FINITE ELEMENT METHOD TO LIMIT ANALYSIS.
            PROC.CONF.COMP. METHODS IN NONLINEAR MECHANICS, AUSTIN,
            TEXAS, 1974.
2060   WILKINSON J.H., THE ALGEBRAIC EIGENVALUE PROBLEM, OXFORD UNIVERSITY
            PRESS, OXFORD, 1965.
2061   WILKINSON J.H., ROUNDING ERRORS IN ALGEBRAIC PROCESSES. H.M.S.O.,
            LONDON, AND PRENTICE HALL, NEW JERSEY, 1963.
2062   WILKINSON J.H., AND REINSCH C., HANDBOOK FOR AUTOMATIC
            COMPUTATION, VOL. II, LINEAR ALGEBRA. SPRINGER, VERLAG,
            BERLIN, 1971.
2063   WILLE F., GALERKINS LOSUNGSNAHERUNGEN BEI MONOTONEN ABBILDUNGEN,
            MATHEMATISCHE ZEIT. 127, 10-16, 1972.
2064   WILLIAMS F.W., COMPARISON BETWEEN SPARSE STIFFNESS MATRIX AND
            SUB-STRUCTURE METHODS. INT.J.NUMER.METH.ENG.5, 383-394,
            1973.

2065 WILLOUGHBY R.A., (ED.), SPARSE MATRIX PROCEEDINGS, RA-1, IBM RESEARCH
          PUBLICATION, 1969.
2066 WILSON, E.L., FINITE ELEMENT ANALYSIS OF TWO DIMENSIONAL STRUCTURES.
          STRUCTURAL ENG. LAB. REPT.63-2, UNIVERSITY OF CALIFORNIA,
          BERKELEY, 1963.
2067 WILSON E.L., AND PARSONS B., FINITE ELEMENT ANALYSIS OF ELASTIC
          CONTACT PROBLEMS USING DIFFERENTIAL DISPLACEMENTS.
          INT.J.NUMER.METH.ENG.2, 387-396, 1970
2068 WILSON E.L., BATHE K.-J., AND DOHERTY W.P., COMPUTER PROGRAM FOR STATIC
          AND DYNAMIC ANALYSIS OF LINEAR STRUCTURAL SYSTEMS. E.E.R.G.
          REPORT NO. 72-10, UNIVERSITY OF CALIFORNIA, BERKELEY, 1972.
2069 WILSON E.L., BATHE K.-J., AND DOHERTY W.P., DIRECT SOLUTION OF LARGE
          SYSTEMS OF LINEAR EQUATIONS. COMPUTERS AND STRUCTURES
          4, 363-372, 1974.
2070 WILSON E.L., BATHE K.J., AND PETERSON F.E., FINITE ELEMENT ANALYSIS
          OF LINEAR AND NONLINEAR HEAT TRANSFER. PAPER L1/4, PROCEEDINGS
          2ND STRUCTURAL MECH . IN REACTOR TECH. CONF., BERLIN, 1973.
2071 WILSON E.L., FARHOOMAND I., AND BATHE K.J., NONLINEAR DYNAMIC
          ANALYSIS OF COMPLEX STRUCTURES. EARTHQUAKE ENG. AND
          STRUCT.DYNAMICS 1, 241-252, 1973.
2072 WILSON E.L. AND NICKELL R.E., APPLICATION OF FINITE ELEMENT METHOD
          TO THE HEAT CONDUCTION EQUATION. NUCLEAR ENG. AND DESIGN 3,
          1-11, 1966.
2073 WINSLOW A.M., NUMERICAL SOLUTION OF THE QUASILINEAR POISSON
          EQUATION IN A NON-UNIFORM TRIANGULAR MESH. J.COMP.PHYS.1,
          149-172, 1967.
2074 WITMER E.A., AND KOTANCHIK J.J., PROGRESS REPORT ON DISCRETE ELEMENT
          ELASTIC AND ELASTIC-PLASTIC ANALYSIS OF SHELLS OF REVOLUTION
          SUBJECTED TO AXISYMMETRIC AND ASYMMETRIC LOADING, PROC.2ND CONF.
          MATRIX METHODS IN STRUCTURAL MECHANICS, WRIGHT-PATTERSON AFB,
          OHIO, AFFDL-TR-68-150, 1968.
2075 WOOD W.L., SIMPLE TEACHING EXAMPLES FOR FINITE ELEMENT STRESS
          ANALYSIS. INT.J.MATH.EDUC.SCI.TECHNOL. 3, 133-146, 1972.
2076 WOODFORD C.H., SMOOTH CURVE INTERPOLATION. B.I.T.9, 69-77, 1969.
2077 WU E.H. AND WITMER E.A., FINITE ELEMENT ANALYSIS OF LARGE
          ELASTIC PLASTIC TRANSIENT DEFORMATIONS OF SIMPLE
          STRUCTURES. AIAA J. 9, 1719-1724, 1971.
2078 WU R.W.-H., AND WITMER E.A., THE DYNAMIC RESPONSES OF CYLINDRICAL
          SHELLS INCLUDING GEOMETRIC AND MATERIAL NONLINEARITIES.
          J.SOLIDS AND STRUCTURES 10, 243-260, 1974.
2079 WU S.T., UNSTEADY MHD DUCT FLOW BY THE FINITE ELEMENT METHOD.
          INT.J.NUMER.METH.ENG.6, 3-10, 1973.
2080 WU S.T., AND JENG D.R., ON STEADY MHD FLOW IN ARBITRARY DUCTS BY
          POINT-MATCHING METHOD. AIAA J.7, 1612-1614, 1969.
2081 YAGAWA G., AND ANDO Y., THREE-DIMENSIONAL FINITE ELEMENT METHOD
          OF THERMOELASTOPLASTICITY WITH CREEP EFFECT. PAPER L4/3,
          PROC. 2ND STRUCTURAL MECH. IN REACTOR TECH. CONF., BERLIN,
          1973.
2082 YAMADA Y., DYNAMIC ANALYSIS OF STRUCTURES. PP.181-200 OF J.T. ODEN,
          R.W. CLOUGH AND Y. YAMOMOTO (EDS.), ADVANCES IN COMPUTATIONAL
          METHODS IN STRUCTURAL MECHANICS AND DESIGN. UNIVERSITY OF
          ALABAMA PRESS, HUNTSVILLE, 1972.
2083 YAMADA Y., INCREMENTAL FORMULATIONS FOR PROBLEMS WITH GEOMETRIC AND
          MATERIAL NONLINEARITIES. PP.325-356 OF J.T. ODEN, R.W. CLOUGH,
          AND Y. YAMOMOTO (EDS.), ADVANCES IN COMPUTATIONAL METHODS IN
          STRUCTURAL MECHANICS AND DESIGN. UNIVERSITY OF ALABAMA PRESS,
          HUNTSVILLE, 1972.
2084 YAMADA Y. AND GALLAGHER R.H. (EDS.), THEORY AND PRACTICE IN FINITE
          ELEMENT STRUCTURAL ANALYSIS. UNIVERSITY OF TOKYO PRESS,1973.
2085 YAMADA Y. AND IWATA K., FINITE ELEMENT ANALYSIS OF THERMOVISCOELASTIC
          PROBLEMS.
2086 YAMADA Y., KAWAI T. AND YOSHIMURA N., ANALYSIS OF THE ELASTIC-PLASTIC
          PROBLEMS BY THE MATRIX DISPLACEMENT METHOD.
          PROC.2ND CONF.MATRIX METHODS IN STRUCTURAL MECHANICS,
          WRIGHT-PATTERSON AFB,OHIO,AFFDL-TR-68-150,1968
2087 YAMADA Y., NAKAGIRI S. AND TAKATSUKA K., ELASTIC-PLASTIC ANALYSIS

OF SAINT-VENANT TORSION PROBLEM BY A HYBRID STRESS MODEL.
INT J.NUMER.METH.ENG.5, 193-207,1972.

2088 YAMADA Y., YASHIMURA N., AND SAKURAI T., STRESS STRAIN MATRIX AND
ITS APPLICATIONS FOR THE SOLUTION OF ELASTIC-PLASTIC PROBLEMS
BY THE FINITE ELEMENT METHOD. INT.J.MECH.SCI.10,343-354, 1968.

2089 YAMADA Y. AND YOKOUCHI Y., INCREMENTAL SOLUTION OF AXISYMMETRIC PLATE
AND SHELL FINITE DEFORMATION.
PROCEEDINGS IUTAM SYMPOSIUM,LIEGE,1970.

2090 YAMOMOTO Y., SOME CONSIDERATIONS OF ROUNDOFF ERRORS OF THE FINITE ELEMENT
METHOD. PP. 69-86 OF J.T. ODEN, R.W. CLOUGH AND Y. YAMOMOTO
(EDS.), ADVANCES IN COMPUTATIONAL METHODS IN STRUCTURAL MECHANICS
AND DESIGN. UNIVERSITY OF ALABAMA PRESS, HUNTSVILLE, 1972.

2091 YAMAMOTO Y., RATE OF CONVERGENCE FOR THE ITERATIVE APPROACH IN
ELASTIC PLASTIC ANALYSIS OF CONTINUA. INT.J.NUMER.METH.ENG.
7, 497-500, 1973.

2092 YAMAMOTO Y., AND KOKURO K., EFFECTS OF IMPERFECTIONS ON BUCKLING
OF SPHERICAL SHELLS. PROC.CONF.COMP.METHODS IN NONLINEAR
MECHANICS, AUSTIN, TEXAS, 1974.

2093 YAMAMOTO Y., AND TOKUDA N., A NOTE ON CONVERGENCE OF FINITE ELEMENT
SOLUTIONS. INT.J. NUMER.METH.ENG.3, 485-494, 1971.

2094 YAMAMOTO Y., AND TOKUDA N., DETERMINATION OF STRESS INTENSITY
FACTORS IN CRACKED PLATES BY THE FINITE ELEMENT METHOD.
INT.J.NUMER.METH.ENG.6, 427-439, 1973.

2095 YANG H.T., A FINITE ELEMENT STRESS ANALYSIS OF THE VERTICAL BUTTRESSES
OF A NUCLEAR CONTAINMENT VESSEL. NUCLEAR ENG. DESIGN 11, 255-268,
1969.

2096 YANG H.T., FLEXIBLE PLATE FINITE ELEMENT ON AN ELASTIC FOUNDATION.
J.STRUCTURAL DIV. A.S.C.E. 96, ST10, 2083-2091, 1970.

2097 YANG T.Y., ELASTIC SNAP THROUGH ANALYSIS OF CURVED PLATES USING DISCRETE
ELEMENTS. AIAA J. 10, 371-372, 1972.

2098 YEO M.F., A MORE EFFICIENT FRONT SOLUTION: ALLOCATING ASSEMBLY LOCATIONS
BY LONGEVITY CONSIDERATIONS. INT.J.NUMER.METH.ENG.7, 570-573,1973.

2099 YETTRAM A.L., AND HIRST M.J.S., THE SOLUTION OF STRUCTURAL
EQUILIBRIUM EQUATIONS BY THE CONJUGATE GRADIENT METHOD WITH
PARTICULAR REFERENCE TO PLANE STRESS ANALYSIS. INT.J.NUMER.
METH.ENG.3, 349-360,1971.

2100 YETTRAM A.L., AND HUSAIN H.M., GENERALISED MATRIX FORCE AND
DISPLACEMENT METHODS FOR LINEAR STRUCTURAL ANALYSIS.
AIAA J.3, 1154-1156, 1965.

2101 YETTRAM A.L., AND HUSAIN H.M., PLANE FRAMEWORK METHODS FOR PLATES
IN EXTENSION. J.ENG.MECH.DIV ASCE 92, 157-168, 1966.

2102 YIN F.C.P., INTERFACE CORE DISCRETE ELEMENT STIFFENED SHELLS OF
REVOLUTION. AIAA J.5, 2270-2273, 1967.

2103 YOKOUCHI Y., YAMADA Y., AND SANBONGI S., FINITE DIFFERENCE SOLUTIONS FOR
LARGE DEFORMATIONS OF CYLINDRICAL SHELLS; A COMPARISON WITH FINITE
ELEMENT SOLUTIONS. PP.107-126 OF J.T. ODEN, R.W. CLOUGH AND Y.
YAMOMOTO (EDS.), ADVANCES IN COMPUTATIONAL METHODS IN STRUCTURAL
MECHANICS AND DESIGN. UNIVERSITY OF ALABAMA PRESS, HUNTSVILLE,
1972.

2104 YOSHIDA Y., EQUIVALENT FINITE ELEMENTS ON DIFFERENT BASES. PP. 133-149
OF J.T. ODEN, R.W. CLOUGH AND Y. YAMOMOTO (EDS.), ADVANCES IN
COMPUTATIONAL METHODS IN STRUCTURAL MECHANICS AND DESIGN,
UNIVERSITY OF ALABAMA PRESS, HUNTSVILLE, 1972.

2105 YOSIDA K., FUNCTIONAL ANALYSIS. ACADEMIC PRESS, NEW YORK, 1965.

2106 YOUNG R.C. AND MOTE C.D., SOLUTION OF MIXED BOUNDARY VALUE
VALUE PROBLEMS WITH LOCAL ERROR BOUND BY THE FINITE
ELEMENT METHOD. COMP.METH.APPL.MECH.ENG.2, 159-184, 1973.

2107 ZANGWILL W., NONLINEAR PROGRAMMING, A UNIFIED APPROACH. PRENTICE
HILL, NEW JERSEY, 1969.

2108 ZARANTONELLO E.H., CONTRIBUTIONS TO NONLINEAR FUNCTIONAL ANALYSIS.
ACADEMIC PRESS, NEW YORK, 1971.

2109 ZEHLEIN H., ASSESSMENT OF COMMERCIALLY AVAILABLE FINITE ELEMENT
CODES FOR THE DYNAMIC ANALYSIS OF STRUCTURES. REPORT
KERNFORSCHUNGSZENTRUM, INSTITUT FUER REAKTORENTWICKLUNG,
KARLSRUHE, 1973.

2110 ZENISEK A., THE CONVERGENCE OF THE FINITE ELEMENT METHOD FOR

BOUNDARY VALUE PROBLEMS OF SYSTEMS OF ELLIPTIC EQUATIONS
(IN CZECH) APLIKACE MAT.14, 355-377, 1969.

2111 ZENISEK A., INTERPOLATION POLYNOMIALS ON THE TRIANGLE.
NUMER.MATH.15, 283-296, 1970.

2112 ZENISEK A., INTERPOLATION POLYNOMIALS ON THE TRIANGLE AND ON THE
TETRAHEDRON AND THE FINITE ELEMENT METHOD. TO APPEAR.

2113 ZENISEK A. AND ZLAMAL M., CONVERGENCE OF A FINITE ELEMENT PROCEDURE
FOR SOLVING BOUNDARY VALUE PROBLEMS OF THE FOURTH ORDER.
INT.J.NUMER.METH.ENG.2,307-310,1971.

2114 ZIENKIEWICZ O.C., ELASTIC TORSIONAL ANALYSIS OF IRREGULAR SHAPES.
J. ENG. MECH. DIV. A.S.C.E. 92,EM4, 78-79, 1966.

2115 ZIENKIEWICZ O.C., THE FINITE ELEMENT METHOD: FROM INTUITION TO
GENERALITY. APPLIED MECHANICS REVIEWS 23, 249-256, 1970.

2116 ZIENKIEWICZ O.C., THE FINITE ELEMENT METHOD IN ENGINEERING SCIENCE,
(2ND ED.), MCGRAW HILL, NEW YORK, 1971.

2117 ZIENKIEWICZ O.C., FINITE ELEMENTS - THE BACKGROUND STORY. PP.
1-35 OF J.R. WHITEMAN (ED.), THE MATHEMATICS OF FINITE
ELEMENTS AND APPLICATIONS. ACADEMIC PRESS, LONDON, 1973.

2118 ZIENKIEWICZ O.C., ISOPARAMETRIC ELEMENT FORMS IN FINITE ELEMENT ANALYSIS.
PP.379-414 OF J.T. ODEN AND E.R. DE A. OLIVEIRA (EDS.), LECTURES ON
FINITE ELEMENT METHODS IN CONTINUUM MECHANICS. UNIVERSITY OF
ALABAMA PRESS, HUNTSVILLE, 1973.

2119 ZIENKIEWICZ O.C., SOME LINEAR AND NONLINEAR PROBLEMS IN FLUID
MECHANICS. FINITE ELEMENT METHOD FORMULATION AND SOLUTION.
NATO SYMPOSIUM, KJELLER, NORWAY, 1973.

2120 ZIENKIEWICZ O.C., CONSTRAINED VARIATIONAL PRINCIPLES AND PENALTY
FUNCTION METHODS IN FINITE ELEMENT ANALYSIS. PP.207-214 OF G.A.
WATSON (ED.), PROCEEDINGS OF CONF. ON NUMERICAL SOLUTION OF
DIFFERENTIAL EQUATIONS. LECTURE NOTES IN MATHEMATICS, NO.363.
SPRINGER-VERLAG,BERLIN,1974.

2121 ZIENKIEWICZ O.C., VISCOPLASTICITY, PLASTICITY AND PLASTIC FLOW.
PROC.CONF.COMP.METHODS IN NONLINEAR MECH., AUSTIN, TEXAS,
1974.

2122 ZIENKIEWICZ O.C., WHY FINITE ELEMENTS. IN VOL.1 OF R.H.GALLAGHER,
J.T.ODEN. C.TAYLOR, AND O.C. ZIENKIEWICS (EDS..), FINITE
ELEMENTS IN FLOW PROBLEMS. WILEY, LONDON, 1974.

2123 ZIENKIEWICZ O.C., ARLETT P.L. AND BAHRANI A.K., SOLUTION OF THREE
DIMENSIONAL FIELD PROBLEMS BY THE FINITE ELEMENT METHOD.
ENGINEER 224, 547-550, 1967.

2124 ZIENKIEWICZ O.C., AND CHEUNG Y.K., THE FINITE ELEMENT METHOD FOR
ANALYSIS OF ELASTIC ISOTROPIC AND ORTHOTROPIC SLABS.
PROC.INST.CIV.ENG.28, 471-488,1964.

2125 ZIENKIEWICZ O.C., AND CHEUNG Y.K., FINITE ELEMENTS IN THE SOLUTION
OF FIELD PROBLEMS. ENGINEER 200, 507-510, 1965.

2126 ZIENKIEWICZ O.C., AND CHEUNG Y.K., FINITE ELEMENT METHOD OF
ANALYSIS OF ARCH DAMS AND COMPARISON WITH
FINITE-DIFFERENCE PROCEDURES. PROC.SYMP.ON THEORY OF
ARCH DAMS, PERGAMON PRESS, OXFORD, 1965.

2127 ZIENKIEWICZ O.C., AND CHEUNG Y.K., THE FINITE ELEMENT METHOD IN
STRUCTURAL AND CONTINUUM MECHANICS. MCGRAW HILL, LONDON 1967.

2128 ZIENKIEWICZ O.C., CHEUNG Y.K., AND STAGG K.G., STRESSES IN ANISOTROPIC
MEDIA OF ROCK MECHANICS. J. STRAIN ANALYSIS 1, 172-182, 1966.

2129 ZIENKIEWICZ O.C., AND CORMEAU I.C., VISCO-PLASTICITY PLASTICITY AND
CREEP IN PLASTIC SOLIDS. A UNIFIED NUMERICAL SOLUTION
APPROACH. INT.J.NUMER.METH.ENG.8, 821-845, 1974.

2130 ZIENKIEWICZ O.C. AND GODBOLE P.N., FLOW OF PLASTIC AND VISCO-PLASTIC
SOLIDS WITH SPECIAL REFERENCE TO EXTRUSION AND FORMING PROCESSES.
INT.J.NUMER.METH.ENG.8, 3-16,1974.

2131 ZIENKIEWICZ O.C., AND HOLISTER G.S. (EDS.), STRESS ANALYSIS.
WILEY, NEW YORK, 1965.

2132 ZIENKIEWICZ O.C., IRONS B.M., ERGATOUDIS J., AND AHMAD S., ISOPARAMETRIC
AND ASSOCIATED ELEMENT FAMILIES FOR TWO AND THREE DIMENSIONAL
ANALYSIS. FINITE ELEMENT ANALYSIS, TAPIR, TRONDHEIM, 1969.

2133 ZIENKIEWICZ O.C., IRONS B., AND NATH P., NATURAL FREQUENCIES OF
COMPLEX FREE OR SUBMERGED STRUCTURES BY THE FINITE ELEMENT
METHOD. SYMP.ON VIBRATION IN CIV.ENG.,INST.CIV.ENG.,

BUTTERWORTH,LONDON, 1965.

2134 ZIENKIEWICZ O.C., AND LEWIS R.W., AN ANALYSIS OF VARIOUS TIME-STEPPING
      SCHEMES FOR INITIAL VALUE PROBLEMS. INT.J.EARTHQUAKE ENG.STRUCT.
      DYN.1, 407-408, 1973.

2135 ZIENKIEWICZ O.C., MAYER P., AND CHEUNG Y.K., SOLUTION OF ANISOTROPIC
      SEEPAGE PROBLEMS BY FINITE ELEMENTS. J.ENG.MECH.DIV.PROC.
      ASCE 92, 111-120, 1966.

2136 ZIENKIEWICZ O.C., AND NAYAK G.C., A GENERAL APPROACH TO PROBLEMS OF
      LARGE DEFORMATION AND PLASTICITY USING ISOPARAMETRIC ELEMENTS.
      PROC.3RD CONF. MATRIX METHODS IN STRUCTURAL MECHANICS, WRIGHT-
      PATTERSON AFB.,OHIO, 1971.

2137 ZIENKIEWICZ O.C., AND NAYLOR D.J., FINITE ELEMENT STUDIES OF SOILS AND
      POROUS MEDIA. PP. 459-494 OF J.T. ODEN AND E.R. DE A. OLIVEIRA
      (EDS.), LECTURES ON FINITE ELEMENT METHODS IN CONTINUUM MECHANICS.
      UNIVERSITY OF ALABAMA PRESS, HUNTSVILLE, 1973.

2138 ZIENKIEWICZ O.C., AND NEWTON R.E., COUPLED VIBRATIONS OF A STRUCTURE
      SUBMERGED IN A COMPRESSIBLE FLUID. INT.SYMP.ON FINITE
      ELEMENT TECHNIQUES IN SHIPBUILDING, STUTTGART,1969.

2139 ZIENKIEWICZ O.C., OWEN D.R.J., AND LEE K.N., LEAST SQUARE-FINITE ELEMENT
      FOR ELASTO-STATIC PROBLEMS. USE OF REDUCED INTEGRATION. INT.J.
      NUMER.METH.ENG.8, 341-358, 1974.

2140 ZIENKIEWICZ O.C., AND PAREKH C.J., TRANSIENT FIELD PROBLEMS - TWO
      AND THREE DIMENSIONAL ANALYSIS BY ISOPARAMETRIC FINITE ELEMENTS.
      INT.J.NUMER.METH.ENG.2,61-71,1970.

2141 ZIENKIEWICZ O.C., AND PHILLIPS D.V., AN AUTOMATIC MESH GENERATION
      SCHEME FOR PLANE AND CURVED SURFACE BY "ISOPARAMETRIC"
      CO-ORDINATES. INT.J.NUMER.METH.ENG.3,519-528,1970.

2142 ZIENKIEWICZ O.C., TAYLOR R.L., AND TOO J.M., REDUCED INTEGRATION
      TECHNIQUE IN GENERAL ANALYSIS OF PLATES AND SHELLS.

2143 ZIENKIEWICZ O.C., AND VALLIAPPAN S., ANALYSIS OF REAL STRUCTURES
      OF CREEP PLASTICITY AND OTHER COMPLEX CONSTITUTIVE LAWS.
      CONF.ON MATERIALS IN CIV.ENG.UNIV.OF SOUTHAMPTON 1969.
      WILEY, LONDON, 1970.

2144 ZIENKIEWICZ O.C., VALLIAPPAN S., AND KING I.P., STRESS ANALYSIS
      OF ROCK AS A N- TENSION METHOD. GEOTECHNIQUE 18, 55-66.

2145 ZIENKIEWICZ O.C., VALLIAPAN S., AND KING I.P., ELASTO-PLASTIC
      SOLUTIONS OF ENGINEERING PROBLEMS. INT.J.NUMER.METH.ENG.
      1, 75-100, 1969.

2146 ZIENKIEWICZ O.C., WATSON M., AND CHEUNG Y.K., STRESS ANALYSIS BY THE
      FINITE ELEMENT METHOD: THERMAL EFFECTS. PP. 357-362 OF M.S.
      UDALL (ED.), PRESTRESSED CONCRETE PRESSURE VESSELS. INST. OF
      CIVIL ENGINEERS, LONDON, 1968.

2147 ZIENKIEWICZ O.C., WATSON M., AND KING I.P., A NUMERICAL METHOD
      OF VISCO-ELASTIC STRESS ANALYSIS. INT.J.MECH.SCI.10,
      807-827, 1968.

2148 ZLAMAL M., ON THE FINITE ELEMENT METHOD. NUMER.MATH.12, 394-409,
      1968.

2149 ZLAMAL M., ON SOME FINITE ELEMENT PROCEDURES FOR SOLVING SECOND
      ORDER BOUNDARY VALUE PROBLEMS. NUMER.MATH.14,42-48, 1969.

2150 ZLAMAL M., A FINITE ELEMENT PROCEDURE OF SECOND ORDER ACCURACY.
      NUMERISCHE MATH.14, 394-402 1970.

2151 ZLAMAL M., CURVED ELEMENTS IN THE FINITE ELEMENT METHOD, I.
      SIAM. J. NUMER. ANAL. 10, 229-240,1973.

2152 ZLAMAL M., THE FINITE ELEMENT METHOD IN DOMAINS WITH CURVED BOUNDARIES.
      INT.J.NUMER.METH.ENG.5, 367-373, 1973.

2153 ZLAMAL M., SOME RECENT ADVANCES IN THE MATHEMATICS OF FINITE
      ELEMENTS. PP. 59-81 OF J.R. WHITEMAN (ED.), THE MATHEMATICS
      OF FINITE ELEMENTS AND APPLICATIONS. ACADEMIC PRESS, LONDON,
      1973.

2154 ZLAMAL M., CURVED ELEMENTS IN THE FINITE ELEMENT METHOD, II. SIAM J.
      NUMER. ANAL.11, 347-362, 1974.

2155 ZLAMAL M., FINITE ELEMENT METHODS FOR PARABOLIC EQUATIONS. PP.215-221
      OF G.A. WATSON (ED.), PROCEEDINGS OF CONF. ON NUMERICAL SOLUTION
      OF DIFFERENTIAL EQUATIONS. LECTURE NOTES IN MATHEMATICS, NO.363.
      SPRINGER-VERLAG,BERLIN,1974.

2156 ZLAMAL M., FINITE ELEMENT METHODS FOR PARABOLIC EQUATIONS.

MATH.COMP.28, 393-404, 1974.

2157 ZLAMAL M., UNCONDITIONALLY STABLE FINITE ELEMENT SCHEMS FOR
          PARABOLIC EQUATIONS. PROC.2ND INT.CONFERENCE ON
          NUMERICAL ANALYSIS, DUBLIN, 1974.

2158 ZUDANS Z., ANALYSES OF ASYMMETRIC STIFFENED SHELL TYPE STRUCTURES BY THE
          FINITE ELEMENT METHOD; I FLAT RECTANGULAR ELEMENTS, NUCL.ENG.
          DESIGN 8, 367-379, 1968.

2159 ZUDANS Z., FINITE ELEMENT INCREMENTAL ELASTIC-PLASTIC ANALYSIS OF PRESSURE
          VESSELS. J.ENG.IND.92, 293-302, 1970.

2160 ZUDANS Z., REDDI M.M., FISHMAN H.M., AND GRAY D., THREE-DIMENSIONAL
          FINITE ELEMENT COMPUTER CODE FOR THE ANALYSIS OF COMPLEX
          STRUCTURES. NUCL.ENG. AND DESIGN 20, 149-167, 1972.

2161 ANON ANON, PROCEEDINGS 1ST CONFERENCE MATRIX METHODS IN STRUCTURAL
          MECHANICS, WRIGHT-PATTERSON AFB, OHIO, AFFDL TR66-80, 1965.

2162 ANON ANON, PROCEEDINGS 2ND CONFERENCE MATRIX METHODS IN STRUCTURAL
          MECHANICS, WRIGHT-PATTERSON AFB, OHIO, AFFDL-TR-68-150,1968.

2163 ANON ANON, PROCEEDINGS 3RD CONFERENCE MATRIX METHODS IN STRUCTURAL
          MECHANICS, WRIGHT-PATTERSON AFB, OHIO, 1971.

2164 ANON ANON, PROCEEDINGS CONFERENCE APPLICATION OF FINITE ELEMENT
          METHODS TO STRESS ANALYSIS PROBLEMS IN NUCLEAR ENGINEERING,
          ISPRA, ITALY, 1971.

2165 ANON ANON, THEORIE DES EQUIVALENCES APPLICATION AU GENIE CIVIL
          CEBTP,PARIS,1972.

2166 ANON ANON, FINITE ELEMENTE IN DER STATIK. WILHELM ERNST,BERLIN,1973.

Key Word Listing

AUTOMATIC

No.	Entry
1453	BERGEN AN AUTOMATIC FINITE ELEMENT MESH GENERATION PROGRAM FOR ARBITRARY ST
658	AUTOMATIC FINITE ELEMENT MESH GENERATION.
1274	MESHGEN, A COMPUTER CODE FOR AUTOMATIC GENERATION OF FINITE ELEMENT MATRICES.
1609	AN AUTOMATIC GENERATION SCHEME FOR PLANE AND CURVED SURFACES BY ISOP
1878	AN AUTOMATIC MESH GENERATION SCHEME FOR FINITE ELEMENT ANALYSIS.
740	TWO DIMENSIONAL AUTOMATIC MESH GENERATION FOR STRUCTURAL ANALYSIS.
1069	AUTOMATIC MESH GENERATION IN TWO AND THREE DIMENSIONAL INTERCONNE
2141	AN AUTOMATIC MESH GENERATION SCHEME FOR PLANE AND CURVED SURFACE BY
1852	OF AUTOMATIC MESH GENERATION SCHEMES BY ISOPARAMETRIC CO-ORDINATES.
40	RESTRICTIONS IN THE APPLICATION OF AUTOMATIC NODE-RELABELLING SCHEME FOR BANDWIDTH MINIMIZATION OF S
503	AN AUTOMATIC RENUMBERING.
424	BANDWIDTH REDUCTION BY AUTOMATIC TRIANGULATION OF ARBITRARY PLANAR DOMAINS FOR THE FINIT
397	AUTOMATIC TRIANGULATION OF TWO DIMENSIONAL REGIONS.

AXI-SYMMETRIC

No.	Entry
900	A CONICAL ELEMENT FOR DISPLACEMENT ANALYSIS OF AXI-SYMMETRIC SHELLS.
818	A CURVED ELEMENT APPROXIMATION IN THE ANALYSIS OF AXI-SYMMETRIC THIN SHELLS

AXISYMMETRIC

No.	Entry
2074	PLASTIC ANALYSIS OF SHELLS OF REVOLUTION SUBJECTED TO AXISYMMETRIC AND ASYMMETRIC LOADING.
853	CREEP ANALYSIS OF AXISYMMETRIC BODIES USING FINITE ELEMENTS.
1143	APPLICATION OF FINITE ELEMENT METHOD TO AXISYMMETRIC BUCKLING OF SHALLOW SPHERICAL SHELLS UNDER EXTERNAL
1681	ANALYSIS OF AXISYMMETRIC COMPOSITE STRUCTURES BY THE FINITE ELEMENT METHOD.
1295	W TRIANGULAR ELEMENT FOR FINITE DIFFERENCE SOLUTION OF AXISYMMETRIC CONDUCTION PROBLEMS IN CYLINDRICAL CO-ORDINATES.
1541	NUMERICAL ANALYSIS OF FINITE AXISYMMETRIC DEFORMATIONS OF INCOMPRESSIBLE ELASTIC SOLIDS OF REV
854	PURE MOMENT LOADING OF AXISYMMETRIC FINITE ELEMENT MODELS.
1712	HICK-WALLED PRESSURE VESSELS USING THE CONSTANT STRAIN AXISYMMETRIC FINITE ELEMENT.
1755	O GENERATE FINITE ELEMENT MESH FOR PROBLEMS OF COMPLEX AXISYMMETRIC GEOMETRY.
1756	GENERATE A FINITE ELEMENT MESH FOR PROBLEMS OF COMPLEX AXISYMMETRIC GEOMETRY.
2089	INCREMENTAL SOLUTION OF AXISYMMETRIC PLATE AND SHELL FINITE DEFORMATION.
29	ELL AND MEMBRANE ELEMENTS WITH PARTICULAR REFERENCE TO AXISYMMETRIC PROBLEMS.
1859	FINITE ELEMENT ANALYSIS OF AXISYMMETRIC ROTORS.
1163	A FINITE ELEMENT PROGRAM PACKAGE FOR AXISYMMETRIC SCALAR FIELD PROBLEMS.
91	IMPROVED LINEAR AXISYMMETRIC SHELL FLUID MODEL FOR LAUNCH VEHICLE LONGITUDAL RESP
851	ANALYSIS OF AXISYMMETRIC SHELLS BY THE DIRECT STIFFNESS METHOD.
831	ANALYSIS OF AXISYMMETRIC SHELLS UNDER ARBITRARY TRANSIENT PRESSURES
666	MIXED FINITE ELEMENT METHOD FOR AXISYMMETRIC SHELLS.
917	DYNAMIC ANALYSIS OF AXISYMMETRIC SOLIDS BY THE FINITE ELEMENT METHOD.
911	NONLINEAR DYNAMIC ANALYSIS OF AXISYMMETRIC SOLIDS WITH MATERIAL AND GEOMETRIC NONLINEARITIES BY
1582	STATIC STRESS ANALYSIS OF AXISYMMETRIC SOLIDS.
1121	FINITE ELEMENT PROCEDURE FOR CONDUCTION IN ANISOTROPIC AXISYMMETRIC SOLIDS.
494	ROCEDURE FOR THE LARGE DEFORMATION DYNAMIC RESPONSE OF AXISYMMETRIC SOLIDS.
870	FINITE ELEMENTS FOR THE REPRESENTATION OF PLANE AND AXISYMMETRIC STRESS ANALYSIS BY FINITE ELEMENT AND FINITE DIFFERE
404	DYNAMIC ELASTIC-PLASTIC ANALYSIS OF AXISYMMETRIC STRUCTURAL REINFORCEMENT.
1435	FINITE ELEMENT STRESS ANALYSIS OF PLANE AND AXISYMMETRIC STRUCTURES BY FINITE ELEMENT METHOD COMPUTER CODE.
405	HYDROELASTIC ANALYSIS OF AXISYMMETRIC STRUCTURES.
882	ESS HYBRID FINITE ELEMENT MODEL FOR THE ANALYSIS OF AN AXISYMMETRIC SYSTEMS BY A FINITE ELEMENT METHOD.
83	FINITE ELEMENT ANALYSIS OF AXISYMMETRIC THERMOVISCOELASTIC
1817	AXISYMMETRIC THICK-WALLED PRESSURE VESSEL.
1669	AXISYMMETRIC TRIANGULAR FINITE ELEMENTS FOR THE SCALAR HELMHOLTZ
	AXISYMMETRIC VIBRATIONS OF LINEARLY TAPERED ANNULAR PLATES.

BANACH

No.	Entry
1837	PROXIMATE METHODS OF SOLVING DIFFERENTIAL EQUATIONS IN BANACH SPACES.
1648	LAGRANGE AND HERMITE INTERPOLATION IN BANACH SPACES.
1178	SCALES OF BANACH SPACES.
380	IATIONAL INEQUALITIES AND MAXIMAL MONOTONE MAPPINGS IN BANACH SPACES.

No.	Entry
263	ON THE QUINTIC TRIANGULAR PLATE BENDING ELEMENT.
1088	AND CONSTANT THICKNESS HIGH PRECISION TRIANGULAR PLATE BENDING ELEMENTS FOR PLATE AND SHELL NETWORKS.
1985	TRIANGULAR THICK PLATE BENDING ELEMENTS IN THE ANALYSIS OF VARIABLE THICKNESS PLATES.
1286	TRIANGULAR PLATE BENDING ELEMENTS PAPER M6/5, PROCEEDINGS 1ST STRUCTURAL MECH.
913	CONFORMING RECTANGULAR AND TRIANGULAR PLATE BENDING ELEMENTS WITH ENFORCED COMPATIBILITY.
1643	STUDIES OF EIGENVALUE SOLUTIONS USING TWO FINITE PLATE BENDING ELEMENTS.
1246	ALGORITHM FOR THE STIFFNESS MATRIX OF TRIANGULAR PLATE BENDING ELEMENTS.
1176	STIFFNESS MATRICES FOR PLATE BENDING ELEMENTS.
973	A NEW FORMULATION FOR PLATE BENDING ELEMENTS.
1023	OMETRIC FUNCTION REPRESENTATIONS FOR RECTANGULAR PLATE BENDING ELEMENTS.
431	TRIANGULAR PLATE BENDING EQUILIBRIUM ANALYSIS.
262	FINITE ELEMENT PLATE BENDING FINITE ELEMENT.
63	A COMPATIBLE TRIANGULAR PLATE BENDING FINITE ELEMENT.
392	A REFINED TRIANGULAR PLATE BENDING FINITE ELEMENTS.
261	TECHNIQUE FOR MESH GRADING APPLIED TO CONFORMING PLATE BENDING FINITE ELEMENTS.
1840	SHEAR IN C0 AND C1 PLATE BENDING FINITE ELEMENTS.
759	ENERGY BALANCING TECHNIQUE IN THE GENERATING OF PLATE BENDING FINITE ELEMENTS.
761	THE CONSTANT BENDING MOMENT PLATE BENDING ELEMENT.
1413	A TRIANGULAR EQUILIBRIUM ELEMENT WITH LINEARLY VARYING BENDING MOMENTS FOR PLATE BENDING PROBLEMS.
1408	S FOR PLATE BENDING WITH CONSTANT AND LINEARLY VARYING BENDING MOMENTS.
50	ON THE THEORY OF BENDING OF CIRCULAR PLATES OF HARDENING MATERIAL.
1641	BENDING OF ELASTIC PLATES.
1698	BENDING OF MICROPOLAR PLATES.
1164	BENDING OF MICROPOLAR PLATES.
834	FINITE ELEMENT APPROACH TO BENDING PROBLEMS BY SIMPLIFIED HYBRID DISPLACEMENT METHOD.
1134	A POLYGONAL FINITE ELEMENT FOR PLATE BENDING PROBLEMS USING THE ASSUMED STRESS APPROACH.
53	RIANGULAR EQUILIBRIUM ELEMENT IN THE SOLUTION OF PLATE BENDING PROBLEMS.
1409	LEMENT WITH LINEARLY VARYING BENDING MOMENTS FOR PLATE BENDING PROBLEMS.
1498	ON THE CONVERGENCE OF A MIXED FINITE ELEMENT FOR PLATE BENDING PROBLEMS.
1058	PIECEWISE RICUBIC METHODS FOR PLATE BENDING PROBLEMS.
994	A CONFORMING TRIANGULAR FINITE ELEMENT PLATE BENDING SOLUTION.
65	TRIANGULAR FINITE ELEMENTS FOR PLATE BENDING USING TRIANGULAR ELEMENTS.
1936	ANALYSIS OF TRIANGULAR FINITE ELEMENT BENDING WITH CONSTANT AND LINEARLY VARYING BENDING MOMENTS.
50	FINITE ELEMENT ANALYSIS OF PLATE BENDING WITH TRANSVERSE SHEAR DEFORMATION.
868	ELEMENT ANALYSIS FOR MODERATELY RECTANGULAR PLATES IN BENDING.
1826	LEMENT ANALYSIS OF THIN RECTANGULAR AND SKEW PLATES IN BENDING.
1827	AN ANNULAR SEGMENT ELEMENT FOR PLATE BENDING.
1750	ANNULAR AND CIRCULAR SECTOR FINITE ELEMENTS FOR PLATE BENDING.
1573	EQUIVALENT MASS MATRICES FOR RECTANGULAR PLATES IN BENDING.
1658	STIFFNESS MATRIX FOR A TRIANGULAR SANDWICH ELEMENT IN BENDING.
1322	A STIFFNESS MATRIX FOR THE ANALYSIS OF THIN PLATES IN BENDING.
1359	A STIFFNESS MATRIX FOR THE ANALYSIS OF THIN PLATES IN BENDING.
1343	FINITE ELEMENT ANALYSIS OF SKEW PLATES IN BENDING.
1405	TWISTED BEAM ELEMENT MATRICES FOR PLATE BENDING.
1208	CONVERGENCE OF A MIXED FINITE ELEMENT SCHEME FOR PLATE BENDING.
1141	MULTIANGULAR ELEMENTS IN PLATE BENDING.
1152	A CONFORMING QUARTIC TRIANGULAR ELEMENT FOR PLATE BENDING.
1018	SIMPLE FINITE ELEMENT MODEL FOR ELASTIC PLASTIC PLATE BENDING.
963	DISTRIBUTED MASS MATRIX FOR PLATE ELEMENT BENDING.
1015	OF NODAL CONNECTIONS IN A STIFFNESS SOLUTION FOR PLATE BENDING.
1025	DISTRIBUTED MASS MATRIX FOR PLATE ELEMENTS IN BENDING.
880	DISTRIBUTED MASS MATRIX FOR PLATE ELEMENTS IN BENDING.
736	AN EQUILIBRIUM MODEL FOR PLATE BENDING.

COMPLEMENTARY VARIATIONAL PRINCIPLES FOR STEADY HEAT CONDUCTION WITH MIXED BOUNDARY CONDITIONS.   71
ON VARIATIONAL APPROACHES TO STEADY STATE HEAT CONDUCTION.   1839
ERGENCE OF THE FINITE ELEMENT METHOD IN NONLINEAR HEAT CONDUCTION.   1337
A MINIMUM PRINCIPLE FOR HEAT CONDUCTION.   391

A NUMERICAL CONFORMAL TRANSFORMATION METHOD FOR HARMONIC MIXED BOUNDARY VALUE   1587
CONFORMAL TRANSFORMATION METHODS FOR THE NUMERICAL SOLUTION OF HA   2056
TREATMENT OF HARMONIC MIXED BOUNDARY PROBLEMS BY CONFORMAL TRANSFORMATION METHODS.   2057

CONFORMING AND NONCONFORMING FINITE ELEMENT METHODS FOR SOLVING T   536
CONFORMING AND NONCONFORMING FINITE ELEMENT METHODS FOR SOLVING T   465
CONFORMING DISPLACEMENT FINITE ELEMENTS.   1285
CONFORMING FINITE ELEMENT FOR PLATE BENDING.   728
CONFORMING FINITE ELEMENT METHODS FOR THE CLAMPED PLATE PROBLEM.   2053
TE BOTH UPPER AND LOWER BOUNDS TO PLATE EIGENVALUES BY A CONFORMING PLATE BENDING FINITE ELEMENT.   1840
A TECHNIQUE FOR MESH GRADING APPLIED TO CONFORMING QUARTIC TRIANGULAR ELEMENT FOR PLATE BENDING.   1018
A CONFORMING RECTANGULAR AND TRIANGULAR PLATE BENDING ELEMENTS.   1643
A CONFORMING TRIANGULAR FINITE ELEMENT PLATE BENDING SOLUTION.   65

A CONICAL ELEMENT FOR DISPLACEMENT ANALYSIS OF AXI-SYMMETRIC SHELLS   900
BOUNDARY PROBLEMS FOR ELLIPTIC EQUATIONS WITH CONICAL OR ANGULAR POINTS.   1160

GENERALIZED CONJUGATE APPROXIMATION FUNCTIONS IN FINITE ELEMENT ANALYSIS.   366
CONJUGATE FUNCTIONS FOR MIXED FINITE ELEMENT APPROXIMATIONS OF BO   1522
HE SOLUTION OF STRUCTURAL EQUILIBRIUM EQUATIONS BY THE CONJUGATE GRADIENT METHOD WITH PARTICULAR REFERENCE TO PLANE STRE   2099
LINEAR ANALYSIS OF STRUCTURES BY THE METHOD OF CONJUGATE GRADIENTS.   269
THEORY OF CONJUGATE PROJECTIONS IN FINITE ELEMENT ANALYSIS.   1530
ON A CONJUGATE SEMI-VARIATIONAL METHOD FOR PARABOLIC EQUATIONS.   961

CONTINUA AND DISCONTINUA.   95
THE ITERATIVE APPROACH IN ELASTIC PLASTIC ANALYSIS OF CONTINUA.   2091
EW DISPLACEMENT HYBRID FINITE ELEMENT MODELS FOR SOLID CONTINUA.   1942
BASIS OF FINITE ELEMENT METHODS FOR SOLID CONTINUA.   1621
FORMULATIONS OF FINITE ELEMENT METHODS FOR SOLID CONTINUA.   1614
VARIATIONAL FORMULATIONS OF NUMERICAL METHODS IN SOLID CONTINUA.   1615
FINITE ELEMENTS OF NONLINEAR CONTINUA.   1521
FORMATION AND IRREVERSIBLE THERMODYNAMICS OF NONLINEAR CONTINUA.   1518
FINITE ELEMENTS FOR COMPRESSIBLE AND INCOMPRESSIBLE CONTINUA.   992
STIC MATRIX DISPLACEMENT ANALYSIS OF THREE-DIMENSIONAL CONTINUA.   96

A CONTINUUM APPROACH TOWARD DYNAMICS OF GRIDWORKS.   1879
CONTINUUM MECHANICS AS AN APPROACH TO ROCK MASS PROBLEMS.   1846
APPLICATION OF FINITE ELEMENT METHOD TO CONTINUUM MECHANICS.   438
THE FINITE ELEMENT METHOD IN STRUCTURAL AND CONTINUUM MECHANICS.   2127
ATIONAL PRINCIPLE FOR LINEAR COUPLED FIELD PROBLEMS IN CONTINUUM MECHANICS.   1742
IPLES FOR BOUNDARY VALUE AND INITIAL VALUE PROBLEMS IN CONTINUUM MECHANICS.   1743
FINITE ELEMENT METHODS IN CONTINUUM MECHANICS.   1623
LECTURES ON FINITE ELEMENT METHODS IN CONTINUUM MECHANICS.   1520
VARIATIONAL PRINCIPLES IN NONLINEAR CONTINUUM MECHANICS.   1524
LECTURES ON FINITE ELEMENT METHODS IN CONTINUUM MECHANICS.   1548
IN CONTINUUM MECHANICS.   326
MENT ELASTIC PLASTIC CREEP ANALYSIS OF TWO DIMENSIONAL CONTINUUM WITH TEMPERATURE DEPENDANT MATERIAL PROPERTIES.   540
REVIEW OF CONTINUUM, FINITE ELEMENT AND HYBRID TECHNIQUES IN THE ANALYSIS O   1677

1073   MESH AND CONTOUR PLOT FOR TRIANGLE AND ISOPARAMETRIC ELEMENTS.
1264   MESH GENERATION CODE AND A CONTOUR PLOTTING ROUTINE.

1255   THE RITZ-GALERKIN PROCEDURE FOR NONLINEAR OPTIMAL CONTROL OF SYSTEMS GOVERNED BY PARTIAL DIFFERENTIAL EQUATIONS.
341   OPTIMAL CONTROL PROBLEMS.
340   L PROPERTIES OF THE RITZ-TREFFTZ ALGORITHM FOR OPTIMAL CONTROL.

280   RATE OF CONVERGENCE CRITERIA FOR ITERATIVE PROCESSES.
354   CONVERGENCE ESTIMATES FOR NON-SELFADJOINT EIGENVALUE APPROXIMATIO
1922   CONVERGENCE ESTIMATES FOR SEMI-DISCRETE GALERKIN METHODS FOR INIT
582   FINITE ELEMENT CONVERGENCE FOR SINGULAR DATA.
1793   METHODS OF GALERKIN TYPE ACHIEVING OPTIMUM L2 RATES OF CONVERGENCE FOR THE FINITE ELEMENT METHOD.
471   MAXIMUM PRINCIPLE AND UNIFORM CONVERGENCE FOR THE FINITE ELEMENT METHOD.
197   THE RATE OF CONVERGENCE FOR THE ITERATIVE APPROACH IN ELASTIC PLASTIC ANALYSI
2091   RATE OF CONVERGENCE IN FINITE ELEMENT METHODS.
463   ORDERS OF CONVERGENCE IN LINEAR SHELL PROBLEMS.
1397   FINITE ELEMENT METHOD OF MIXED TYPE AND ITS CONVERGENCE IN THE FINITE ELEMENT METHOD.
1562   COMPLETENESS AND CONVERGENCE IN THE THEORY OF SHELLS.
1568   THE ROLE OF A CONVERGENCE INVESTIGATION OF THE DIRECT STIFFNESS METHOD.
1119   CONVERGENCE OF A FINITE ELEMENT PROCEDURE FOR SOLVING BOUNDARY VA
2113   CONVERGENCE OF A FINITE ELEMENT SOLUTION OF LAPLACE'S EQUATION.
1695   CONVERGENCE OF A FINITE-ELEMENT METHOD BASED ON ASSUMED STRESS DI
1948   CONVERGENCE OF A GALERKIN METHOD TO SOLVE THE INITIAL PROBLEM OF
1685   ON THE CONSTRUCTION AND THE CONVERGENCE OF A MIXED FINITE ELEMENT SCHEME FOR PLATE BENDING.
1058   ON THE CONVERGENCE OF A MIXED FINITE ELEMENT SCHEME FOR PLATE BENDING.
1141   CONVERGENCE OF A RITZ APPROXIMATION FOR THE STEADY STATE HEAT FLO
894   CONVERGENCE OF A RITZ APPROXIMATION TO A PLANE STRAIN ELASTICITY
893   FOURTH ORDER CONVERGENCE OF BIVARIATE SPLINES.
1305   ON THE VARIATIONAL CHARACTERIZATION AND L-INFINITY CONVERGENCE OF COLLOCATION AND GALERKIN APPROXIMATIONS TO LINEAR
425   CONVERGENCE OF CONSISTENTLY DERIVED TIMOSHENKO BEAM FINITE ELEMEN
1469   CONVERGENCE OF DIFFERENCE SCHEMES IN CASE OF IMPROVED APPROXIMATI
1554   CONVERGENCE OF FINITE ELEMENT APPROXIMATION.
2013   A NOTE ON ACCURACY AND CONVERGENCE OF FINITE ELEMENT APPROXIMATIONS.
1535   ACCURACY AND CONVERGENCE OF FINITE ELEMENT GALERKIN APPROXIMATIONS OF TIME DEP
2031   THE CONVERGENCE OF FINITE ELEMENT METHOD IN SOLVING LINEAR ELASTIC PR
1619   ON THE CONVERGENCE OF FINITE ELEMENT SCHEMES BASED ON NON-CONFORMING AND
1142   CONVERGENCE OF FINITE ELEMENT SCHEMES BASED ON NONCONFORMING AND
1139   CONVERGENCE OF FINITE ELEMENT SOLUTIONS IN VISCOUS FLOW PROBLEMS.
1567   CONVERGENCE OF FINITE ELEMENT SOLUTIONS REPRESENTED BY A NON-CONF
1396   A NOTE ON THE CONVERGENCE OF FINITE ELEMENT SOLUTIONS.
2093   COMMENTS ON THE CONVERGENCE OF FINITE ELEMENT SOLUTIONS.
1429   EXPERIENCE WITH THE PATCH TEST FOR CONVERGENCE OF FINITE ELEMENTS.
1028   ON THE CONVERGENCE OF FINITE ELEMENTS.
1043   CONVERGENCE OF FINITE-DIFFERENCE APPROXIMATIONS TO ONE-DIMENSIONA
2058   CONVERGENCE OF FINITE-DIFFERENCE TECHNIQUES FOR A HARMONIC MIXED
1140   CONVERGENCE OF LUMPED FINITE ELEMENT SCHEMES FOR SELECTED INITIAL
1486   CONVERGENCE OF NONCONFORMING METHODS.
711   ON THE CONVERGENCE OF SOR ITERATIONS FOR FINITE ELEMENT APPROXIMATIONS T
654   ON THE RATE OF CONVERGENCE OF THE BUBNOV- GALERKIN METHOD.
2110   THE CONVERGENCE OF THE FINITE ELEMENT METHOD FOR BOUNDARY VALUE PROBL
1337   CONVERGENCE OF THE FINITE ELEMENT METHOD IN NONLINEAR HEAT CONDUC
1947   THE CONVERGENCE OF THE FINITE ELEMENT METHOD IN SOLVING LINEAR ELASTI
1059   THE CONVERGENCE OF THE FINITE ELEMENT METHOD IN THE THEORY OF ELASTIC

COMPUTER PREDICTION OF FATIGUE CRACK PROPAGATION UNDER RANDOM LOADING.
FINITE ELEMENTS FOR DETERMINATION OF CRACK TIP ELASTIC STRESS INTENSITY FACTORS.
CRACK TIP FINITE ELEMENTS ARE UNNECESSARY.
CALCULATION OF STRESS INTENSITY FACTORS AT CRACK TIPS USING SPECIAL FINITE ELEMENTS.

S STATE CRITERION FOR FAILURE IN THE VICINITY OF SHARP CRACKS.

CREEP ANALYSIS OF AXISYMMETRIC BODIES USING FINITE ELEMENTS.
CREEP ANALYSIS OF PIPELINES BY FINITE ELEMENTS.
CREEP ANALYSIS OF PIPELINES BY FINITE ELEMENTS.
ELASTIC-PLASTIC AND CREEP ANALYSIS OF TWO DIMENSIONAL CONTINUUM WITH TEMPERATURE DEPE
ELASTIC-PLASTIC AND CREEP ANALYSIS WITH THE ASKA PROGRAM SYSTEM.
FINITE ELEMENT ELASTIC PLASTIC CREEP AND SWELLING FOR LMF REACTOR DESIGN.
ELASTOPLASTIC AND CREEP EFFECT.
FOR THE ANALYSIS OF THERMAL CREEP, IRRADIATION INDUCED CREEP IN ELASTIC SOLIDS.
L FINITE ELEMENT METHOD OF THERMOELASTOPLASTICITY WITH CREEP EFFECT.
VISCO-PLASTICITY PLASTICITY AND CREEP IN STRUCTURES.
ANALYSIS OF REAL STRUCTURES OF CREEP PLASTICITY AND OTHER COMPLEX CONSTITUTIVE LAWS.
FINITE ELEMENT FRACTURE MECHANICS ANALYSIS OF CREEP RUPTURE OF FUEL ELEMENT CLADDING.
THE ANALYSIS OF THERMAL CREEP, IRRADIATIATION INDUCED CREEP, ABD SWELLING FOR LMFBR DESIGN.
ANALYSIS OF TIME-DEPENDENT STRESS PROBLEMS, INCLUDING CREEP, BY THE FINITE ELEMENT
OF FINITE ELEMENT METHODS FOR THE ANALYSIS OF THERMAL CREEP, IRRADIATIATION INDUCED CREEP, ABD SWELLING FOR LMFBR REACTOR DES
OF FINITE ELEMENT METHODS FOR THE ANALYSIS OF THERMAL CREEP, IRRADIATION INDUCED CREEP AND SWELLING FOR LMF REACTOR DES
HER-ORDER NUMERICAL METHODS FOR PROBLEMS OF NON-LINEAR CREEP, VISCOELACTICITY AND ELASTO-PLASTICITY.

NUMERICAL QUADRATURE AND CUBATURE.

NATURAL CUBIC AND BICUBIC SPLINE INTERPOLATION.
PIECEWISE CUBIC INTERPOLATION AND DEFERRED CORRECTION.
PIECEWISE CUBIC INTERPOLATION AND TWO-POINT BOUNDARY PROBLEMS.
PIECEWISE CUBIC INTERPOLATION AND TWO-POINT BOUNDARY VALUE PROBLEMS.
CUBIC SPLINE FUNCTION AND DIFFERENCE METHOD.
CUBIC SPLINE INTERPOLATION TO HARMONIC FUNCTIONS.
CUBIC SPLINE SOLUTIONS TO TWO-POINT BOUNDARY VALUE PROBLEMS.
A CUBIC SPLINE TECHNIQUE FOR THE ONE DIMENSIONAL HEAT CONDUCTION EQ
CUBIC SPLINES.
L SOLUTION OF TWO-POINT BOUNDARY VALUE PROBLEMS USING CUBIC SPLINES.
A CURVED TRIANGULAR FINITE ELEMENT FOR POTENTIAL FLOW PROBLEMS.

GENERALIZED STIFFNESS MATRIX OF A CURVED BEAM ELEMENT.
A COMPARISON OF CURVED BEAM FINITE ELEMENTS WHEN USED IN VIBRATION PROBLEMS
THE COMBINED EFFECT OF CURVED BOUNDARIES AND NUMERICAL INTEGRATION IN ISOPARAMETRIC FINI
THE FINITE ELEMENT METHOD IN DOMAINS WITH CURVED BOUNDARIES.
THE USE OF PARABOLIC ARCS IN MATCHING CURVED BOUNDARIES.
SION TRIANGULAR PLATE-BENDING ELEMENT TO PROBLEMS WITH CURVED BOUNDARIES.
CURVED BOUNDARY ELEMENTS, GENERAL FORMS OF POLYNOMIAL MAPPINGS.
CURVED BOX-GIRDER BRIDGE.
FINITE ELEMENT ANALYSIS OF SKEW, CURVED CUBIC TRIANGULAR FINITE ELEMENT FOR POTENTIAL FLOW PROBLEM
A CURVED CYLINDRICAL SHELL FINITE ELEMENT.
A CURVED CYLINDRICAL SHELL FINITE ELEMENT.
FINITE ELEMENT ANALYSIS OF PLATES WITH CURVED EDGES.
A CURVED ELEMENT APPROXIMATION IN THE ANALYSIS OF AXI-SYMMETRIC THI
ETHODS: BLENDING FUNCTION INTERPOLATION OVER ARBITRARY CURVED ELEMENT DOMAINS.
A CURVED ELEMENT FOR THIN ELASTIC SHELLS.
A REFINED CURVED ELEMENT FOR THIN SHELLS OF REVOLUTION.

1394
1954
941
328

2008

853
952
950
540
224
1881
2081
2129
965
2143
452
1882
1054
1882
1881
257

676

892
1732
304
1731
1734
1588
41
1586
567
1034

1217
1726
470
2152
1340
444
1275
451
1434
1728
408
445
818
844
649
1642

885 RIGID - BODY DISPLACEMENTS OF CURVED ELEMENTS IN THE ANALYSIS OF SHELLS BY THE MATRIX DISPLACEM

1387 CURVED ELEMENTS IN THE FINITE ELEMENT METHOD.

1339 BASIS FUNCTIONS FOR CURVED ELEMENTS IN THE FINITE ELEMENT METHOD.

1341 THE CONSTRUCTION OF BASIS FUNCTIONS FOR CURVED ELEMENTS IN THE FINITE ELEMENT METHOD. I'.

2151 CURVED ELEMENTS IN THE FINITE ELEMENT METHOD. II.

2154 CURVED ELEMENTS WITH APPLICATIONS TO FINITE ELEMENT METHODS.'

469 S METHOD OF ANALYSIS OF SHELLS OF REVOLUTION UTILIZING INTERPOLATION THEORY OVER CURVED ELEMENTS.

1064 ANALYSIS OF THICK AND THIN SHELL STRUCTURES BY GENERAL CURVED ELEMENTS.

30 F SHELLS OF REVOLUTION BY MATRIX DISPLACEMENT METHOD (CURVED ELEMENTS).

1872 CURVED FINITE ELEMENTS IN THE ANALYSIS OF SHELL STRUCTURES BY THE

561 FURTHER STUDIES IN THE APPLICATION OF CURVED FINITE ELEMENTS TO CIRCULAR ARCHES.

167 LIMITATIONS OF CERTAIN CURVED FINITE ELEMENTS WHEN APPLIED TO ARCHES.'

165 IMPLICIT RIGID BODY MOTION IN CURVED FINITE ELEMENTS.

1346 RIGID BODY MOTIONS IN CURVED FINITE ELEMENTS.

406 CURVED ISOPARAMETRIC QUADRILATERAL ELEMENTS IN FINITE ELEMENT ANA

681 THE APPLICATION OF A CURVED MIXED TYPE SHELL ELEMENT.

1991 VIBRATION ANALYSIS OF CANTILEVERED CURVED PLATES USING A NEW CYLINDRICAL SHELL FINITE ELEMENT.

1572 ELASTIC SNAP THROUGH ANALYSIS OF CURVED PLATES USING DISCRETE ELEMENTS.

2097 PLANAR AND CURVED SHELL ELEMENTS.

1348 CURVED SIDES.

2000 A RATIONAL BASIS FOR FUNCTION APPROXIMATION; CURVED STRUCTURES USING QUADRILATERAL FINITE ELEMENTS.

1606 VIBRATION OF CURVED STRUCTURES.

1265 THE LARGE DEFLECTION GEOMETRICALLY NONLINEAR PLANE AND CURVED SURFACE BY (ISOPARAMETRIC) CO-ORDINATES.

2141 AN AUTOMATIC MESH GENERATION SCHEME FOR PLANE AND CURVED SURFACES BY ISOPARAMETRIC COORDINATES.

1609 AN AUTOMATIC GENERATION SCHEME FOR FLAT AND CURVED SURFACES.

1608 TECHNIQUES IN TRIANGULAR MESH GENERATION FOR FLAT AND CURVED SURFACES.

29 CURVED THICK SHELL AND MEMBRANE ELEMENTS WITH PARTICULAR REFERENC

1185 ANALYSIS OF THICK- WALLED VESSEL-NOZZLE JUNCTIONS WITH CURVED TRANSITIONS.

1184 NT ANALYSIS OF THICK WALLED PIPE-NOZZLE JUNCTIONS WITH CURVED TRANSITIONS.

600 NUMERICAL ANALYSIS OF THIN SHELLS BY CURVED TRIANGULAR ELEMENTS BASED ON DISCRETE-KIRCHOFF HYPOTHESIS.

339 CURVED TRIANGULAR ELEMENTS FOR THE ANALYSIS OF SHELLS.

1574 DYNAMIC ANALYSIS OF SHALLOW SHELLS WITH A DOUBLY CURVED TRIANGULAR FINITE ELEMENT.

28 VIBRATION OF THICK, CURVED, SHELLS WITH PARTICULAR REFERENCE TO TURBINE BLADES.

811 THERMO-ELASTIC-PLASTIC CYCLIC ANALYSIS BY FINITE ELEMENT METHOD.

277 OF CRACK PROPAGATION IN THREE-DIMENSIONAL SOLIDS UNDER CYCLIC LOADING.

153 DS FOR THE PLASTIC ANALYSIS OF STRUCTURES SUBJECTED TO CYCLIC LOADING.

225 FINITE ELEMENT ANALYSIS OF A THICK WALLED MICROPOLAR CYLINDER LOADED AXISYMETRICALLY.

1295 ERENCE SOLUTION OF AXISYMMETRIC CONDUCTION PROBLEMS IN CYLINDRICAL CO-ORDINATES.

548 FINITE ELEMENTS FOR FIELD PROBLEMS IN CYLINDRICAL CO-ORDINATES.

333 FINITE DEFLECTION STRUCTURAL ANALYSIS USING PLATE AND A CYLINDRICAL SHELL DISCRETE ELEMENT, AIAA JOURNAL 5, 745-750, 1967

1766 CYLINDRICAL SHELL DISCRETE ELEMENTS.

168 A CORRECTED ASSESSMENT OF THE CYLINDRICAL SHELL FINITE ELEMENT BASED ON SIMPLE INDEPENDENT STRA

1724 THE SHALLOW TO THE NON-SHALLOW STIFFNESS MATRIX FOR A CYLINDRICAL SHELL FINITE ELEMENT OF BOGNER, FOX AND SCHMIT WHEN A

1728 ION ANALYSIS OF CANTILEVERED CURVED PLATES USING A NEW CYLINDRICAL SHELL FINITE ELEMENT.

1572 CYLINDRICAL SHELL FINITE ELEMENT.

1247 A HIGH-PRECISION TRIANGULAR CYLINDRICAL SHELL FINITE ELEMENT.

940 A NEW HYBRID CYLINDRICAL SHELL FINITE ELEMENT.

408 A CURVED CYLINDRICAL SHELL FINITE ELEMENT.

375 NASTRAN BUCKLING ANALYSIS OF A LARGE STIFFENED CYLINDRICAL SHELL WITH A CUTOUT.

2078 THE DYNAMIC RESPONSES OF CYLINDRICAL SHELLS INCLUDING GEOMETRIC AND MATERIAL NONLINEARITIE

613 FINITE ELEMENT ANALYSIS OF SANDWICH PLATES AND CYLINDRICAL SHELLS WITH LAMINATED FACES.

1430 ION OF THE FINITE ELEMENT METHOD TO THE CALCULATION OF CYLINDRICAL SHELLS WITH RECTANGULAR OPENINGS.

76 TION OF STRESS CONCENTRATIONS AROUND ELLIPTIC HOLES IN CYLINDRICAL SHELLS.

2103 E FINITE ELEMENT METHOD TO THE ANALYSIS OF FRACTURE OF CYLINDRICAL SHELLS.

1102 FINITE DIFFERENCE SOLUTIONS FOR LARGE DEFORMATIONS OF CYLINDRICAL SHELLS; A COMPARISON WITH FINITE ELEMENT SOLUTIONS;

ON THE STATES OF STRESS AND DEFORMATION OF CYLINDRICAL SPECIMENS OF BRITTLE MATERIAL UNDER UNIAXIAL COMPRES

483 THE STRESS DISTRIBUTION OF NORFOLK DAM.

691 EFFECTS OF TIME AND DAMPING ON FINITE ELEMENT ANALYSIS OF RESPONSE OF STRUCTURES.

1959 STIFFNESS AND DEFLECTION ANALYSIS OF COMPLEX STRUCTURES,

1629 INCREMENTAL LARGE DEFLECTION ANALYSIS OF ELASTIC STRUCTURES.

281 FINITE ELEMENT LARGE DEFLECTION ANALYSIS OF PLATES, AND SHALLOW SHELLS USING THE FINITE

1427 EFFECT OF INITIAL DISPLACEMENT ON PROBLEM OF LARGE DEFLECTION ANALYSIS OF PLATES.

1308 VATION OF STIFFNESS MATRICES FOR THE ANALYSIS OF LARGE DEFLECTION AND STABILITY PROBLEMS.

1970 -ELEMENT STIFFNESS MATRICES ASSOCIATED WITH POLYHEDRAL DEFLECTION DISTRIBUTIONS.

1729 LARGE DEFLECTION GEOMETRICALLY NON-LINEAR FINITE ELEMENT ANALYSIS OF C

1265 ALGORITHM FOR THE LARGE DEFLECTION GEOMETRICALLY NONLINEAR PLANE AND CURVED STRUCTURES.

1103 ANALYSIS OF LARGE DEFLECTION OF PLATES BY THE FINITE ELEMENT METHOD.

1960 LARGE DEFLECTION OF STRUCTURES SUBJECT TO HEATING AND EXTERNAL LOADS.

172 INITE ELEMENT MODEL FOR INCREMENTAL ANALYSIS OF LARGE DEFLECTION PROBLEMS.

1766 FINITE DEFLECTION STRUCTURAL ANALYSIS USING PLATE AND CYLINDRICAL SHELL

1765 FINITE DEFLECTION STRUCTURAL ANALYSIS USING PLATE AND SHELL DISCRETE ELE

888 NAPLAS A FINITE ELEMENT PROGRAM FOR THE DYNAMIC, LARGE DEFLECTION, ELASTIC-PLASTIC ANALYSIS OF STIFFENED SHELLS OF REVOL

301 A LARGE DEFORMATION ANALYSIS OF CRYSTALLINE ELASTIC-VISCOPLASTIC MATERIAL

1989 THE FINITE ELEMENT METHOD IN DEFORMATION AND HEAT CONDUCTION PROBLEMS.

1518 FINITE ELEMENT FORMULATION OF PROBLEMS OF FINITE DEFORMATION AND IRREVERSIBLE THERMODYNAMICS OF NONLINEAR CONTINUA

2136 A GENERAL APPROACH TO PROBLEMS OF LARGE DEFORMATION AND PLASTICITY USING ISOPARAMETRIC ELEMENTS.

1806 PRE-SEISMIC CRUSTAL DEFORMATION CAUSED BY AN UNDERTHRUSTING OCEANIC PLATE IN EASTERN

1121 A FINITE ELEMENT PROCEDURE FOR THE LARGE DEFORMATION DYNAMIC RESPONSE OF AXISYMMETRIC SOLIDS.

499 IMPROVED DEFORMATION FUNCTIONS FOR FINITE ELEMENT ANALYSIS OF BEAM SYSTEMS

1233 ELASTIC AND PLASTIC INTERLAMINAR SHEAR DEFORMATION IN LAMINATED COMPOSITES UNDER GENERALIZED PLANE STRES

1544 A NOTE ON THE FINITE DEFORMATION OF A THICK ELASTIC SHELL OF REVOLUTION.

1102 ON THE STATES OF STRESS AND DEFORMATION OF CYLINDRICAL SPECIMENS OF BRITTLE MATERIAL UNDER U

3 ELASTIC DEFORMATION OF POLYCRYSTALLINE METALS.

845 A FINITE ELEMENT ANALYSIS OF GENERAL DEFORMATION OF SHEET METALS.

1277 ISON OF FINITE ELEMENT AND EXPERIMENTAL STUDIES ON THE DEFORMATION OF ZIRCONIUM NOTCHED BEND SPECIMENS.

2089 MENTAL SOLUTION OF AXISYMMETRIC PLATE AND SHELL DEFORMATION.

868 LEMENT ANALYSIS OF PLATE BENDING WITH TRANSVERSE SHEAR DEFORMATION.

1543 IONS OF THE INCREMENTAL STIFFNESS RELATIONS FOR FINITE DEFORMATIONS OF COMPRESSIBLE AND INCOMPRESSIBLE FINITE ELEMENTS.

2103 FINITE DIFFERENCE SOLUTIONS FOR LARGE DEFORMATIONS OF CYLINDRICAL SHELLS; A COMPARISON WITH FINITE ELE

1542 ANALYSIS OF FINITE DEFORMATIONS OF ELASTIC SOLIDS BY THE FINITE ELEMENT METHOD.

1525 APPROXIMATIONS AND NUMERICAL ANALYSIS OF FINITE DEFORMATIONS OF ELASTIC SOLIDS.

1916 ELASTIC-PLASTIC DEFORMATIONS OF FLEXURALLY STIFF FRAMEWORKS FROM SECOND ORDER STR

1541 NUMERICAL ANALYSIS OF FINITE AXISYMMETRIC DEFORMATIONS OF INCOMPRESSIBLE ELASTIC SOLIDS OF REVOLUTION.

1268 THE DETERMINATION OF STRESSED AND DEFORMATIONS OF REINFORCED CONCRETE AFTER CRACKING.

2077 TE ELEMENT ANALYSIS OF LARGE ELASTIC PLASTIC TRANSIENT DEFORMATIONS OF SIMPLE STRUCTURES.

1988 F NON-STATIONARY TEMPERATURE DISTRIBUTION AND THERMAL DEFORMATIONS.

1009   FINITE ELEMENT ANALYSIS OF SOIL DEFORMATIONS.
73   N AND PLATE THEORY TO PROBLEMS OF REACTOR FUEL ELEMENT DEFORMATIONS.

1713   THER LIMITS ON GENERAL HEXAHEDRON FINITE ELEMENTS WITH DERIVATIVE DEGREES OF FREEDOM.
1689   PROGRAMME FOR TRIANGULAR BENDING ELEMENT WITH DERIVATIVE SMOOTHING.

188   COMPUTATION OF DERIVATIVES IN THE FINITE ELEMENT METHOD.
717   SOME INEQUALITIES FOR THE SOLUTION AND ITS DERIVATIVES OF AN EQUATION OF ELLIPTIC TYPE IN L PROC.
327   ON THE SUMMABILITY OF HIGHEST ORDER DERIVATIVES OF THE SOLUTION OF THE DIRICHLET PROBLEM IN A DOMAIN
323   HERMITE INTERPOLATION ERRORS FOR DERIVATIVES.

544   NUMERICAL SOLUTION OF DIELECTRIC LOADED WAVEGUIDES.

1884   CREATION AND COMPARISON OF FINITE DIFFERENCE ANALOGS OF SOME FINITE ELEMENT SCHEMES.
1887   THE RELATIVE EFFIENCY OF FINITE- DIFFERENCE AND FINITE ELEMENT METHODS.
300   ENERGY APPROACHES TO FINITE DIFFERENCE AND FINITE ELEMENT METHODS.
534   THE FINITE DIFFERENCE AND LOCALIZED RITZ METHODS.
1080   RDER FINITE ELEMENT PROCEDURES AND A LOW ORDER FINITE DIFFERENCE APPROXIMATION PROCEDURE FOR THE NUMERICAL SOL
1190   FINITE DIFFERENCE APPROXIMATIONS FOR EIGENVALUES OF UNIFORMLY ELLIPTIC O
1592   NORMAL GRADIENT BOUNDARY CONDITIONS IN FINITE DIFFERENCE CALCULATIONS.
870   SYMMETRIC STRESS ANALYSIS BY FINITE ELEMENT AND FINITE DIFFERENCE COMPUTER TECHNIQUES.
642   A FACTORIZATION PROCEDURE FOR THE SOLUTION OF ELLIPTIC DIFFERENCE EQUATIONS
1924   ERRELAXATION TECHNIQUE FOR SOLVING SECOND ORDER FINITE DIFFERENCE EQUATIONS AND INTERIOR REGULARITY.
1939   DIFFERENCE EQUATIONS FOR TRIANGULAR MESHES.
1106   DIFFERENCE EQUATIONS ON A MESH ARISING FROM A GENERAL TRIANGULATI
1107   DIFFERENCE EQUATIONS ON A TRIANGULATION.
303   RITZ DIFFERENCE EQUATION.
174   FINITE DIFFERENCE FORMULAE FOR THE SQUARE LATTICE.
822   OR LINEAR ELLIPTIC OPERATORS BY GALLERKIN"S AND FINITE DIFFERENCE MET
1737   THEORY OF A FINITE DIFFERENCE METHOD ON IRREGULAR NETWORKS.
1734   HE USE OF WHICH IS EQUIVALENT TO THE USE OF THE FINITE DIFFERENCE METHOD.
1886   CUBIC SPLINE FUNCTION AND DIFFERENCE METHODS.
1888   RATIVE EFFICIENCY OF CERTAIN FINITE ELEMENT AND FINITE DIFFERENCE METHODS FOR A HYPERBOLIC PROBLEM.
456   RATIVE EFFICIENCY OF CERTAIN FINITE ELEMENT AND FINITE DIFFERENCE METHODS FOR A HYPERBOLIC PROBLEM.
1710   A COMPARISON OF SOME FINITE ELEMENT AND FINITE DIFFERENCE METHODS FOR A SIMPLE SLOSH PROBLEM.
719   DIFFERENCE METHODS FOR INITIAL VALUE PROBLEMS.
2052   LAGRANGIAN FINITE ELEMENT AND FINITE DIFFERENCE METHODS FOR PARTIAL DIFFERENTIAL EQUATIONS.
1127   COMPARISON OF FINITE ELEMENT AND FINITE DIFFERENCE METHODS FOR POISSON PROBLEMS.
533   L BOUNDARY CONDITIONS IN THE FINITE ELEMENT AND FINITE DIFFERENCE METHODS.
1918   DISCRETE INTERIOR SCHAUDER ESTIMATES FOR ELLIPTIC DIFFERENCE OPERATORS.
1000   A SELF-ADJUSTING GRID FOR FINITE- DIFFERENCE PROGRAMS.
1383   GALERKIN APPROXIMATIONS AND THE OPTIMIZATION OF DIFFERENCE SCHEMES FOR BOUNDARY VALUE PROBLEMS.
1556   INVESTIGATION OF THE CONVERGENCE RATE OF VARIATIONAL- DIFFERENCE SCHEMES FOR ELLIPTIC SECOND ORDER EQUATIONS IN A TWO D
1706   CONVERGENT FINITE DIFFERENCE SCHEMES FOR NONLINEAR PARABOLIC EQUATIONS.
1555   VARIATIONAL DIFFERENCE SCHEMES FOR SECOND ORDER LINEAR ELLIPTIC EQUATIONS IN
1919   SPLINE APPROXIMATIONS AND DIFFERENCE SCHEMES FOR THE HEAT EQUATIONS.
1554   CONVERGENCE OF DIFFERENCE SCHEMES IN CASE OF IMPROVED APPROXIMATION OF THE BOUND
1295   A NEW TRIANGULAR ELEMENT FOR FINITE DIFFERENCE SOLUTION OF AXISYMMETRIC CONDUCTION PROBLEMS IN CYLIND
2103   DIFFERENCE SOLUTIONS OF CYLINDRICAL SHELLS
673   FOR DETERMINING THE EFFECT OF SINGULARITIES IN FINITE DIFFERENCE SOLUTIONS ILLUSTRATED BY APPLICATION TO PLANE ELASTIC

1635   SUR UNE INTERPOLATION DE LA METHODE DES DIFFERENCES FINIES QUI PEUT FOURNIR DES BORNES SUPERIEURES OU IN
1762   FINITE ELEMENTS VERSUS FINITE DIFFERENCES, A COMPARISON OF THE TWO METHODS FOR THE SOLUTION OF

PENALTY GALERKIN METHODS FOR PARTIAL DIFFERENTIAL EQUATIONS. 580

RKIN FINITE ELEMENT APPROXIMATIONS OF ELLIPTIC PARTIAL DIFFERENTIAL EQUATIONS. 595

PENALTY GALERKIN METHODS FOR PARTIAL DIFFERENTIAL EQUATIONS. 581

TE ELEMENT METHOD FOR THE GENERAL SOLUTION OF ORDINARY DIFFERENTIAL EQUATIONS. 542

ERMITEAN METHODS FOR INITIAL VALUE PROBLEMS IN PARTIAL DIFFERENTIAL EQUATIONS. 501

NUMERICAL SOLUTION OF PARTIAL DIFFERENTIAL EQUATIONS. 347

IN ONE AND TWO VARIABLES WITH APPLICATIONS TO PARTIAL DIFFERENTIAL EQUATIONS. 324

PARTIAL DIFFERENTIAL EQUATIONS. 295

EXISTENCE THEOREMS IN PARTIAL DIFFERENTIAL EQUATIONS. 294

NUMERICAL SOLUTION OF PARTIAL DIFFERENTIAL EQUATIONS. 201

NUMERICAL PROCESSES IN PARTIAL DIFFERENTIAL EQUATIONS. 209

MAXIMUM AND MINIMUM PRINCIPLES FOR A CLASS OF PARTIAL DIFFERENTIAL EQUATIONS. 164

THE FINITE ELEMENT METHOD FOR ELLIPTIC DIFFERENTIAL EQUATIONS. 190

THE FINITE ELEMENT METHOD FOR ELLIPTIC DIFFERENTIAL EQUATIONS. 196

N METHOD SOLUTIONS TO SELECTED ELLIPTIC AND PARABOLIC DIFFERENTIAL EQUATIONS. 86

A FINITE ELEMENT COLLOCATION SOLUTION OF DIFFERENTIAL EQUATIONS. 34

TH RESPECT TO BASIC PROBLEMS OF THE THEORY OF PARTIAL DIFFERENTIAL EQUATIONS, ESPECIALLY WITH RESPECT TO 184

ND THEIR APPLICATIONS TO BOUNDARY PROBLEMS FOR PARTIAL DIFFERENTIAL EQUATIONS, LENINGR. 1822

LECTURE SERIES IN DIFFERENTIAL EQUATIONS, VOL. 183

ON SOME NONLINEAR ELLIPTIC DIFFERENTIAL FUNCTIONAL EQUATIONS. 908

BOUNDS FOR EIGENVALUES OF SOME DIFFERENTIAL OPERATORS BY THE RAYLEIGH-RITZ METHOD. 2035

DIFFERENTIAL OPERATORS OF MATHEMATICAL PHYSICS. 932

LINEAR DIFFERENTIAL OPERATORS. 1198

LINEAR PARTIAL DIFFERENTIAL OPERATORS. 984

NUMERISCHE LOSUNG NICHTLINEARER PARTIALLER DIFFERENTIAL UND INTEGRODIFFERENTIALGLEICHUNGEN. 81

OXIMATIONS OF TIME DEPENDENT PROBLEMS WITH EMPHASIS ON DIFFUSION AND CONVECTION. 2031

CATION OF FINITE ELEMENT SOLUTION TECHNIQUE TO NEUTRON DIFFUSION AND TRANSPORT EQUATIONS. 1558

NUMERICAL SOLUTION OF THE NEUTRON DIFFUSION EQUATION IN THE PRESENCE OF CORNERS AND INTERFACES. 206

OF THE FINITE ELEMENT METHOD TO THE MULTIGROUP NEUTRON DIFFUSION EQUATION. 1797

COMPARISON OF THE TWO METHODS FOR THE SOLUTION OF THE DIFFUSION EQUATION. 1762

REACTOR MATHEMATICS; NUMERICAL SOLUTION OF THE NEUTRON DIFFUSION EQUATION. 1079

TECHNIQUES WITH HIGHER ORDER FINITE ELEMENT MODELS FOR DIFFUSION EQUATIONS WITH PIECEWISE CONTINUOUS MATERIAL PROPERTIES 936

NUMERICAL SOLUTION OF STEADY STATE DIFFUSION PROBLEMS CONTAINING SINGULARITIES. 2051

IN THE FINITE ELEMENT APPROXIMATION OF TWO DIMENSIONAL DIFFUSION PROBLEMS. 937

FINITE ELEMENT METHOD APPLIED TO NEUTRON DIFFUSION PROBLEMS. 586

CATION OF THE FINITE ELEMENT METHOD TO TWO-DIMENSIONAL DIFFUSION PROBLEMS. 585

E ELEMENT METHOD TO TWO-DIMENSIONAL MULTIGROUP NEUTRON DIFFUSION. 1796

THE DIMENSION OF PIECEWISE POLYNOMIAL SPACES AND ONE-SIDED APPROXIMAT 1863

METHOD OF ORTHOGONAL PROJECTION AND DIRICHLET BOUNDARY VALUE PROBLEM IN A GRAIN DOMAIN. 454

FINIS DE LAGRANGE D"ORDRE UN ET DEUX, DU PROBLEME DE DIRICHLET POUR L"OPERATEUR BIHARMONIQUE. 828

TY OF HIGHEST ORDER DERIVATIVES OF THE SOLUTION OF THE DIRICHLET PROBLEM IN A DOMAIN WITH PIECEWISE SMOOTH BOUNDARY. 327

ON THE DISCRETIZATION ERROR OF THE DIRICHLET PROBLEM IN A PLANE REGION WITH CORNERS. 1193

SOME REMARKS ON THE DIRICHLET PROBLEM IN PIECEWISE SMOOTH DOMAINS. 897

ERROR ANALYSIS OF GALERKIN METHODS FOR THE SOLUTION OF DIRICHLET PROBLEMS CONTAINING BOUNDARY SINGULARITIES. 241

ON ERROR OF DISCRETE APPROXIMATIONS A PROJECTION METHOD FOR DIRICHLET PROBLEMS IN A DOMAIN WITH CORNERS. 1192

NAL ANALOGUES OF VARIATIONAL AND QUASI-LINEAR ELLIPTIC DIRICHLET PROBLEMS USING SUBSPACES WITH NEARLY ZERO BOUNDARY CON 1484

ON THE SMOOTHNESS OF SOLUTIONS OF UNILATERAL DIRICHLET PROBLEMS. 1853

A GENERALISED RITZ-LEAST-SQUARES METHOD FOR DIRICHLET PROBLEMS. 1421

DIRICHLET PROBLEMS. 352

Keyword-in-context index (reference number followed by entry). Truncated words at line starts are reproduced as printed.

507 E CONSISTENT AND SIMPLIFIED FINITE ELEMENT ANALYSES OF EIGENVALUE PROBLEMS WITH A MIXED PLATE ELEMENT.
1984 AN EXTENDED KANTOROVICH METHOD FOR THE SOLUTION OF EIGENVALUE PROBLEMS.
1114 VARIATIONAL METHODS FOR EIGENVALUE PROBLEMS.
849 RECENT ADVANCES IN NUMERICAL ANALYSIS OF STRUCTURAL EIGENVALUE PROBLEMS.
877 ER ACCURACY FOR NONLINEAR BOUNDARY VALUE PROBLEMS III: EIGENVALUE PROBLEMS III:
476 RGENCE RESULTS FOR THE RAYLEIGH-RITZ METHOD APPLIED TO EIGENVALUE PROBLEMS. I, ESTIMATES RELATING RAYLEIGH-RITZ AND GALE
1625 RGENCE RESULTS FOR THE RAYLEIGH-RITZ METHOD APPLIED TO EIGENVALUE PROBLEMS;2 ERROR BOUNDS FOR EIGENFUNCTIONS.
1626 STRUCTURAL EIGENVALUE PROBLEMS; ELIMINATION OF UNWANTED VARIABLES.
1014 CONVERGENCE STUDIES OF STRUCTURAL EIGENVALUE SOLUTIONS USING TWO FINITE PLATE BENDING ELEMENTS.
1246 STRUCTURAL EIGENVALUE SOLUTIONS USING TWO FINITE PLATE BENDING ELEMENTS.
707 RRORS IN FINITE ELEMENT APPROXIMATION OF STEADY STATE, EIGENVALUE, AND PARABOLIC PROBLEMS.

1285 ETHOD TO GENERATE BOTH UPPER AND LOWER BOUNDS TO PLATE EIGENVALUES BY CONFORMING DISPLACEMENT FINITE ELEMENTS.
317 ACCURATE EIGENVALUES COMPUTATIONS FOR ELLIPTIC PROBLEMS.
353 APPROXIMATION OF STEKLOV EIGENVALUES OF NON-SELFADJOINT SECOND ORDER ELLIPTIC OPERATORS.
714 BOUNDS AND APPROXIMATIONS FOR EIGENVALUES OF RECTANGULAR POLYGONAL MEMBRANES.
705 FINITE DIFFERENCE APPROXIMATIONS FOR EIGENVALUES OF SELF-ADJOINT BOUNDARY VALUE PROBLEMS.
2035 EIGENVALUES OF SOME DIFFERENTIAL OPERATORS BY THE RAYLEIGH-RITZ M
1190 EIGENVALUES OF UNIFORMLY ELLIPTIC OPERATORS.
1183 ASSESSMENT OF ACCURACIES OF FINITE ELEMENT EIGENVALUES.

415 ETICAL FORMULATION OF FINITE ELEMENT METHODS IN LINEAR ELASTIC ANALYSIS OF CRACKED THIN SHELLS BY THE FINITE ELEMENT MET
173 APPLICATION OF MATRIX DISPLACEMENT METHOD TO LINEAR ELASTIC ANALYSIS OF GENERAL SHELLS.
1602 ELASTIC ANALYSIS OF SHELLS OF REVOLUTION.
1006 EE-DIMENSIONAL FINITE ELEMENT PROGRAM FOR SMALL STRAIN ELASTIC ANALYSIS.
2074 ELASTIC AND ELASTIC-PLASTIC ANALYSIS OF SHELLS OF REVOLUTION SUB
1233 PROGRESS REPORT ON DISCRETE ELEMENT ELASTIC AND PLASTIC INTERLAMINAR SHEAR DEFORMATION IN LAMINATED C
947 ELASTIC AND TORSIONAL ANALYSIS OF IRREGULAR SHAPES.
1464 NUMERICAL METHODS OF ANALYSIS IN BARS, PLATES AND ELASTIC BODIES.
1328 VIBRATION OF AN ELASTIC BODY IMMERSED IN A FLUID.
953 FINITE ELEMENT ELASTIC BUCKLING ANALYSES.
2067 DIRECT FINITE ELEMENT ANALYSIS OF ELASTIC CONTACT PROBLEMS USING DIFFERENTIAL DISPLACEMENTS.
802 FINITE ELEMENTS FOR THREE-DIMENSIONAL ELASTIC CONTACT PROBLEMS.
1624 ELASTIC CRACK ANALYSIS BY A FINITE ELEMENT HYBRID METHOD.
1955 ELASTIC CRACK ANALYSIS.
3 ELASTIC DEFORMATION OF POLYCRYSTALLINE METALS.
730 A NEW VARIATIONAL PRINCIPLE FOR FINITE ELASTIC DISPLACEMENTS.
1804 THREE-DIMENSIONAL ELASTIC FINITE ELEMENT ANALYSIS USING GRADIENT DEGREES OF FREEDO
1549 PPROXIMATE METHOD FOR COMPUTING CONSISTENT STRESSES IN ELASTIC FINITE ELEMENTS.
2096 FLEXIBLE PLATE FINITE ELEMENT ON AN ELASTIC FOUNDATION.
1880 ELEMENT ANALYSIS OF FINITE SIZED PLATES BOUNDED TO AN ELASTIC HALF SPACE.
799 THE FINITE ELEMENT METHOD FOR ANALYSIS OF ELASTIC INSTABILITY PREDICTIONS FOR DOUBLY CONNECTED REGIONS.
2124 ELASTIC ISOTROPIC AND ORTHOTROPIC SLABS.
100 THREE DIMENSIONAL ANISOTROPIC AND INHOMOGENEOUS ELASTIC MEDIA MATRIX ANALYSIS FOR SMALL AND LARGE DISPLACEMENTS.
1744 FINITE ELEMENT ANALYSIS OF SEEPAGE IN ELASTIC MEDIA.
813 FLOW OF COMPRESSIBLE FLUID IN POROUS ELASTIC MEDIA.
1551 FINITE STRAINS AND DISPLACEMENTS OF ELASTIC MEMBRANES BY THE FINITE ELEMENT METHOD.
1545 NONLINEAR DYNAMICS OF ELASTIC MEMBRANES BY THE FINITE ELEMENT METHOD.
656 A THEORETICAL AND NUMERICAL COMPARISON OF ELASTIC NONLINEAR FINITE ELEMENT METHODS.
2091 RATE OF CONVERGENCE FOR THE ITERATIVE APPROACH IN ELASTIC PLASTIC ANALYSIS OF CONTINUA.
540 FINITE ELEMENT ELASTIC PLASTIC CREEP ANALYSIS OF TWO DIMENSIONAL CONTINUUM WITH
963 A SIMPLE FINITE ELEMENT MODEL FOR ELASTIC PLASTIC PLATE BENDING.
268 FINITE ELEMENT METHOD FOR ELASTIC PLASTIC PLATES.

```
 903   CONTAINING CAVITIES AND INCLUSIONS WITH REFERENCE TO FATIGUE CRACK INITIATION.
1394                         COMPUTER PREDICTION OF FATIGUE CRACK PROPAGATION UNDER RANDOM LOADING.
1424   ITE ELEMENT ELASTIC-PLASTIC STRESS ANALYSIS TO NOTCHED FATIGUE SPECIMEN BEHAVIOUR.
1735            AN EXAMPLE OF THE EVALUATION OF FATIGUE STRENGTH OF A HIGH TEMPERATURE STRUCTURE BY ASME CODE.
1963   AN APPROACH TO INVESTIGATE NOTCH EFFECT ON LOW-CYCLE FATIGUE WITH FINITE ELEMENTS.

1493   THE TENSIBLE BEHAVIOUR OF TWO-PHASE ALLOY CONTAINING A FIBRE UNDER TENSIBLE LOAD.

 386         FINITE ELEMENT COMPUTER CODES FOR STRESS ANALYSIS AND FIELD PROBLEMS.
1161                            LINEAR ACCELERATOR CAVITY FIELD CALCULATION BY THE FINITE ELEMENT METHOD.
 322   UBIC SPLINE INTERPOLATION AS A METHOD FOR TREATMENT OF FIELD DATA.
2020                             AN EQUILIBRIUM STRESS FIELD MODEL FOR FINITE ELEMENT SOLUTIONS OF TWO-DIMENSIONAL ELAST
1664   ONAL PROJECTION METHOD FOR THE NUMERICAL SOLUTION OF A FIELD PROBLEM.
2140                                             TRANSIENT FIELD PROBLEMS - TWO AND THREE DIMENSIONAL ANALYSIS BY ISOPARAMET
2123                       SOLUTION OF THREE DIMENSIONAL FIELD PROBLEMS BY THE FINITE ELEMENT METHOD.
1742             A VARIATIONAL PRINCIPLE FOR LINEAR COUPLED FIELD PROBLEMS IN CONTINUUM MECHANICS.
 548                            FINITE ELEMENTS FOR FIELD PROBLEMS IN CYLINDRICAL CO-ORDINATES.
 906   ER-ORDER TRIANGULAR FINITE ELEMENT FOR THE SOLUTION OF FIELD PROBLEMS IN ORTHOTROPIC MEDIA.
2125                   FINITE ELEMENTS IN THE SOLUTION OF FIELD PROBLEMS.
1816      FINITE ELEMENT SOLUTION OF TWO-DIMENSIONAL EXTERIOR FIELD PROBLEMS.
1589          A FINITE ELEMENT SOLUTION OF TIME DEPENDENT FIELD PROBLEMS.
1163   FINITE ELEMENT PROGRAM PACKAGE FOR AXISYMMETRIC SCALAR FIELD PROBLEMS.
1162       SCALAR FINITE ELEMENT PACKAGE FOR TWO DIMENSIONAL FIELD PROBLEMS.
1145      AN ITERATIVE APPROACH TO THE FINITE ELEMENT METHOD IN FIELD PROBLEMS.
 384     ELEMENT WEIGHTED RESIDUAL SOLUTION TO ONE-DIMENSIONAL FIELD PROBLEMS.
1401                                 FINITE ELEMENT STRESS FIELD SOLUTION OF THE PROBLEM OF SAINT VENANT TORSION.
 612   ARIATIONAL PRINCIPLE OF HAMILTONIAN TYPE FOR CLASSICAL FIELD THEORY.

1503   FINEL: UNIVERSELLES PROGRAMSYSTEM ZUR BERECHNUNG DER ELASTISCHEN

 671            THE USE OF SINGULARITY PROGRAMMING IN FINITE-DIFFERENCE AND FINITE-ELEMENT COMPUTATIONS OF TEMPERATURE.
1043                                 ON THE CONVERGENCE OF FINITE-DIFFERENCE APPROXIMATIONS TO ONE-DIMENSIONAL SINGULAR BOUN
2204                                  THE CONNECTION BETWEEN FINITE-DIFFERENCE LIKE METHODS AND THE METHODS BASED ON INITIAL
2126   NT METHOD OF ANALYSIS OF ARCH DAMS AND COMPARISON WITH A FINITE-DIFFERENCE PROCEDURES.
 768                                                         A FINITE-DIFFERENCE SCHEME FOR GENERALISED NEUMANN PROBLEMS.
2049                                                           FINITE-DIFFERENCE TECHNIQUES FOR A HARMONIC MIXED BOUNDARY PROBLE
2058                                          CONVERGENCE OF FINITE-DIFFERENCE TECHNIQUES FOR A HARMONIC MIXED BOUNDARY VALUE

 337                                FLEXIBILITY AND STIFFNESS MATRICES FOR AN OPEN-TUBE WARPING CONST
 792                                 DIRECT FLEXIBILITY FINITE ELEMENT FLASTOPLASTIC ANALYSIS.

1800                         THE FINITE ELEMENT METHOD FOR FLEXURE OF SLABS WHEN STRESS DISTRIBUTIONS ARE ASSUMED.
1747                      A SECTOR ELEMENT FOR THIN PLATE FLEXURE.
 393        A STUDY OF FINITE ELEMENTS APPLIED TO PLATE FLEXURE.

 137                                          POTENTIAL FLOW ANALYSIS BY FINITE ELEMENTS.
1094                                            STEADY FLOW ANALYSIS OF INCOMPRESSIBLE VISCOUS FLUID BY THE FINITE ELEME
1945                                                   FLOW ANALYSIS.
  13               THE FINITE ELEMENT METHOD IN FLUID FLOW ANALYSIS.
1455              THE USE OF FINITE METHODS IN HEAT FLOW AND HEAT TRANSFER IN AN EXTRUDER CHANNEL.
1965   ICATIONS OF FINITE ELEMENT METHODS TO UNSTEADY VISCOUS FLOW AROUND A BOX-GIRDER BRIDGE OSCILLATING IN UNIFORM FLOW.
2079                                      UNSTEADY MHD DUCT FLOW BY THE FINITE ELEMENT METHOD.
1221                             LINEARIZED COMPRESSIBLE FLOW BY THE FINITE ELEMENT METHOD.
```

A NOVEL, HERMITE TYPE, **INTERPOLATION** PROCEDURE. — 182

ERROR BOUNDS FOR A BIVARIATE SPLINE **INTERPOLATION** SCHEME. — 1782

INTERPOLATION TECHNIQUES FOR VARIATIONAL METHODS. — 409

PIECEWISE **INTERPOLATION** THEORY OVER CURVED ELEMENTS WITH APPLICATIONS TO FI — 469

SPLINE-BLENDED **INTERPOLATION** THEORY. — 862

INTERPOLATION THROUGH CURVE NETWORKS. — 840

INTERPOLATION TO BOUNDARY DATA IN TRIANGLES. — 312

INTERPOLATION TO BOUNDARY DATA ON N-SIMPLICES. — 235

CUBIC SPLINE **INTERPOLATION** TO HARMONIC FUNCTIONS. — 1588

APPLICATION OF THE SMOOTH-SURFACE **INTERPOLATION** TO THE FINITE ELEMENT METHOD. — 563

DISCRETE DATA SMOOTHING BY SPLINE **INTERPOLATION** WITH APPLICATION TO INITIAL GEOMETRY OF CARLF NETS — 1150

SMOOTH **INTERPOLATION** WITHOUT TWIST CONSTRAINTS. — 861

SMOOTH **INTERPOLATION** WITHOUT TWIST CONSTRAINTS. — 859

SMOOTH CURVE **INTERPOLATION**. — 2076

ERROR BOUNDS FOR SPLINE **INTERPOLATION**. — 1975

ERROR BOUNDS FOR SPLINE AND L-SPLINE **INTERPOLATION**. — 1885

ON HERMITE-BIRKHOFF **INTERPOLATION**. — 1768

ON A CLASS OF FINITE ELEMENTS GENERATED BY LAGRANGE **INTERPOLATION**. — 1471

ON GENERAL HERMITE TRIGONOMETRIC **INTERPOLATION**. — 1179

ERROR BOUNDS FOR SPLINE **INTERPOLATION**. — 890

NATURAL CUBIC AND BICUBIC SPLINE **INTERPOLATION**. — 892

AN ALGORITHM FOR HERMITE-BIRKHOFF **INTERPOLATION**. — 698

BICUBIC SPLINE **INTERPOLATION**. — 565

ON BOUNDING SPLINE **INTERPOLATION**. — 570

ERROR BOUNDS FOR BICUBIC SPLINE **INTERPOLATION**. — 413

SMOOTH SURFACE **INTERPOLATION**. — 320

FRROR BOUNDS FOR SPLINE **INTERPOLATION**. — 314

ASS OF LINFAR FUNCTIONALS WITH APPLICATIONS TO HERMITE **INTERPOLATION**. — 351

ACES WITH APPLICATION TO FOURIER TRANSFORMS AND SPLINE **INTERPOLATION**. — 350

ERROR BOUNDS FOR POLYNOMIAL SPLINE **INTERPOLATION**, MATH. — 1780

RATIONALE **INTERPOLATION**, NORMALITAT UND MONOSPLINES. — 346

TRICUBIC POLYNOMIAL **INTERPOLATION**, PROC. — 309

SMOOTH-CURVE **INTERPOLATION**: PROCEDURE. — 826

HE APPLICATION OF AUTOMATIC MESH GENERATION SCHEMES BY **ISOPARAMETRIC** AND ASSOCIATED ELEMENT FAMILIES FOR TWO AND THREE D — 2132

TIC GENERATION SCHEME FOR PLANE AND CURVED SURFACES BY **ISOPARAMETRIC** CO-ORDINATES. — 1852

ISOPARAMETRIC COORDINATES. — 1609

ISOPARAMETRIC ELEMENT FORMS IN FINITE ELEMENT ANALYSIS. — 2118

LASTIC-PLASTIC STRESS ANALYSIS, A GENERALISATION USING **ISOPARAMETRIC** ELEMENTS AND VARIOUS CONSTITUTIVE RELATIONS. — 1449

A TECHNIQUE FOR DEGENERATING BRICK-TYPE **ISOPARAMETRIC** ELEMENTS USING HIERARCHICAL MIDSIDE NODES. — 1024

TO PROBLEMS OF LARGE DEFORMATION AND PLASTICITY USING **ISOPARAMETRIC** ELEMENTS. — 2136

MESH AND CONTOUR PLOT FOR TRIANGLE AND **ISOPARAMETRIC** ELEMENTS. — 1073

FINITE ELEMENT MESH GENERATOR PROGRAM FOR EIGHT NODE **ISOPARAMETRIC** ELEMENTS. — 1008

MORE ON REDUCED INTEGRATION AND **ISOPARAMETRIC** ELEMENTS. — 509

ISOPARAMETRIC FINITE ELEMENT METHODS FOR THE NEUTRON TRANSPORT EQ — 1230

FECT OF CURVED BOUNDARIES AND NUMERICAL INTEGRATION IN **ISOPARAMETRIC** FINITE ELEMENT METHODS. — 470

THE **ISOPARAMETRIC** F'NITE ELEMENT SYSTEM - A NEW CONCEPT IN FINITE ELE — 1031

THE PSEUDO **ISOPARAMETRIC** FINITE ELEMENTS FOR SHELL AND PLATE ANALYSIS. — 27

FIELD PROBLEMS - TWO AND THREE DIMENSIONAL ANALYSIS BY **ISOPARAMETRIC** FINITE ELEMENTS. — 2140

REPRESFNTATION OF SINGULARITIES WITH CURVED **ISOPARAMETRIC** FINITE ELEMENTS. — 271

CURVED **ISOPARAMETRIC** QUADRILATERAL ELEMENTS IN FINITE ELEMENT ANALYSIS. — 681

VIBRATION PROBLEMS USING **ISOPARAMETRIC** SHELL ELEMENTS. — 970

BASIS FOR **ISOPARAMETRIC** STRESS ELEMENTS. — 1714

223 SHAPE FUNCTION SUBROUTINE FOR AN ISOPARAMETRIC THIN PLATE ELEMENT.
2141 ESH GENERATION SCHEME FOR PLANE AND CURVED SURFACE BY)ISOPARAMETRIC) CO-ORDINATES.

2124 THE FINITE ELEMENT METHOD FOR ANALYSIS OF ELASTIC ISOTROPIC AND ORTHOTROPIC SLABS.

766 DIRECT ITERATION METHOD FOR THE INCORPORATION OF PHASE CHANGE IN FINITE
1276 OF AN UNSUPPORTED VIBRATING STRUCTURE BY SIMULTANEOUS ITERATION.

1973 MATRIX ITERATIVE ANALYSIS.
1054 THE COMPARISON OF ITERATIVE AND DIRECT SOLUTION TECHNIQUES IN THE ANALYSIS OF TIME-
2091 RATE OF CONVERGENCE FOR THE ITERATIVE APPROACH IN ELASTIC PLASTIC ANALYSIS OF CONTINUA.
1145 AN ITERATIVE APPROACH TO THE FINITE ELEMENT METHOD IN FIELD PROBLEMS
1355 ITERATIVE LOSUNG NICHTLINEARER GLEICHUNGS-SYSTEME UND DISKRETISIE
1605 ITERATIVE METHODS FOR THE SOLUTION OF LINEAR OPERATOR EQUATIONS I
749 DIRECT AND ITERATIVE METHODS IN FINITE ELEMENT ANALYSIS.
280 MORE ON GRADIENT ITERATIVE PROCESSES.
1115 CONVERGENCE CRITERIA FOR ITERATIVE PROCESSES.
1437 A NEW ITERATIVE SCHEME FOR THE SOLUTION OF PARTIAL DIFFERENTIAL EQUATIO
1998 ITERATIVE SOLUTIONS FOR THE FINITE ELEMENT METHOD.
ITERATIVE SOLUTIONS OF ELLIPTIC SYSTEMS.

288 KERNEL FUNCTIONS AND ELLIPTIC DIFFERENTIAL EQUATIONS IN MATHEMATI
234 SARD KERNEL THEOREMS ON TRIANGULAR AND RECTANGULAR DOMAINS WITH EXTENS
236 THE EXTENSION AND APPLICATION OF SARD KERNEL THEOREMS TO COMPUTE FINITE ELEMENT ERROR BOUNDS.

1209 L-INFINITY BOUNDS FOR MULTIVARIATE LAGRANGE APPROXIMATION.
425 L-INFINITY CONVERGENCE OF COLLOCATION AND GALERKIN APPROXIMATIONS

1885 ERROR BOUNDS FOR SPLINE AND L-SPLINE INTERPOLATION.

1787 L-SPLINES.

1648 LAGRANGE AND HERMITE INTERPOLATION IN BANACH SPACES.
468 GENERAL LAGRANGE AND HERMITE INTERPOLATION IN RN WITH APPLICATIONS TO FIN
1209 LAGRANGE APPROXIMATION.
828 APPROXIMATIONS EXTERNES, PAR ELEMENTS FINIS DE LAGRANGE D"ORDRE UN ET DEUX. DU PROBLEME DE DIRICHLET POUR L"OPE
1472 ON A CLASS OF FINITE ELEMENTS GENERATED BY LAGRANGE INTERPOLATION II.
1471 ON A CLASS OF FINITE ELEMENTS GENERATED BY LAGRANGE INTERPOLATION.

2052 LAGRANGIAN FINITE ELEMENT AND FINITE DIFFERENCE METHODS FOR POISS
1506 A LAGRANGIAN FINITE ELEMENT SOLUTION TO UNSTEADY FLOW.
1470 ON LAGRANGIAN INTERPOLATION IN N VARIABLES.
198 THE FINITE ELEMENT METHOD WITH LAGRANGIAN MULTIPLIERS.
199 THE FINITE ELEMENT METHOD WITH LAGRANGIAN MULTIPLIERS.
1203 ELEMENT METHOD IN TWO-DIMENSIONAL HYDRODYNAMICS USING LAGRANGIAN VARIABLES.

1593 NON-POLYNOMIAL LAGRANGIANS.

1501 NONLINEAR FINITE ELEMENT ANALYSIS OF LAMINATED COMPOSITE SHELLS.
1233 ELASTIC AND PLASTIC INTERLAMINAR SHEAR DEFORMATION IN LAMINATED COMPOSITES UNDER GENERALIZED PLANE STRESS.
228 THREE DIMENSIONAL FINITE ELEMENT ANALYSIS OF LAMINATED COMPOSITES.
1404 NALYSIS OF SANDWICH PLATES AND CYLINDRICAL SHELLS WITH LAMINATED FACES.
1654 INCLUDING TRANSVERSE SHEAR EFFECTS FOR APPLICATIONS TO LAMINATED PLATES.
1159 ICIENT FINITE ELEMENT ANALYSIS OF RECTANGULAR AND SKEW LAMINATED PLATES.

562 NONLINEAR HYBRID STRESS FINITE ELEMENT ANALYSIS OF LAMINATED SHELLS.

1194 ESSES AROUND HOLES IN STIFFENED COMPOSITE PANELS USING LAURENT-SERIES AND FINITE ELEMENT METHODS.

2139 A COMPUTATIONAL INVESTIGATION OF LEAST SQUARE-FINITE ELEMENT FOR ELASTO-STATIC PROBLEMS.
1799 A COMPARISON OF LEAST SQUARES AND OTHER PROJECTION METHODS FOR THE APPROXIMATE S
578 LEAST SQUARES AND VARIATIONAL METHODS FOR THE SOLUTION OF DIFFERE
357 ERICAL SOLUTION OF ELLIPTIC BOUNDARY VALUE PROBLEMS BY LEAST SQUARES APPROXIMATION OF THE DATA.
1287 USE OF THE LEAST SQUARES CRITERION IN THE FINITE ELEMENT FORMULATION.
1283 USE OF THE LEAST SQUARES CRITERION IN THE FINITE ELEMENT FORMULATION.
1282 LEAST SQUARES FINITE ELEMENT ANALYSIS OF LAMINAR BOUNDARY LAYER F
1281 A LEAST SQUARES FINITE ELEMENT ANALYSIS OF NONLINEAR BIOPHYSICAL DI
221 SIMPLIFIED PROOFS OF ERROR ESTIMATES FOR THE LEAST SQUARES METHOD FOR DIRICHLET"S PROBLEM.
1284 FINITE ELEMENTS FORMULATED BY THE WEIGHTED DISCRETE LEAST SQUARES METHOD.
957 HING OF DISCONTINUOUS FINITE ELEMENT FUNCTIONS USING A LEAST SQUARES METHOD.
37 FINITE ELEMENT APPLICATIONS OF THE LEAST SQUARES METHOD.
356 LEAST SQUARES METHODS FOR 2MTH ORDER ELLIPTIC BOUNDARY-VALUE PROB
1703 UTION OF THE FIRST BIHARMONIC PROBLEM BY THE METHOD OF LEAST SQUARES ON THE BOUNDARY.
1237 AN APPLICATION OF LEAST SQUARES TO ONE DIMENSIONAL TRANSIENT PROBLEMS.
35 A LEAST SQUARES-FINITE ELEMENT SOLUTION OF NONLINEAR OPERATORS.

1463 DISCRETE LEGENDRE ORTHOGONAL POLYNOMIALS.

370 S EQUATIONS ET INEQUATIONS NON LINEAIRES DANS LE ESPACES VECTORIELS EN DUALITE.
1809 S EQUATIONS AUX DERIVEES PARTIELLES NON LINEAIRES, DE TYPE MONOTONE.
1254 S METHODES DE RESOLUTION DES PROBLEMES AUX LIMITES NON LINEAIRES.

425 VERGENCE OF COLLOCATION AND GALERKIN APPROXIMATIONS TO LINEAR TWO-POINT PARABOLIC PROBLEMS.
1161 LINEAR ACCELERATOR CAVITY FIELD CALCULATION BY THE FINITE ELEMENT
1794 COMPUTER PROGRAMS FOR THE SOLUTION OF SYSTEMS OF LINEAR ALGEBRAIC EQUATIONS
718 COMPUTER SOLUTION OF LINEAR ALGEBRAIC SYSTEMS.
269 LINEAR ANALYSIS OF STRUCTURES BY THE METHOD OF CONJUGATE GRADIENT
1864 THE FINITE ELEMENT METHOD, LINEAR AND NONLINEAR APPLICATIONS.
2070 FINITE ELEMENT ANALYSIS OF LINEAR AND NONLINEAR HEAT TRANSFER.
1532 FINITE ELEMENT APPROXIMATIONS OF LINEAR AND NONLINEAR OPERATORS.
2119 SOME LINEAR AND NONLINEAR PROBLEMS IN FLUID MECHANICS.
1052 R THE GENERATION OF WEAK-WEAK FINITE APPROXIMATIONS OF LINEAR AND NONLINEAR SPLINE FUNCTIONS.
692 MINIMIZATION PROBLEMS AND LINEAR AND NONLINEAR TWO-DIMENSIONAL STRUCTURES.
1849 REFINED FINITE ELEMENT ANALYSIS OF LINEAR APPROXIMATION FORMULAS IN TWO VARIABLES.
1745 THE REMAINDER OF CERTAIN LINEAR APPROXIMATION.
91 IMPROVED LINEAR AXISYMMETRIC SHELL FLUID MODEL FOR LAUNCH VEHICLE LONGITUD
1742 A VARIATIONAL PRINCIPLE FOR LINEAR COUPLED FIELD PROBLEMS IN CONTINUUM MECHANICS.
1198 LINEAR DIFFERENTIAL OPERATORS.
173 THEORETICAL FORMULATION OF FINITE ELEMENT METHODS IN LINEAR ELASTIC ANALYSIS OF GENERAL SHELLS.
1602 APPLICATION OF MATRIX DISPLACEMENT METHOD TO LINEAR ELASTIC ANALYSIS OF SHELLS OF REVOLUTION.
1619 THE CONVERGENCE OF FINITE ELEMENT METHOD IN SOLVING LINEAR ELASTIC PROBLEMS.
1947 HE CONVERGENCE OF THE FINITE ELEMENT METHOD IN SOLVING LINEAR ELASTIC PROBLEMS.
272 T COMPUTER PROGRAM THAT CALCULATES THE INTENSITIES OF LINEAR ELASTIC SINGULARITIES.
1857 TRANSIENT RESPONSE OF LINEAR ELASTIC STRUCTURES DETERMINED BY THE MATRIX EXPONENTIAL ME
1334 THE THEORY OF THE DISPLACEMENT METHOD OF ANALYSIS FOR LINEAR ELASTICITY.
982 A REMARK ON LINEAR ELLIPTIC DIFFERENTIAL EQUATIONS OF SECOND ORDER.
1555 VARIATIONAL DIFFERENCE SCHEMES FOR SECOND ORDER LINEAR ELLIPTIC EQUATIONS IN A TWO-DIMENSIONAL REGION WITH A PIE
174 E APPROXIMATE SOLUTIONS OF BOUNDARY VALUE PROBLEMS FOR LINEAR ELLIPTIC OPERATORS BY GALLERKIN"S AND FINITE DIFFERENCE ME

```
                              ANALYSIS OF  LOADING  AND THERMAL EFFECTS ON FUELLED GRAPHITE BRICKS FOR A HIGH     655
                             PURE MOMENT  LOADING  OF AXISYMMETRIC FINITE ELEMENT MODELS.                         854
                 BUCKLING UNDER COMBINED  LOADING  OF THIN FLAT-WALLED STRUCTURES BY A COMPLEX  FINITE STRIP      1633
   OF REVOLUTION SUBJECTED TO AXISYMMETRIC AND ASYMMETRIC  LOADING.                                               2074
   R PREDICTION OF FATIGUE CRACK PROPAGATION UNDER RANDOM  LOADING.                                               1394
   OF REVOLUTION UNDER ASYMMETRIC MECHANICAL AND THERMAL  LOADING.                                                402
   K PROPAGATION IN THREE-DIMENSIONAL SOLIDS UNDER CYCLIC  LOADING.                                               277
   THE PLASTIC ANALYSIS OF STRUCTURES SUBJECTED TO CYCLIC  LOADING.                                               153

   FLECTION OF STRUCTURES SUBJECT TO HEATING AND EXTERNAL  LOADS.                                                 1960

                    GENERALISED MATRIX INVERSE TECHNIQUES FOR  ON LOCAL  APPROXIMATION PROPERTIES OF L2 PROJECTION ON SPLINE SUBSPAC   1491
                 GENERALISED MATRIX INVERSE TECHNIQUES FOR  LOCAL  APPROXIMATIONS OF OPERATOR EQUATIONS.          543
   ION OF SMOOTHING AND INTERPOLATING NATURAL SPLINES VIA  LOCAL  BASES.                                          1279
                      THE FINITE ELEMENT METHOD AND  LOCAL  BOUNDS FOR BOUNDARY VALUE PROBLEMS OF ELASTIC STRUCTURES.   1772
   SOLUTION OF MIXED BOUNDARY VALUE PROBLEMS WITH  LOCAL  ERROR BOUND BY THE FINITE ELEMENT METHOD.               2106
                             LOCAL  GLOBAL SMOOTHING OF DISCONTINUOUS FINITE ELEMENT FUNCTIONS   957
                    MATRIX ANALYSIS OF  LOCAL  INSTABILITY IN PLATES STIFFENED PANELS AND COLUMNS   1660
                             LOCAL  MESH REFINEMENT WITH FINITE ELEMENTS FOR ELLIPTIC PROBLEMS.   863
   A METHOD FOR SIMULATING CONCENTRATED FORCES AND  LOCAL  REINFORCEMENTS IN STRESS COMPUTATION.                  1934

                           THE  LUMINA  ELEMENT FOR THE MATRIX DISPLACEMENT METHOD.                               127

                             L2  AND L INFINITY CONDITION NUMBERS OF THE FINITE ELEMENT STIFFNE   758
                             L2  ERROR BOUNDS FOR THE RAYLEIGH-RITZ-GALERKIN METHOD.              1781
                             L2  ERROR ESTIMATES FOR GALERKIN APPROXIMATIONS TO PARABOLIC PARTI   2042
                      A PRIORI  L2  ERROR ESTIMATES FOR PARABOLIC GALERKIN METHODS.               644
                         SOME  L2  OPTIMALITY AND WEIGHTED H1 PROJECTIONS INTO PIECEWISE POLYNOMI   648
   ON LOCAL APPROXIMATION PROPERTIES OF  L2  PROJECTION ON SPLINE SUBSPACES.                      1491
   TWO METHODS OF GALERKIN TYPE ACHIEVING OPTIMUM  L2  RATES OF CONVERGENCE FOR FIRST ORDER HYPERBOLICS.   582

   THE BEHAVIOUR OF DIFFERENTIABLE FUNCTIONS ON NONSMOOTH  MANIFOLDS.                             297

   3 A FINITE ELEMENT PROGRAM TO DETERMINE STIFFNESS AND  MASS  MATRICES OF RING STIFFENED SHELLS OF REVOLUTION.   889
   HE GENRATION OF INTERELEMENT COMPATIABLE STIFFNESS AND  MASS  MATRICES BY USE OF INTERPOLATION FORMULAE.   332
             DEVELOPMENT OF EXPLICIT STIFFNESS AND  MASS  MATRICES FOR A TRIANGULAR PLATE ELEMENT.   1823
             DERIVATION OF GEOMETRIC STIFFNESS AND  MASS  MATRICES FOR FINITE ELEMENT HYBRID MODELS.   1946
                        EQUIVALENT  MASS  MATRICES FOR RECTANGULAR PLATES IN BENDING.   1658
             REDUCTION OF STIFFNESS AND  MASS  MATRICES.                                          881
   CONDITION NUMBERS OF THE FINITE ELEMENT STIFFNESS AND  MASS  MATRICES, AND THE POINTWISE CONVERGENCE OF THE METHOD.   758
                        CONSISTENT  MASS  MATRIX FOR DISTRIBUTED MASS SYSTEMS.                    89
                        DISTRIBUTED  MASS  MATRIX FOR PLATE ELEMENT BENDING.                      1015
                        DISTRIBUTED  MASS  MATRIX FOR PLATE ELEMENTS IN BENDING.                  889
   PREDICTION OF PLATE VIBRATIONS USING A CONSISTENT  MASS  MATRIX.                               1881
   CONTINUUM MECHANICS AS AN APPROACH TO ROCK  MASS  PROBLEMS.                                    1846
             CONSISTENT MASS MATRIX FOR DISTRIBUTED  MASS  SYSTEMS.                               89

   FINITE ELEMENT ANALYSIS OF COMBINED PROBLEMS OF  MATERIAL  AND GEOMETRIC BEHAVIOUR.            1309
   STATIC STRESS ANALYSIS OF AXISYMMETRIC SOLIDS WITH  MATERIAL  AND GEOMETRIC NONLINEARITIES.    911
   UCTION OF A FINITE ELEMENT APPROXIMATION WHICH CROSSES  MATERIAL  INTERFACES.                  1210
   ESPONSES OF CYLINDRICAL SHELLS INCLUDING GEOMETRIC AND  MATERIAL  NONLINEARITIES.              2078
   CREMENTAL FORMULATIONS FOR PROBLEMS WITH GEOMETRIC AND  MATERIAL  NONLINEARITIES.              2083
   MENT ANALYSIS, WITH SPECIAL REFERENCE TO GEOMETRIC AND  MATERIAL  NONLINEARITIES.              1310
```

505 — STIFFNESS MATRIX FOR SHALLOW RECTANGULAR SHELL ELEMENT.
1725 — A STIFFNESS MATRIX FOR SHALLOW SHELL FINITE ELEMENTS.
1343 — A STIFFNESS MATRIX FOR THE ANALYSIS OF THIN PLATES IN BENDING.
1359 — A STIFFNESS MATRIX FOR THE ANALYSIS OF THIN PLATES IN BENDING.
2100 — GENERALISED MATRIX FORCE AND DISPLACEMENT METHODS FOR LINEAR STRUCTURAL ANALY
1656 — TRIANGULAR PLATE ELEMENTS IN THE MATRIX FORCE METHOD OF STRUCTURAL ANALYSIS
1657 — TETRAHEDRON ELEMENTS IN THE MATRIX FORCE METHOD OF STRUCTURAL ANALYSIS.
90 — CONSISTENT MATRIX FORMULATION FOR STRUCTURAL ANALYSIS USING FINITE ELEMENT T
616 — ALYSIS OF RADIATION DAMAGE STRESSES IN THE GRAPHITE OF MATRIX FUEL ELEMENTS.
874 — ELEMENT STIFFNESS MATRIX GENERATION.
543 — GENERALISED MATRIX INVERSE TECHNIQUES FOR LOCAL APPROXIMATIONS OF OPERATOR EQ
1973 — MATRIX ITERATIVE ANALYSIS.
584 — A MATRIX METHOD OF STRUCTURAL ANALYSIS.
1262 — MATRIX METHODS IN STRUCTURAL ANALYSIS.
134 — MATRIX METHODS IN STRUCTURAL ANALYSIS, A PRECIS OF RECENT DEVELOP
2161 — PROCEEDINGS 1ST CONFERENCE MATRIX METHODS IN STRUCTURAL MECHANICS.
2162 — PROCEEDINGS 2ND CONFERENCE MATRIX METHODS IN STRUCTURAL MECHANICS.
2163 — PROCEEDINGS 3RD CONFERENCE MATRIX METHODS IN STRUCTURAL MECHANICS.
798 — INTRODUCTION IN MATRIX METHODS OF STRUCTURAL ANALYSIS, AND DESIGN.
1321 — INTRODUCTION TO MATRIX METHODS OF STRUCTURAL ANALYSIS.
725 — MATRIX METHODS OF STRUCTURAL ANALYSIS.
93 — MATRIX METHODS OF STRUCTURAL ANALYSIS.
52 — MATRIX METHODS OF STRUCTURAL ANALYSIS.
94 — MATRIX METHODS OF STRUCTURAL ANALYSIS.
1217 — RECENT ADVANCES IN MATRIX METHODS OF STRUCTURAL ANALYSIS.
1176 — GENERALIZED STIFFNESS MATRIX OF A CURVED BEAM ELEMENT.
2065 — A SIMPLE ALGORITHM FOR THE STIFFNESS MATRIX OF TRIANGULAR PLATE BENDING ELEMENTS.
1869 — SPARSE MATRIX PROCEEDINGS, RA=1,
778 — INTEGRATION OF AREA COORDINATES IN MATRIX STRUCTURAL ANALYSIS.
724 — A CORRELATION STUDY OF METHODS OF MATRIX STRUCTURAL ANALYSIS.
1659 — UPPER AND LOWER BOUNDS IN MATRIX STRUCTURAL ANALYSIS.
1968 — THEORY OF MATRIX STRUCTURAL ANALYSIS.
1081 — CIT EXPRESSIONS FOR TRIANGULAR TORUS ELEMENT STIFFNESS MATRIX.
876 — PREDICTION OF PLATE VIBRATIONS USING A CONSISTENT MASS MATRIX.
N OF FRAMES AND OTHER STRUCTURES WITH BANDED STIFFNESS MATRIX.

1006 — MATUS A THREE-DIMENSIONAL FINITE ELEMENT PROGRAM FOR SMALL STRAIN

164 — MAXIMUM AND MINIMUM PRINCIPLES FOR A CLASS OF PARTIAL DIFFERENTIA
471 — MAXIMUM PRINCIPLE AND UNIFORM CONVERGENCE FOR THE FINITE ELEMENT

1993 — HREE DIMENSIONAL ANISOTROPIC AND INHOMOGENEOUS ELASTIC MEDIA BY FINITE ELEMENTS.
100 — NONLINEAR FLOW IN POROUS MEDIA MATRIX ANALYSIS FOR SMALL AND LARGE DISPLACEMENTS.
2128 — STRESSES IN ANISOTROPIC MEDIA OF ROCK MECHANICS.
2137 — FINITE ELEMENT STUDIES OF SOILS AND POROUS MEDIA.
1744 — FINITE ELEMENT ANALYSIS OF SEEPAGE IN ELASTIC MEDIA.
1238 — COUPLED ELECTROKINETIC AND HYDRODYNAMIC FLOW IN POROUS MEDIA.
1049 — THE FINITE ELEMENT METHOD TO TRANSIENT FLOW IN POROUS MEDIA.
906 — MENT FOR THE SOLUTION OF FIELD PROBLEMS IN ORTHOTROPIC MEDIA.
989 — FINITE ELEMENT ANALYSIS OF NONLINEAR SOIL MEDIA.
813 — FLOW OF COMPRESSIBLE FLUID IN POROUS ELASTIC MEDIA.
746 — M, COUPLED STEFAN PROBLEM, FROST PROPAGATION IN POROUS MEDIA.
383 — NT SOLUTION OF UNSTEADY AND UNSATURATED FLOW IN POROUS MEDIA.

478   RDER ACCURACY FOR NONLINEAR BOUNDARY VALUE PROBLEMS V: MONOTONE OPERATORS.

1811   QUATIONS AUX DERIVEES PARTIELLES NON LINEARRES DE TYPE MONOTONE.

1809   QUATIONS AUX DERIVEES PARTIELLES NON LINEAIRES DE TYPE MONOTONE.

371   HOD D''APPROXIMATION ET D''ITERATION POUR LES OPERATEURS MONOTONES'.

1964   SOLUTION OF MULTI-DIMENSIONAL NEUTRON TRANSPORT EQUATION BY FINITE ELEMENT ME

291   THE CALCULATION OF MULTIDIMENSIONAL AND GRAM-CHARIER COEFFICIENTS.

1778   RAYLEIGH-RITZ-GALERKIN METHODS FOR MULTIDIMENSIONAL PROBLEMS.

313   MULTIVARIATE APPROXIMATION BY LOCALLY BLENDED UNIVARIATE INTERPOL

1775   LINFINITY MULTIVARIATE APPROXIMATION THEORY.

841   BLENDING FUNCTION METHODS FOR BIVARIATE AND MULTIVARIATE APPROXIMATION.

839   DISTRIBUTIVE LATTICES AND APPROXIMATION OF MULTIVARIATE FUNCTIONS.

1209   L-INFINITY BOUNDS FOR MULTIVARIATE LAGRANGE APPROXIMATION.

342   NUMERICAL PROPERTIES OF A MULTIVARIATE RITZ-TREFFTZ METHOD.

1457   MULTIVARIATE SMOOTHING AND INTERPOLATING SPLINES.

1777   MULTIVARIATE SPLINE FUNCTIONS AND ELLIPTIC PROBLEMS.

1776   APPROXIMATION THEORY OF MULTIVARIATE SPLINE FUNCTIONS AND ELLIPTIC PROBLEMS, SIAM J'.

1783   MULTIVARIATE SPLINE FUNCTIONS IN SOBOLEV SPACES.

375   NASTRAN BUCKLING ANALYSIS OF A LARGE STIFFENED CYLINDRICAL SHELL

1292   THE NASTRAN COMPUTER PROGRAM FOR STRUCTURAL ANALYSIS.

48   NASTRAN EXPERIENCES OF FORT WORTH OPERATION.

821   RE FOR EFFICIENTLY GENERATING, CHECKING AND DISPLAYING NASTRAN INPUT AND OUTPUT DATA FOR ANALYSIS OF AEROSPACE VEHICLE

1331   THE NASTRAN PROGRAM FOR STRUCTURAL ANALYSIS.

1291   SOME ORGANIZATIONAL ASPECTS OF NASTRAN.

535   AN INTERACTIVE COMPUTER GRAPHICS PROGRAM FOR NASTRAN.

515   AUTOMATED INPUT DATA PREPARATION FOR NASTRAN.

533   THE TREATMENT OF NATURAL BOUNDARY CONDITIONS IN THE FINITE ELEMENT AND FINITE DIFF

751   SOME ASPECTS OF THE NATURAL CO-ORDINATE SYSTEM IN THE FINITE-ELEMENT METHOD

38   NATURAL CO-ORDINATE SYSTEMS.

661   ON FINITE ELEMENT INTEGRATION IN NATURAL CO-ORDINATES.

892   NATURAL CUBIC AND BICUBIC SPLINE INTERPOLATION'.

2133   NATURAL FREQUENCIES OF COMPLEX FREE OR SUBMERGED STRUCTURES BY TH

459   N OF FINITE ELEMENT ANALYSIS TO UNDERGROUND STORAGE OF NATURAL GAS.

132   NATURAL GEOMETRY OF SURFACES WITH SPECIFIC REFFRENCE TO THE MATRI

141   SOME GENERAL CONSIDERATIONS ON THE NATURAL MODE TECHNIQUE; I: SMALL DISPLACEMENTS, II: LARGE DISPLAC

1276   ON THE COMPUTATION OF NATURAL MODES OF AN UNSUPPORTED VIBRATING STRUCTURE BY SIMULTANEO

1278   L PROCEDURES FOR COMPUTING SMOOTHING AND INTERPOLATING NATURAL SPLINES

1279   COMPUTATION OF SMOOTHING AND INTERPOLATING NATURAL SPLINES VIA LOCAL BASES'.

125   LARGE NATURAL STRAINS AND SOME SPECIAL DIFFICULTIES DUE TO NON-LINEARIT

514   BODY ORIENTED (NATURAL) CO-ORDINATES FOR GENERATING THREE DIMENSIONAL MESHES.

441   NUMERICAL SOLUTION OF THE NAVIER STOKES EQUATIONS BY THE FINITE ELEMENT METHOD'

1907   A NUMERICAL SOLUTION OF THE NAVIER STOKES EQUATIONS USING THE FINITE ELEMENT TECHNIQUE.

1685   IN METHOD TO SOLVE THE INITIAL PROBLEM OF A STABILIZED NAVIER-STOKES EQUATION.

1045   NUMERICAL SOLUTION OF THE STATIONARY NAVIER-STOKES EQUATIONS BY FINITE ELEMENT METHODS.

1558   APPLICATION OF FINITE ELEMENT SOLUTION TECHNIQUE TO NEUTRON DIFFUSION AND TRANSPORT EQUATIONS.

THE METHOD OF ORTHOGONAL PROJECTION AND DIRICHLET BOUNDARY VALUE PROBLEM IN A G 454
THE METHOD OF ORTHOGONAL PROJECTION IN POTENTIAL THEORY. 2041
THE ORTHOGONAL PROJECTION METHOD FOR THE NUMERICAL SOLUTION OF A FIEL 1664
ORTHOGONAL RIEHENENTWICKLUNG NACH LINEAREN SPLINE FUNKTIONEN. 1480

ORTHOGONALITY PROPERTIES OF SPLINE FUNCTIONS. 20
THE SOLUTION OF DUAL COSINE SERIES BY THE USE OF ORTHOGONALITY RELATIONS. 1500
SOLUTION OF DUAL TRIGONOMETRICAL SERIES USING ORTHOGONALITY RELATIONS. 1499

AN OPTIMAL SUCCESSIVE OVERRELAXATION TECHNIQUE FOR SOLVING SECOND ORDER FINITE DIFFEREN 1939
EXTENSIONS OF THE SUCCESSIVE OVERRELAXATION THEORY WITH APPLICATIONS TO FINITE ELEMENT APPROXI 1980
QUADRATIC PROGRAMMING FOR SUCCESSIVE OVERRELAXATION. 657

PAFEC 70+, THE MANUAL OF THE PAFEC-SYSTEM. 938

ENT GALERKIN METHOD SOLUTIONS TO SELECTED ELLIPTIC AND PARABOLIC DIFFERENTIAL EQUATIONS. 86
FOR GALERKIN METHODS FOR ONE-DIMENSIONAL SECOND ORDER PARABOLIC AND HYPERBOLIC EQUATIONS. 2043
ALTERNATING-DIRECTION GALERKIN METHODS FOR PARABOLIC AND HYPERBOLIC PROBLEMS ON RECTANGULAR POLYGONS. 583
THE USE OF PARABOLIC ARCS IN MATCHING CURVED BOUNDARIES. 1340
SEMIDISCRETE LEAST SQUARES METHODS FOR A PARABOLIC BOUNDARY VALUE PROBLEM. 358
QUASILINEAR PARABOLIC BOUNDARY VALUE PROBLEMS. APPROXIMATE SOLUTIONS AND ERROR 448
ATION OF DIRECT VARIATIONAL METHODS TO THE SOLUTION OF PARABOLIC BOUNDARY VALUE PROBLEMS OF ARBITRARY ORDER IN THE SPACE 1702
APPROXIMATE SOLUTION OF ELLIPTIC AND PARABOLIC BOUNDARY VALUE PROBLEMS. 329
ON A SEMI-VARIATIONAL METHOD FOR PARABOLIC EQUATIONS I. 959
ON A SEMI VARIATIONAL METHOD FOR PARABOLIC EQUATIONS II. 960
COLLOCATION METHODS FOR PARABOLIC EQUATIONS IN A SINGLE SPACE VARIABLE. 630
A SURVEY OF DISCRETE GALERKIN METHODS FOR PARABOLIC EQUATIONS IN ONE SPACE VARIABLE. 687
GALERKIN METHODS FOR PARABOLIC EQUATIONS WITH NONLINEAR BOUNDARY CONDITIONS. 627
FINITE ELEMENT METHODS FOR PARABOLIC EQUATIONS. 2156
UNCONDITIONALLY STABLE FINITE FINITE ELEMENT SCHEMS FOR PARABOLIC EQUATIONS. 2157
FINITE ELEMENT METHODS FOR PARABOLIC EQUATIONS. 2155
ERROR BOUNDS FOR THE GALERKIN METHOD FOR NONLINEAR PARABOLIC EQUATIONS. 1785
CONVERGENT FINITE DIFFERENCE SCHEMES FOR NONLINEAR PARABOLIC EQUATIONS. 1706
ICAL INTEGRATION IN FINITE ELEMENT METHODS FOR SOLVING PARABOLIC EQUATIONS. 1687
ERROR BOUNDS FOR THE GALERKIN METHOD FOR LINEAR PARABOLIC EQUATIONS. 1784
ON A CONJUGATE SEMI-VARIATIONAL METHODS FOR PARABOLIC EQUATIONS. 961
THREE LEVEL GALERKIN METHODS FOR PARABOLIC EQUATIONS. 647
A FINITE ELEMENT COLLOCATION METHOD FOR QUASILINEAR PARABOLIC EQUATIONS. 628
A FINITE ELEMENT COLLOCATION METHOD FOR QUASILINEAR PARABOLIC EQUATIONS. 626
GALERKIN METHODS FOR PARABOLIC EQUATIONS. 624
GALERKIN METHODS FOR PARABOLIC EQUATIONS. 621
GALERKIN METHODS FOR PARABOLIC GALERKIN METHODS, 689
RICHARDSON EXTRAPOLATION FOR PARABOLIC GALERKIN METHODS. 644
SOME L2 ERROR ESTIMATES FOR PARABOLIC PARTIAL DIFFERENTIAL EQUATIONS. 2442
IORI L2 ERROR ESTIMATES FOR GALERKIN APPROXIMATIONS TO PARABOLIC PARTIAL DIFFERENTIAL EQUATIONS. 1661
E GALERKIN APPROXIMATIONS FOR SECOND ORDER NON-LINEAR PARABOLIC PARTIAL DIFFERENTIAL EQUATIONS. 1441
NUMERICAL SOLUTION OF PARABOLIC PROBLEMS BY THE GENERALIZED CRANK NICHOLSON SCHEME. 1651
OR BOUNDS FOR SEMIDISCRETE GALERKIN APPROXIMATIONS FOR LINEAR PARABOLIC PROBLEMS WITH APPLICATIONS TO PETROLEUM RESERVOIR MECHA 1979
CHEBYSHEV SEMI-DISCRETE APPROXIMATIONS FOR LINEAR PARABOLIC PROBLEMS. 707
ELEMENT APPROXIMATION OF STEADY STATE, EIGENVALUE, AND PARABOLIC PROBLEMS. 425
ATION AND GALERKIN APPROXIMATIONS TO LINEAR TWO-POINT PARABOLIC PROBLEMS. 763
PARTIAL DIFFERENTIAL EQUATIONS OF PARABOLIC TYPE.

MEMBRANE PARALLELOGRAM ELEMENT WITH LINEARLY VARYING EDGE STRAIN FOR THE M 104

```
1737  A RECTANGULAR PLATE BENDING ELEMENT THE USE OF WHICH IS EQUIVALENT TO THE USE O
1990  A REFINED MIXED TYPE PLATE BENDING ELEMENT.
1413  THE CONSTANT BENDING MOMENT PLATE BENDING ELEMENT.
 559  FOR THE RECTANGULAR PLATE BENDING ELEMENT.
 525  ND DYNAMIC APPLICATIONS OF A HIGH-PRECISION TRIANGULAR PLATE BENDING ELEMENT.
 263  FORMULATION OF A NEW TRIANGULAR PLATE BENDING ELEMENT.
1985  ON THE QUINTIC TRIANGULAR PLATE BENDING ELEMENTS IN THE ANALYSIS OF VARIABLE THICKNESS PLAT
1286  IABLE AND CONSTANT THICKNESS HIGH PRECISION TRIANGULAR PLATE BENDING ELEMENTS PAPER M6/5, PROCEEDINGS 1ST STRUCTURAL MEC
 913  TRIANGULAR THICK PLATE BENDING ELEMENTS WITH ENFORCED COMPATIBILITY.
1643  TRIANGULAR PLATE BENDING ELEMENTS.
1176  CONFORMING RECTANGULAR AND TRIANGULAR PLATE BENDING ELEMENTS.
1246  IMPLE ALGORITHM FOR THE STIFFNESS MATRIX OF TRIANGULAR PLATE BENDING ELEMENTS.
1023  GENCE STUDIES OF EIGENVALUE SOLUTIONS USING TWO FINITE PLATE BENDING ELEMENTS.
 973  A NEW FORMULATION FOR PLATE BENDING ELEMENTS.
 431  STIFFNESS MATRICES FOR RECTANGULAR PLATE BENDING ELEMENTS.
 262  TRIGONOMETRIC FUNCTION REPRESENTATIONS FOR RECTANGULAR PLATE BENDING ELEMENTS.
  63  TRIANGULAR PLATE BENDING ELEMENTS.
 392  FINITE ELEMENT PLATE BENDING EQUILIBRIUM ANALYSIS.
 261  A COMPATIBLE TRIANGULAR PLATE BENDING FINITE ELEMENT.
1840  A REFINED TRIANGULAR PLATE BENDING FINITE ELEMENT.
 761  A TECHNIQUE FOR MESH GRADING APPLIED TO CONFORMING PLATE BENDING FINITE ELEMENTS.
1134  SIDUAL ENERGY BALANCING TECHNIQUE IN THE GENERATING OF PLATE BENDING FINITE ELEMENTS.
  53  FINITE ELEMENT SOLUTIONS FOR PLATE BENDING PROBLEMS BY SIMPLIFIED HYBRID DISPLACEMENT METHOD.
1499  A POLYGONAL FINITE ELEMENT IN THE SOLUTION OF PLATE BENDING PROBLEMS USING THE ASSUMED STRESS APPROACH.
1408  THE TRIANGULAR EQUILIBRIUM ELEMENT IN THE SOLUTION OF PLATE BENDING PROBLEMS.
1058  RIUM ELEMENT WITH LINEARLY VARYING BENDING MOMENTS FOR PLATE BENDING PROBLEMS.
 994  ON THE CONVERGENCE OF A MIXED FINITE ELEMENT FOR PLATE BENDING PROBLEMS.
  65  PIECEWISE RICUBIC METHODS FOR PLATE BENDING PROBLEMS.
1936  A CONFORMING TRIANGULAR FINITE ELEMENT PLATE BENDING SOLUTION.
  58  ANALYSIS OF PLATE BENDING USING TRIANGULAR ELEMENTS.
 868  TRIANGULAR FINITE ELEMENTS FOR PLATE BENDING WITH CONSTANT AND LINEARLY VARYING BENDING MOMENTS.
1750  FINITE ELEMENT ANALYSIS OF PLATE BENDING WITH TRANSVERSE SHEAR DEFORMATION.
1573  AN ANNULAR SEGMENT ELEMENT FOR PLATE BENDING.
1141  ANNULAR AND CIRCULAR SECTOR FINITE ELEMENTS FOR PLATE BENDING.
1152  CONVERGENCE OF A MIXED FINITE ELEMENT SCHEME FOR PLATE BENDING.
1018  MULTIANGULAR ELEMENTS IN PLATE BENDING.
1025  A CONFORMING QUARTIC TRIANGULAR ELEMENT FOR PLATE BENDING.
 963  QUACY OF NODAL CONNECTIONS IN A STIFFNESS SOLUTION FOR PLATE BENDING.
 729  A SIMPLE FINITE ELEMENT MODEL FOR ELASTIC PLASTIC PLATE BENDING.
 736  AN EQUILIBRIUM MODEL FOR PLATE BENDING.
 491  AN EQUILIBRIUM MODEL FOR ANALYSIS OF PLATE BENDING.
 495  A REFINED QUADRILATERAL ELEMENT FOR ANALYSIS OF PLATE BENDING.
  82  FINITE ELEMENT STIFFNESS MATRICES FOR THE ANALYSIS OF PLATE BENDING.
 421  BICUBIC FUNDAMENTAL SPHERES IN PLATE BENDING.
1997  A FINITE ELEMENT TENSOR APPROACH TO PLATE BUCKLING ANALYSIS USING A FULLY COMPATIBLE FINITE ELEMENT.
1136  SIMPLIFIED HYBRID DISPLACEMENT METHOD APPLIED TO PLATE BUCKLING AND POSTBUCKLING INT.
  51  FINITE ELEMENT ANALYSIS OF PLATE BUCKLING PROBLEMS.
 978  THE FINITE ELEMENT METHOD IN PLATE BUCKLING USING A MIXED VARIATIONAL PRINCIPLE.
1583  ANISOTROPIC PARAMETRIC PLATE BENDING DISCRETE ELEMENTS.
1285  A METHOD TO GENERATE BOTH UPPER AND LOWER BOUNDS TO PLATE BENDING EIGENVALUES BY CONFORMING DISPLACEMENT FINITE ELEMENTS.
1015  DISTRIBUTED MASS MATRIX FOR PLATE BENDING ELEMENT BENDING.
1871  A RAPIDLY CONVERGING TRIANGULAR PLATE BENDING ELEMENT.
```

NALYSIS OF SHELLS OF REVOLUTION BY MINIMIZATION OF THE GENERAL POTENTIAL ENERGY AND COMPLEMENTARY ENERGY MODELS BASED ON STRESS   788
POTENTIAL ENERGY FUNCTIONAL.   847
POTENTIAL FLOW ANALYSIS BY FINITE ELEMENTS.   137
A MIXED CUBIC TRIANGULAR FINITE ELEMENT FOR USE IN POTENTIAL FLOW PROBLEMS,   1356
THE APPLICATION OF THE FINITE ELEMENT TECHNIQUE TO POTENTIAL FLOW PROBLEMS,   1034
APPLICATION OF THE FINITE ELEMENT TECHNIQUE TO POTENTIAL FLOW PROBLEMS, 1 AND 2.   598
FINITE ELEMENT TECHNIQUE TO BOUNDARY VALUE PROBLEMS OF POTENTIAL FLOW.   597
THREE DIMENSIONAL ANALYSIS OF COMPRESSIBLE POTENTIAL FLOWS WITH THE FINITE ELEMENT METHOD.   614
UAL MINIMUM THEOREMS TO THE FINITE ELEMENT SOLUTION OF POTENTIAL PROBLEMS WITH SPECIAL REFERENCE TO SEEPAGE.   1603
HIGH-ORDER POLYNOMIAL TRIANGULAR FINITE ELEMENTS FOR POTENTIAL PROBLEMS.   1637
POTENTIAL PROBLEMS.   1815
INTEGRAL EQUATION METHODS IN POTENTIAL THEORY AND ELASTOSTATICS.   1048
THE METHOD OF ORTHOGONAL PROJECTION IN POTENTIAL THEORY.   2041
MIXED BOUNDARY VALUE PROBLEMS IN POTENTIAL THEORY.   1831
FOUNDATIONS OF POTENTIAL THEORY.   1105

PRESSURE VESSEL ANALYSIS BY FINITE ELEMENT METHOD.   1684
STRESS ANALYSIS OF A PRESTRESSED CONCRETE NUCLEAR PRESSURE VESSEL BY THE FINITE ELEMENT METHOD USING VARIABLE STRAI   1189
APPLICATION OF THE FINITE ELEMENT METHOD TO PRESTRESSED CONCRETE REACTOR PRESSURE VESSEL DESIGN (JAPANESE).   1096
PRESTRESSED CONCRETE REACTOR PRESSURE VESSEL FAILURE ANALYSIS.   969
STRESS ANALYSIS OF BWR PRESSURE VESSEL WITH BERSAFE AND FLHE,   670
STRESS ANALYSIS OF BWR PRESSURE VESSEL WITH BERSAFE AND FLHE.   669
AVIGUR OF A LONGITUDINAL SEMIELLIPTIC CRACK IN A THICK PRESSURE VESSEL.   1312
VIOUR OF A LONGITUDINAL SEMI-ELLIPTIC CRACK IN A THICK PRESSURE VESSEL.   1314
MODEL FOR THE ANALYSIS OF AN AXISYMMETRIC THICK-WALLED PRESSURE VESSEL.   83
NT ANALYSIS OF PERFORATED END CAPS FOR NUCLEAR REACTOR PRESSURE VESSELS BY THE FINITE ELEMENT METHOD.   1851
ANALYSIS OF PRESTRESSED CONCRETE PRESSURE VESSELS FOR NUCLEAR REACTORS IN CZECHOSLOVAKIA.   557
FINITE ELEMENT INCREMENTAL ELASTIC-PLASTIC ANALYSIS OF PRESSURE VESSELS USING THE CONSTANT STRAIN AXISYMMETRIC FINITE EL   1712
ELEMENT FOR THE THREE-DIMENSIONAL STRESS ANALYSIS OF PRESSURE VESSELS.   2159
THREE-DIMENSIONAL ANALYSIS OF REACTOR PRESSURE VESSELS.   1675
E ELEMENT METHODS IN THE STRESS ANALYSIS OF THIN SHELL PRESSURE VESSELS.   1683
ADVANCES IN THE ANALYSIS OF PRESTRESSED CONCRETE PRESSURE VESSELS.   1591
OF CALCULATED AND MEASURED STRAINS IN OLDBURY CONCRETE PRESSURE VESSELS.   1236
TE ELEMENT METHOD FOR STRESS CONCENTRATION PROBLEMS IN PRESSURE VESSELS.   1933
HE ANALYSIS OF PODDED BOILER TYPE PRESTRESSED CONCRETE PRESSURE VESSELS.   819
L CALCULATION METHODS FOR PRESTRESSED CONCRETE REACTOR PRESSURE VESSELS.   414
FRACTURE AND SAFETY ANALYSIS OF NUCLEAR PRESSURE VESSELS.   363
CALCULATIONS FOR PRESTRESSED CONCRETE REACTOR PRESSURE VESSELS.   249
PRESSURE VESSELS.   149
HE FINITE ELEMENT METHOD TO THE CALCULATION OF REACTOR PRESSURE VESSELS.   256
IC BUCKLING OF SHALLOW SPHERICAL SHELLS UNDER EXTERNAL PRESSURE.   1143

BERGEN AN AUTOMATIC FINITE ELEMENT MESH GENERATION PROGRAM FOR ARBITRARY STRUCTURES.   1453
A FINITE ELEMENT MESH GENERATION PROGRAM FOR ARBITRARY TWO-AND THREE-DIMENSIONAL STRUCTURES,   1452
A FINITE ELEMENT MESH GENERATION PROGRAM FOR ARBITRARY TWO-AND THREE-DIMENSIONAL STRUCTURES.   1454
H-3D A THREE-DIMENSIONAL FINITE ELEMENT MESH GENERATOR PROGRAM FOR EIGHT NODE ISOPARAMETRIC ELEMENTS.   1008
A FRONTAL SOLUTION PROGRAM FOR FINITE ELEMENT ANALYSIS,   1020
MESH GENERATION PROGRAM FOR HIGHWAY EXCAVATION CUTS.   1068
AN INTERACTIVE COMPUTER GRAPHICS PROGRAM FOR NASTRAN.   535
NONSAP - A GENERAL FINITE ELEMENT PROGRAM FOR NONLINEAR DYNAMIC ANALYSIS OF, COMPLEX STRUCTURES.   251
MATUS A THREE-DIMENSIONAL FINITE ELEMENT PROGRAM FOR SMALL STRAIN ELASTIC ANALYSIS.   1006
COMPUTER PROGRAM FOR SOLUTION OF LARGE, SPARSE, UNSYMMETRIC SYSTEMS OF LIN   1118
COMPUTER PROGRAM FOR STATIC AND DYNAMIC ANALYSIS OF LINEAR STRUCTURAL SYST   2068

APACHE A THREE-DIMENSIONAL FINITE ELEMENT PROGRAM FOR STEADY-STATE OR TRANSIENT HEAT CONDUCTION ANALYSIS. 1007

THE NASTRAN PROGRAM FOR STRUCTURAL ANALYSIS. 1331

THE NASTRAN COMPUTER PROGRAM FOR STRUCTURAL ANALYSIS. 1292

DYNAPLAS A FINITE ELEMENT PROGRAM FOR THE DYNAMIC, LARGE DEFLECTION, ELASTIC-PLASTIC ANALYS 888

A FINITE ELEMENT PROGRAM PACKAGE FOR AXISYMMETRIC SCALAR FIELD PROBLEMS. 1163

OF THE IBM-360 VERSION OF THE FINITE ELEMENT COMPUTER PROGRAM SAFE 3D. 1327

ELASTOPLASTIC AND CREEP ANALYSIS WITH THE ASKA PROGRAM SYSTEM. 224

CHILES A FINITE ELEMENT COMPUTER PROGRAM THAT CALCULATES THE INTENSITIES OF LINEAR ELASTIC SINGUL 272

SAMMSOR 3 A FINITE ELEMENT PROGRAM TO DETERMINE STIFFNESS AND MASS MATRICES OF RING STIFFEN 889

EURCYL, A COMPUTER PROGRAM TO GENERATE FINITE ELEMENT MESHES FOR CYLINDER-CYL 599

A GENERAL HIGH-ORDER FINITE-ELEMENT WAVEGUIDE ANALYSIS PROGRAM. 1814

A SELF-ORGANIZING MESH GENERATION PROGRAM. 1062

TRIANGULAR MESH GENERATION PROGRAM. 1144

SESAM-69, A GENERAL PURPOSE FINITE ELEMENT METHOD PROGRAM. 659

ON AND FUNCTIONAL PURPOSE OF A FINITE ELEMENT COMPUTER PROGRAM. 639

ANALYSIS AND DESIGN CAPABILITIES OF THE STRUDL PROGRAM. 455

ORNL USER'S MANUAL FOR CREEP- PLAST COMPUTER PROGRAM. 481

METHOD OF ORTHOGONAL PROJECTION AND DIRICHLET BOUNDARY VALUE PROBLEM IN A GRAIN DOMAIN 454

THE METHOD OF ORTHOGONAL PROJECTION IN POTENTIAL THEORY. 2041

A PROJECTION METHOD FOR DIRICHLET PROBLEMS USING SUBSPACES WITH NEA 1484

A HIGH ORDER PROJECTION METHOD FOR NONLINEAR TWO POINT BOUNDARY VALUE PROBLEMS 1273

THE ORTHOGONAL PROJECTION METHOD FOR THE NUMERICAL SOLUTION OF A FIELD PROBLEM. 1664

COMPUTATIONAL INVESTIGATION OF LEAST SQUARES AND OTHER PROJECTION METHODS FOR THE APPROXIMATE SOLUTION OF BOUNDARY VALU 1799

INTERIOR ERROR ESTIMATES OF PROJECTION METHODS. 1485

ON LOCAL APPROXIMATION PROPERTIES OF L2 PROJECTION ON SPLINE SUBSPACES. 1491

THE METHOD OF PROJECTIONS AS APPLIED TO THE NUMERICAL SOLUTION OF TWO-POINT BO 567

THEORY OF CONJUGATE PROJECTIONS IN FINITE ELEMENT ANALYSIS. 1530

L2 OPTIMALITY AND WEIGHTED H1 PROJECTIONS INTO PIECEWISE POLYNOMIAL SPACES. 648

THE PUBA FAMILY OF PLATE ELEMENTS FOR THE MATRIX DISPLACEMENT METHOD. 128

QUADRATIC FINITE ELEMENT IN NEUTRON TRANSPORT. 1380

THE PROBLEM OF THE MINIMUM OF A QUADRATIC FUNCTIONAL. 1374

QUADRATIC INTERPOLATING SPLINES: THEORY AND APPLICATIONS. 1071

QUADRATIC MATRIX EQUATIONS FOR FREE VIBRATION ANALYSIS OF STRUCTU 878

SOLUTION OF QUADRATIC PROGRAMMING APPROACH TO THE FINITE ELEMENT METHOD. 1895

THE QUADRATIC PROGRAMMING. 657

NT METHOD FOR FINITE ELEMENT ELASTOPLASTIC ANALYSIS BY QUADRATIC PROGRAMMING. 575

FINITE ELEMENT ELASTOPLASTIC ANALYSIS BY QUADRATIC PROGRAMMING: THE MULTISTAGE METHOD. 576

TWO TYPES OF PIECEWISE QUADRATIC SPACES AND THEIR ORDER OF ACCURACY FOR POISSON'S EQUATI 284

NUMERICAL QUADRATURE AND CUBATURE. 676

NUMERICAL QUADRATURE AND SOLUTION OF ORDINARY DIFFERENTIAL EQUATIONS. 1875

EFFECTS OF QUADRATURE ERRORS IN FINITE ELEMENT APPROXIMATION OF STEADY STATE 707

ON THE EFFECTS OF QUADRATURE ERRORS IN THE FINITE ELEMENT METHOD. 708

THE EFFECT OF QUADRATURE ERRORS IN THE NUMERICAL SOLUTION OF TWO-DIMENSIONAL RO 943

GAUSSIAN QUADRATURE FORMULAS. 524

NONINTERPOLATORY QUADRATURE FORMULAS. 680

GAUSSIAN QUADRATURE FORMULAS. 1877

QUADRATURE RULES FOR BRICK BASED FINITE ELEMENTS. 1022

QUADRATURE SCHEMES FOR THE NUMERICAL SOLUTION OF BOUNDARY VALUE P 942

NUMERICAL QUADRATURE USED FOR RADIATIVE HEAT-TRANSFER COMPUTATIONS. — 1267
EMS WITH APPLICATION TO MECHANICAL DIFFERENTIATION AND QUADRATURE. — 306

SOME PLANE QUADRILATERAL [HYBRID] FINITE ELEMENTS. — 512
A REFINED QUADRILATERAL ELEMENT FOR ANALYSIS OF PLATE BENDING. — 491
MIXED QUADRILATERAL ELEMENTS FOR BENDING. — 376
CURVED ISOPARAMETRIC QUADRILATERAL ELEMENTS IN FINITE ELEMENT ANALYSIS. — 681
QUADRILATERAL FINITE ELEMENTS IN ELASTIC-PLASTIC PLANE STRESS ANA — 1601
VIBRATION OF CURVED STRUCTURES USING QUADRILATERAL FINITE ELEMENTS. — 1606
ARBITRARY QUADRILATERAL SPAR WEBS FOR THE MATRIX DISPLACEMENT METHOD. — 103

A CONFORMING QUARTIC TRIANGULAR ELEMENT FOR PLATE BENDING. — 1018

APPROXIMATE SOLUTIONS AND ERROR BOUNDS FOR QUASILINEAR ELLIPTIC BOUNDARY VALUE PROBLEMS. — 447
QUASILINEAR PARABOLIC BOUNDARY VALUE PROBLEMS APPROXIMATE SOLUTIO — 448
A FINITE ELEMENT COLLOCATION METHOD FOR QUASILINEAR PARABOLIC EQUATIONS. — 628
A FINITE ELEMENT COLLOCATION METHOD FOR QUASILINEAR PARABOLIC EQUATIONS. — 626
NUMERICAL SOLUTION OF THE QUASILINEAR POISSON EQUATION IN A NON-UNIFORM TRIANGULAR MESH. — 2073

ON THE QUINTIC TRIANGULAR PLATE BENDING ELEMENT. — 263

CHEBYSHEV RATIONAL APPROXIMATIONS TO EXP(-X) IN [0, + INFINITY) AND APPLICA — 498
A RATIONAL BASIS FOR FUNCTION APPROXIMATION. — 1999
A RATIONAL BASIS FOR FUNCTION APPROXIMATION; CURVED SIDES. — 2000

N ANALYSIS OF THE LARGE DEFLECTIONS OF BEAMS USING THE RAYLEIGH RITZ FINITE ELEMENT METHOD. — 2007
RAYLEIGH RITZ GALERKIN METHODS FOR SYMMETRIC POSITIVE DIFFERENTIA — 1227

APPLIED TO EIGENVALUE PROBLEMS, I, ESTIMATES RELATING RAYLEIGH-RITZ AND GALERKIN APPROX — 1625
SCHWARZ" INEQUALITY AND THE METHODS OF RAYLEIGH-RITZ AND TREFFTZ. — 602
ORDERS OF CONVERGENCE OF THE RAYLEIGH-RITZ AND WEINSTEIN-BAZLEY METHODS. — 704
RAYLEIGH-RITZ APPROXIMATION BY PIECEWISE POLYNOMIALS. — 316
RAYLEIGH-RITZ APPROXIMATION BY TRIGONOMETRIC POLYNOMIALS. — 318
HIGHER-ORDER RAYLEIGH-RITZ APPROXIMATIONS. — 706
IMPROVEMENT OF RAYLEIGH-RITZ EIGENFUNCTIONS. — 1154
HIGHER ORDER CONVERGENCE RESULTS FOR THE RAYLEIGH-RITZ METHOD APPLIED TO EIGENVALUE PROBLEMS, I, ESTIMATES — 1625
HIGHER ORDER CONVERGENCE RESULTS FOR THE RAYLEIGH-RITZ METHOD APPLIED TO EIGENVALUE PROBLEMS;2 ERROR BOUND — 1626
A MODIFICATION OF THE RAYLEIGH-RITZ METHOD FOR STRESS CONCENTRATION PROBLEMS IN ELASTOP — 1410
A FINITE ELEMENT ANALOGUE OF THE MODIFIED RAYLEIGH-RITZ METHOD FOR VIBRATION PROBLEMS. — 1646
FOR EIGENVALUES OF SOME DIFFERENTIAL OPERATORS BY THE MODIFIED RAYLEIGH-RITZ METHOD. — 2035
A FINITE ELEMENT APPLICATION OF THE MODIFIED RAYLEIGH-RITZ METHOD. — 1411
REMARKS ABOUT THE RAYLEIGH-RITZ METHOD. — 520
LYNOMIAL SOLUTIONS TO THE HELMHOLTZ EQUATION USING THE RAYLEIGH-RITZ METHOD. — 388
CKLING PROBLEMS OF ELASTIC PLATES WITH COMBINED USE OF RAYLEIGH-RITZ PROCEDURE IN THE FINITE ELEMENT METHO — 1101

ERROR-BOUNDS FOR THE RAYLEIGH-RITZ-GALERKIN METHOD. — 1774
ELLIPTIC SPLINE FUNCTIONS AND THE RAYLEIGH-RITZ-GALERKIN METHOD. — 1779
L2 ERROR BOUNDS FOR THE RAYLEIGH-RITZ-GALERKIN METHODS. — 1781
RAYLEIGH-RITZ-GALERKIN METHODS FOR DIRICHLET'S PROBLEM USING SUBS — 355
RAYLEIGH-RITZ-GALERKIN METHODS FOR MULTIDIMENSIONAL PROBLEMS. — 1778
ON RAYLEIGH-RITZ-GALERKIN PROCEDURES FOR NONLINEAR TWO POINT BOUNDAR — 1730

FINITE ELEMENT METHODS FOR REACTOR ANALYSIS. — 1075

BENDING AND VIBRATION OF MULTILAYER **SANDWICH** BEAMS AND PLATES. — 1130

STIFFNESS MATRIX FOR A TRIANGULAR **SANDWICH** ELEMENT IN BENDING. — 1322

A **SANDWICH** PLATE FINITE ELEMENT. — 231

FINITE ELEMENT ANALYSIS OF **SANDWICH** PLATES AND CYLINDRICAL SHELLS WITH LAMINATED FACES. — 1404

TWO HYBRID ELEMENTS FOR ANALYSIS OF THICK THIN AND **SANDWICH** PLATES. — 510

STATIC AND DYNAMIC FINITE ELEMENT ANALYSIS OF **SANDWICH** STRUCTURES. — 4

THE EXTENSION AND APPLICATION OF **SARD** KERNEL THEOREMS ON TRIANGULAR AND RECTANGULAR DOMAINS WITH E — 234

SARD KERNEL THEOREMS TO COMPUTE FINITE ELEMENT ERROR BOUNDS. — 236

VARIATIONAL FORMULATION OF DYNAMICS OF FLUID **SATURATED** POROUS ELASTIC SOLIDS. — 812

A FINITE ELEMENT PROGRAM PACKAGE FOR AXISYMMETRIC **SCALAR** FIELD PROBLEMS. — 1163

SCALAR FINITE ELEMENT PACKAGE FOR TWO DIMENSIONAL FIELD PROBLEMS. — 1162

AXISYMMETRIC TRIANGULAR FINITE ELEMENTS FOR THE **SCALAR** HELMHOLTZ EQUATION. — 1817

A **SECTOR** ELEMENT FOR THIN PLATE FLEXURE. — 1747

SECTOR ELEMENTS FOR MATRIX DISPLACEMENT ANALYSIS. — 1668

STIFFNESS MATRICES FOR **SECTOR** ELEMENTS. — 1573

ANNULAR AND CIRCULAR **SECTOR** FINITE ELEMENTS FOR PLATE BENDING. — 552

SECTOR FINITE ELEMENTS IN THE THEORY OF PLANE ELASTICITY. — 703

SOME FREE SURFACE TRANSIENT FLOW PROBLEMS OF **SEEPAGE** AND IRROTATIONAL FLOW. — 1906

FINITE ELEMENT ANALYSIS OF **SEEPAGE** IN ELASTIC MEDIA. — 1744

SOLUTION OF ANISOTROPIC **SEEPAGE** PROBLEMS BY FINITE ELEMENTS. — 2135

NUMERICAL ANALYSIS OF FREE SURFACE **SEEPAGE** PROBLEMS. — 739

NUMERICAL ANALYSIS OF FREE SURFACE **SEEPAGE** PROBLEMS. — 738

FINITE ELEMENT ANALYSIS OF **SEEPAGE** THROUGH DAMS. — 1244

FINITE ELEMENT ANALYSIS OF **SEEPAGE** THROUGH DAMS, J. — 1637

LUTION OF POTENTIAL PROBLEMS WITH SPECIAL REFERENCE TO **SEEPAGE**. — 558

RESPONSE OF STRUCTURES TO **SEISMIC** EXCITATION. — 457

TURE INTERACTION PROBLEM FOR NUCLEAR POWER PLANT UNDER **SEISMIC** EXCITATION. — 1937

SEISMIC INTERACTION OF SOIL AND POWER PLANTS. — 1426

EFFECT OF GEOLOGIC IRREGULARITIES ON **SEISMIC** RESPONSE. — 278

SESAM A PROGRAMMING SYSTEM FOR FINITE ELEMENT PROBLEMS. — 659

SESAM-69, A GENERAL PURPOSE FINITE ELEMENT METHOD PROGRAM. — 112

A GENERAL METHOD FOR THE **SHAPE** FINDING OF LIGHTWEIGHT TENSION STRUCTURES. — 1029

SHAPE FUNCTION FORMULATIONS FOR ELEMENTS OTHER THAN DISPLACEMENT — 223

SHAPE FUNCTION SUBROUTINE FOR AN ISOPARAMETRIC THIN PLATE ELEMENT — 1909

ON THE COMPLETENESS OF **SHAPE** FUNCTIONS FOR FINITE ELEMENT ANALYSIS. — 1898

ENERGY METHOD OF NETWORKS OF ARBITRARY **SHAPE** IN PROBLEMS OF THE THEORY OF ELASTICITY. — 1761

THE ENERGY METHOD OF NETWORK OF ARBITRARY **SHAPE** IN PROBLEMS OF THEORY OF ELASTICITY. — 1511

TRICES FOR FINITE ELEMENTS OF THIN SHELLS OF ARBITRARY **SHAPE**. — 523

HIGH-PRECISION FINITE ELEMENT FOR SHELLS OF ARBITRARY **SHAPE**. — 527

A SHALLOW SHELL FINITE ELEMENT OF TRIANGULAR **SHAPE**. — 1233

ELASTIC AND PLASTIC INTERLAMINAR **SHEAR** DEFORMATION IN LAMINATED COMPOSITES UNDER GENERALIZED PLANE — 868

NITE ELEMENT ANALYSIS OF PLATE BENDING WITH TRANSVERSE **SHEAR** DEFORMATION. — 1654

A FINITE ELEMENT ANALYSIS INCLUDING TRANSVERSE **SHEAR** EFFECTS FOR APPLICATIONS TO LAMINATED PLATES.

FINITE ELEMENT CONVERGENCE FOR SINGULAR DATA. — 1793
ON THE USE OF SINGULAR ELEMENTS IN VARIATIONAL AND FINITE ELEMENT TRANSPORT COM — 8
NUMERICAL METHODS OF HIGHER-ORDER ACCURACY FOR SINGULAR FUNCTIONS WITH FINITE ELEMENT APPROXIMATION. — 710
SINGULAR NONLINEAR BOUNDARY VALUE PROBLEMS. — 467

STRESS CONCENTRATIONS AND SINGULARITIES AT INTERFACE CORNERS. — 1676
HIGH ORDER SINGULARITIES DUE TO RE-ENTRANT BOUNDARIES IN ELLIPTIC PROBLEMS. — 242
CORNER SINGULARITIES FOR INTERFACE PROBLEMS. — 1108
SINGULARITIES IN ELLIPTIC PROBLEMS BY FINITE ELEMENT METHODS. — 2005
A NUMERICAL TECHNIQUE FOR DETERMINING THE EFFECT OF SINGULARITIES IN FINITE DIFFERENCE SOLUTIONS ILLUSTRATED BY APPL — 673
TREATMENT OF SINGULARITIES IN HARMONIC MIXED BOUNDARY VALUE PROBLEM BY DUAL S — 2047
SINGULARITIES IN THE FINITE ELEMENT APPROXIMATION OF TWO DIMENSIO — 937
SINGULARITIES IN TRANSMISSION LINES. — 549
SINGULARITIES OF ELLIPTIC PROBLEMS. — 310
ANGULAR REPRESENTATION OF SINGULARITIES WITH ISOPARAMETRIC FINITE ELEMENTS. — 271

FOR ELLIPTIC MIXED BOUNDARY VALUE PROBLEMS CONTAINING SINGULARITIES. — 2055
SOLUTION OF STEADY STATE DIFFUSION PROBLEMS CONTAINING SINGULARITIES. — 2051
ELEMENTS FOR THE ANALYSIS OF STRESS CONCENTRATIONS AND SINGULARITIES. — 1677
FINITE ELEMENT ANALYSIS OF STRESS CONCENTRATIONS AND SINGULARITIES. — 1667
SOLUTION OF BOUNDARY VALUE PROBLEMS WITH NON-REMOVABLE SINGULARITIES. — 1414
RAM THAT CALCULATES THE INTENSITIES OF LINEAR ELASTIC SINGULARITIES. — 272
KIN METHODS FOR DIRICHLET PROBLEMS CONTAINING BOUNDARY SINGULARITIES. — 241
SOLUTION OF PROBLEMS WITH INTERFACES AND SINGULARITIES. — 202

EVALUATION OF THE USE OF A SINGULARITY IN THE FINITE ELEMENT ANALYSIS OF A CENTRE-CRACKED PL — 1365
FUNCTIONS NEAR TERMINATION OR INTERSECTION OF GRADIENT SINGULARITY LINES: NUMERICAL METHOD. — 258
THE USE OF SINGULARITY PROGRAMMING IN FINITE-DIFFERENCE AND FINITE-ELEMENT C — 671
BEST FINITE ELEMENTS DISTRIBUTION AROUND A SINGULARITY. — 762

THE ANALYSIS OF SKEW BRIDGE DECKS; A NEW FINITE ELEMENT APPROACH. — 1751
E EFFICIENT FINITE ELEMENT ANALYSIS OF RECTANGULAR AND SKEW LAMINATED PLATES. — 1159
IN THE FINITE ELEMENT ANALYSIS OF THIN RECTANGULAR AND SKEW PLATES IN BENDING. — 1827
FINITE ELEMENT ANALYSIS OF SKEW PLATES IN BENDING. — 1405
FINITE ELEMENT ANALYSIS OF SKEW, CURVED BOX-GIRDER BRIDGE. — 451

STRAIN ENERGY BOUNDS IN FINITE ELEMENT ANALYSIS BY SLAB ANALYSIS. — 737
L EXPERIMENTATION ON THE FINITE ELEMENT METHOD IN BARE SLAB CRITICALITY CALCULATIONS. — 1630
FINITE ELEMENT ANALYSIS OF A BROAD ARCH BRIDGE WITH A SLAB DECK. — 1727

HOD FOR ANALYSIS OF ELASTIC ISOTROPIC AND ORTHOTROPIC SLABS WHEN STRESS DISTRIBUTIONS ARE ASSUMED. — 1800
THE FINITE ELEMENT METHOD FOR FLEXURE OF SLABS. — 2124
MIT ANALYSIS SOLUTIONS FOR CERTAIN REINFORCED CONCRETE SLABS. — 1344
GALERKIN APPROXIMATIONS TO FLOWS WITHIN SLABS, SPHERES AND CYLINDERS. — 1576

SLADE D. — 1126
SLADE. — 1123

ITE ELEMENT AND FINITE DIFFERENCE METHODS FOR A SIMPLE SLOSH PROBLEM. — 456

A NOTE ON SOME VARIATIONAL STATEMENTS FOR THE SLOW FLOW OF A NAVIER-POISSON FLUID. — 300
FINITE ELEMENT ANALYSIS OF SLOW INCOMPRESSIBLE VISCOUS FLUID MOTION. — 136
FINITE ELEMENT ANALYSIS OF SLOW NON-NEWTONIAN CHANNEL FLOW. — 1584
FINITE ELEMENT METHOD FOR INCOMPRESSIBLE SLOW VISCOUS FLOW WITH A FREE SURFACE. — 1926

Permuted (KWIC) index — keyword: **STIFFNESS**

1959	SAMMSOR 3 A FINITE ELEMENT PROGRAM TO DETERMINE	STIFFNESS	AND DEFLECTION ANALYSIS OF COMPLEX STRUCTURES.
889	THE GENERATION OF INTERELEMENT COMPATIABLE	STIFFNESS	AND MASS MATRICES OF RING STIFFENED SHELLS OF REVOLUTI
332	DEVELOPMENT OF EXPLICIT	STIFFNESS	AND MASS MATRICES BY USE OF INTERPOLATION FORMULAE.
1823	DERIVATION OF GEOMETRIC	STIFFNESS	AND MASS MATRICES FOR A TRIANGULAR PLATE ELEMENT.
1946	REDUCTION OF	STIFFNESS	AND MASS MATRICES FOR FINITE ELEMENT HYBRID MODELS.
881	AND L INFINITY CONDITION NUMBERS OF THE FINITE ELEMENT	STIFFNESS	AND MASS MATRICES.
758	ESIGN OF NUCLEAR REACTOR BUILDING, STRESS ANALYSIS AND	STIFFNESS	EVALUATION OF THE ENTIRE BUILDING BY THE FINITE ELEMENT
1957	BEHAVIOUR OF TRIANGULAR SHELL-ELEMENT	STIFFNESS	MATRICES ASSOCIATED WITH POLYHEDRAL DEFLECTION DISTRIBU
1970	DERIVATION OF ELEMENT	STIFFNESS	MATRICES BY ASSUMED STRESS DISTRIBUTIONS.
1611	COMMENTS ON DERIVATION OF ELEMENT	STIFFNESS	MATRICES BY T.
780	FLEXIBILITY AND	STIFFNESS	MATRICES FOR AN OPEN-TUBE WARPING CONSTRAINT FINITE ELE
337	ELEMENT	STIFFNESS	MATRICES FOR BOUNDARY COMPATIBILITY AND FOR PRESCRIBED
1612	CALCULATION OF GEOMETRIC	STIFFNESS	MATRICES FOR COMPLEX STRUCTURES.
1509	CALCULATION OF	STIFFNESS	MATRICES FOR FINITE ELEMENTS OF THIN SHELLS OF ARBITRAR
1511		STIFFNESS	MATRICES FOR PLATE BENDING ELEMENTS.
973	DERIVATION OF	STIFFNESS	MATRICES FOR PROBLEMS OF PLANE ELASTICITY BY THE GALERK
1893		STIFFNESS	MATRICES FOR SECTOR ELEMENTS.
1670	DERIVATION OF	STIFFNESS	MATRICES FOR THE ANALYSIS OF LARGE DEFLECTION AND STABI
1320	FINITE ELEMENT	STIFFNESS	MATRICES FOR THE ANALYSIS OF PLATE BENDING.
495		STIFFNESS	MATRICES FOR THIN TRIANGULAR ELEMENTS OF NON-ZERO GAUSS
1967	EFFECTIVE USE OF THE INCREMENTAL	STIFFNESS	MATRICES IN NONLINEAR GEOMETRIC ANALYSIS.
652	A FORMULA FOR CERTAIN TYPES OF	STIFFNESS	MATRICES OF STRUCTURAL ELEMENTS.
298	ON DERIVATION OF	STIFFNESS	MATRICES WITH ROTATION FIELDS FOR PLATES AND SHELLS.
1969	COMPUTATION OF STRESS RESULTANTS FROM ELEMENT	STIFFNESS	MATRICES.
1868	DERIVATION OF ELEMENT	STIFFNESS	MATRICES.
1610	ING ANALYSIS OF SHELLS OF REVOLUTION USING INCREMENTAL	STIFFNESS	MATRICES.
1557	CRITERION FOR SELECTING	STIFFNESS	MATRICES.
1128	COMPARISON AND EVALUATION OF ELEMENT	STIFFNESS	MATRICES.
1129	TECHNIQUES FOR THE DERIVATIVATION OF ELEMENT	STIFFNESS	MATRICES.
779	A METHOD OF COMPUTING NUMERICALLY INTEGRATED	STIFFNESS	MATRICES.
875	NODE-RELABELLING SCHEME FOR BANDWIDTH MINIMIZATION OF	STIFFNESS	MATRICES.
40		STIFFNESS	MATRICES.
2064	COMPARISON BETWEEN SPARSE	STIFFNESS	MATRIX AND SUB-STRUCTURE METHODS.
1620	RATIONALIZATION IN DRIVING ELEMENT	STIFFNESS	MATRIX BY ASSUMED STRESS APPROACH.
1724	AN EXTENSION OF THE SHALLOW TO THE NON-SHALLOW	STIFFNESS	MATRIX FOR A CYLINDRICAL SHELL FINITE ELEMENT.
1322		STIFFNESS	MATRIX FOR A TRIANGULAR SANDWICH ELEMENT IN BENDING.
781		STIFFNESS	MATRIX FOR SHALLOW RECTANGULAR SHELL ELEMENT.
505	A	STIFFNESS	MATRIX FOR SHALLOW RECTANGULAR SHELL ELEMENT.
1725	A	STIFFNESS	MATRIX FOR SHALLOW SHELL FINITE ELEMENTS.
1359	A	STIFFNESS	MATRIX FOR THE ANALYSIS OF THIN PLATES IN BENDING.
1343	A	STIFFNESS	MATRIX FOR THE ANALYSIS OF THIN PLATES IN BENDING.
874	ELEMENT	STIFFNESS	MATRIX GENERATION.
1217	GENERALIZED	STIFFNESS	MATRIX OF A CURVED BEAM ELEMENT.
1176	A SIMPLE ALGORITHM FOR THE	STIFFNESS	MATRIX OF TRIANGULAR PLATE BENDING ELEMENTS.
1968	EXPLICIT EXPRESSIONS FOR TRIANGULAR TORUS ELEMENT	STIFFNESS	MATRIX.
876	VIBRATION OF FRAMES AND OTHER STRUCTURES WITH BANDED	STIFFNESS	MATRIX.
1450	NOTE ON THE ALPHA-CONSTANT	STIFFNESS	METHOD FOR THE ANALYSIS OF NON-LINEAR PROBLEMS.
1061	A GENERALIZATION OF THE DIRECT	STIFFNESS	METHOD OF ANALYSIS OF SHELLS OF REVOLUTION UTILIZING CU
1119	A CONVERGENCE INVESTIGATION OF THE DIRECT	STIFFNESS	METHOD OF STRUCTURAL ANALYSIS.
1016	ENGINEERING APPLICATION OF NUMERICAL INTEGRATION IN THE DIRECT	STIFFNESS	METHOD.
851	ANALYSIS OF AXISYMMETRIC SHELLS BY THE DIRECT	STIFFNESS	METHOD.
1613	FINITE ELEMENT	STIFFNESS	METHODS BY DIFFERENT VARIATIONAL PRINCIPLES IN ELASTICI

The following is a permuted (KWIC) index listing. Each entry is shown as its printed context, keyword phrase, and reference number.

1894	STIFFNESS METHODS FOR PLATES BY GALERKIN"S METHOD.
331	ON SOME GENERALIZATIONS OF THE INCREMENTAL STIFFNESS OF RECTANGULAR PLATES BY THE FINITE ELEMENT METHOD.
1543	INADEQUACY OF NODAL CONNECTIONS IN A STIFFNESS RELATIONS FOR FINITE DEFORMATIONS OF COMPRESSIBLE AND I
1025	ROUNDOFF CRITERIA IN DIRECT STIFFNESS SOLUTION FOR PLATE BENDING.
1017	STIFFNESS SOLUTIONS.
1151	IS OF THE CORNEO-SCLERAL SHELL BY THE METHOD OF DIRECT STIFFNESS SOLUTIONS.
951	A FINITE ELEMENT FORMULATION FOR PROBLEMS OF LARGE STRAIN AND LARGE DISPLACEMENT.
2008	QUANTITATIVE STRAIN AND STRESS STATE CRITERION FOR FAILURE IN THE VICINITY OF
1712	IS OF THICK-WALLED PRESSURE VESSELS USING THE CONSTANT STRAIN AXISYMMETRIC FINITE ELEMENT.
453	PLANE STRAIN CONSOLIDATION BY FINITE ELEMENTS.
147	ALYSIS OF SHELLS AND PLATES INCLUDING TRANSVERSE SHEAR STRAIN EFFECTS.
144	TE ELEMENT THEORY OF PLATES AND SHELLS INCLUDING SHEAR STRAIN EFFECTS.
1006	S A THREE-DIMENSIONAL FINITE ELEMENT PROGRAM FOR SMALL STRAIN ELASTIC ANALYSIS.
893	H ORDER CONVERGENCE OF A RITZ APPROXIMATION TO A PLANE STRAIN ELASTICITY PROBLEM.
971	LARGE STRAIN FLASTO-PLASTIC FINITE ELEMENT ANALYSIS.
1189	URE VESSEL BY THE FINITE ELEMENT METHOD USING VARIABLE STRAIN ELEMENTS.
737	NLINEAR FINITE ELEMENT ANALYSIS INCLUDING HIGHER-ORDER STRAIN ENERGY BOUNDS IN FINITE ELEMENT ANALYSIS BY SLAB ANALYSIS.
886	APPLICATION OF FINITE ELEMENT METHOD TO PLANE STRAIN ENERGY TERMS.
830	LARGE STRAIN EXTRUSION PROCESSES.
104	BRANE PARALLELOGRAM ELEMENT WITH LINEARLY VARYING EDGE STRAIN FOR THE MATRIX DISPLACEMENT METHOD.
101	TETRAHEDRON ELEMENTS WITH LINEARLY VARYING STRAIN FOR THE MATRIX DISPLACEMENT METHOD.
102	TRIANGULAR ELEMENTS WITH LINEARLY VARYING STRAIN FOR THE MATRIX DISPLACEMENT METHOD.
166	RICAL SHELL FINITE ELEMENT BASED ON SIMPLE INDEPENDENT STRAIN FUNCTIONS.
1550	ICAL SOLUTION OF A CLASS OF PROBLEMS IN A LINEAR FIRST STRAIN GRADIENT THEORY OF ELASTICITY.
2088	STRESS STRAIN MATRIX AND ITS APPLICATIONS FOR THE SOLUTION OF ELASTIC-PL
1512	FINITE PLANE STRAIN OF INCOMPRESSIBLE ELASTIC SOLIDS BY THE FINITE ELEMENT MET
508	FINITE ELEMENT ANALYSIS OF STRAIN RESULTANTS IN CERTAIN FINITE ELEMENTS.
964	STRAIN SOFTENING CLAY.
1598	SOME PROPERTIES OF LINEAR STRAIN TRIANGLES AND OPTIMAL FINITE ELEMENT MODELS.
1035	IC ANALYSIS OF STRUCTURES IN A STATE OF MODIFIED PLANE STRAIN. VISCOELASTIC NUMERICAL ANALYSIS BY MEANS OF FINITE ELEMEN
1092	ON THE UNITY OF THE CONSTANT STRAIN/CONSTANT MOMENT FINITE ELEMENT METHODS
926	SMALL SCALE YIELDING NEAR A CRACK IN PLAIN STRAIN; A FINITE ELEMENT ANALYSIS.
1235	ED FIELDS OF TRIANGULAR ELEMENTS WITH LINEARLY VARYING STRAIN; EFFECTS OF INITIAL STRAINS.
99	LARGE STRAIN
564	STRAP A COMPUTER CODE FOR THE STATIC AND DYNAMIC ANALYSIS OF REAC
1040	ANALYSIS OF BRANCH PIPE STRESS (JAPANESE),
2025	ANISOTROPIC ELASTIC-PLASTIC THERMAL STRESS ANALYSES OF SOLID STRUCTURES.
386	TWO-DIMENSIONAL FINITE ELEMENT COMPUTER CODES FOR STRESS ANALYSIS AND FIELD PROBLEMS.
1957	ASEISMIC DESIGN OF NUCLEAR REACTOR BUILDING, STRESS ANALYSIS AND STIFFNESS EVALUATION OF THE ENTIRE BUILDING B
870	THERMOVISCOELASTIC AXISYMMETRIC STRESS ANALYSIS BY FINITE ELEMENT AND FINITE DIFFERENCE COMPUTER
403	STRESS ANALYSIS BY THE FINITE ELEMENT DISPLACEMENT METHOD.
775	HEAT CONDUCTION AND THERMAL STRESS ANALYSIS BY THE FINITE ELEMENT METHOD.
2146	STRESS ANALYSIS BY THE FINITE ELEMENT METHOD; THERMAL EFFECTS.
1436	CODE FOR STRESS ANALYSIS BY THE THREE DIMENSIONAL FINITE ELEMENT METHOD (J
1234	THREE-DIMENSIONAL ELASTIC-PLASTIC STRESS ANALYSIS FOR FRACTURE MECHANICS.
1398	NONLINEAR THERMAL STRESS ANALYSIS FOR NUCLEAR POWER PLANT BY THE FINITE ELEMENT MET
797	STRESS ANALYSIS IN HEATED COMPLEX SHAPES.
2022	STRESS ANALYSIS IN VISCOELASTIC BODIES USING FINITE ELEMENTS AND
1116	STRESS ANALYSIS OF A CLAMPED RECTANGULAR PLATE.
1189	APPLICATION OF THE EXTENDED KANTOROVICH METHOD TO THE STRESS ANALYSIS OF A PRESTRESSED CONCRETE NUCLEAR PRESSURE VESSEL

LICATION OF FINITE ELEMENT METHODS FOR THE ANALYSIS OF THERMAL CREEP, IRRADIATION INDUCED CREEP AND SWELLING FOR LMF REA   1881
NATION OF NON-STATIONARY TEMPERATURE DISTRIBUTION AND THERMAL DEFORMATIONS.   1988
STRESS ANALYSIS BY THE FINITE ELEMENT METHOD; THERMAL EFFECTS.   655
HODS FOR NONLINEAR EIGENVALUE PROBLEMS ASSOCIATED WITH THERMAL IGNITION.   2146
N SHELLS OF REVOLUTION UNDER ASYMMETRIC MECHANICAL AND THERMAL LOADING.   2002
ANISOTROPIC ELASTIC-PLASTIC THERMAL STRESS ANALYSES OF SOLID STRUCTURES.   402
HEAT CONDUCTION AND THERMAL STRESS ANALYSIS BY THE FINITE ELEMENT METHOD.   2025
NONLINEAR THERMAL STRESS ANALYSIS FOR NUCLEAR POWER PLANT BY THE FINITE ELE   775
ION OF THREE DIMENSIONAL TEMPERATURE DISTRIBUTIONS AND THERMAL STRESSES USING FINITE ELEMENT METHODS.   1398
THE COMPUTATION OF TEMPERATURE DISTRIBUTIONS AND THERMAL STRESSES USING FINITE ELEMENTS TECHNIQUES.   1433
                                                                                                            776

THERMO-ELASTIC-PLASTIC CYCLIC ANALYSIS BY FINITE ELEMENT METHOD.   811

ANALYSIS OF THICK AND THIN SHELL STRUCTURES BY GENERAL CURVED ELEMENTS.   30
A NOTE ON THE FINITE DEFORMATION OF A THICK PLATE SHELL OF REVOLUTION.   1544
TRIANGULAR THICK PLATE BENDING ELEMENTS PAPER M6/5.   1286
C BEHAVIOUR OF A LONGITUDINAL SEMI-ELLIPTIC CRACK IN A THICK PRESSURE VESSEL.   1314
IC BEHAVIOUR OF A LONGITUDINAL SEMIELLIPTIC CRACK IN A THICK PRESSURE VESSEL.   1312
CURVED THICK SHELL AND MEMBRANE ELEMENTS WITH PARTICULAR REFERENCE TO AX   29
IMPROVED NUMERICAL INTEGRATION OF TRIANGULAR THICK SHELL FINITE ELEMENT FOR THE THREE-DIMENSIONAL STRESS ANAL   1675
APPROXIMATE ELASTICITY SOLUTION FOR MODERATELY THICK SHELLS OF REVOLUTION.   1596
TWO HYBRID ELEMENTS FOR ANALYSIS OF THICK THIN AND SANDWICH PLATES.   1903
FINITE ELEMENT ANALYSIS OF A THICK WALLED MICROPOLAR CYLINDER LOADED AXISYMMETRICALLY.   510
THREE DIMENSIONAL FINITE ELEMENT ANALYSIS OF THICK WALLED PIPE-NOZZLE JUNCTIONS WITH CURVED TRANSITIONS.   225
VIBRATION OF THICK, CURVED, SHELLS WITH PARTICULAR REFERENCE TO TURBINE BLADES   1184
                                                                                                            28

TWO HYBRID ELEMENTS FOR ANALYSIS OF THICK THIN AND SANDWICH PLATES.   510
FINITE ELEMENT POST BUCKLING ANALYSIS OF THIN ELASTIC PLATES.   1428
A CURVED ELEMENT FOR THIN ELASTIC SHELLS.   649
BUCKLING UNDER COMBINED LOADING OF THIN FLAT-WALLED STRUCTURES BY A COMPLEX FINITE STRIP METHOD.   1633
SHAPE FUNCTION SUBROUTINE FOR AN ISOPARAMETRIC THIN PLATE FLEMENT.   223
A SECTOR ELEMENT FOR THIN PLATE FLEXURE.   1747
A STIFFNESS MATRIX FOR THE ANALYSIS OF THIN PLATES IN BENDING.   29
A STIFFNESS MATRIX FOR THE ANALYSIS OF THIN PLATES IN BENDING.   1359
NT LINEAR PROGRAMMING METHOD FOR THE LIMIT ANALYSIS OF THIN PLATES.   1343
E NODAL CONTINUITIES IN THE FINITE ELEMENT ANALYSIS OF THIN RECTANGULAR AND SKEW PLATES IN BENDING.   686
NS OF FINITE ELEMENT METHODS IN THE STRESS ANALYSIS OF THIN SHELL PRESSURE VESSELS.   1827
ANALYSIS OF THICK AND THIN SHELL STRUCTURES BY GENERAL CURVED ELEMENTS.   1591
ELEMENT APPROXIMATION IN THE ANALYSIS OF AXI-SYMMETRIC THIN SHELLS   30
NUMERICAL ANALYSIS OF THIN SHELLS BY CURVED TRIANGULAR ELEMENTS BASED ON DISCRETE-KIRCH   818
THE TRANSIENT DYNAMIC ANALYSIS OF THIN SHELLS BY THE FINITE ELEMENT METHOD.   600
THE ANALYSIS OF THIN SHELLS BY THE FINITE ELEMENT METHOD.   1120
ELASTIC ANALYSIS OF CRACKED THIN SHELLS BY THE FINITE ELEMENT METHOD.   1124
THE TRANSIENT DYNAMIC ANALYSIS OF THIN SHELLS IN THE FINITE ELEMENT METHOD.   415
LCULATION OF STIFFNESS MATRICES FOR FINITE ELEMENTS OF THIN SHELLS OF ARBITRARY SHAPE.   1125
CALCULATION OF TEMPERATURE DISTRIBUTIONS IN THE THIN SHELLS OF REVOLUTION BY THE FINITE ELEMENT METHOD.   1511
A DYNAMIC ELASTOPLASTIC ANALYSIS OF THIN SHELLS OF REVOLUTION UNDER ASYMMETRIC MECHANICAL AND THERMAL   11
A REFINED CURVED ELEMENT FOR THIN SHELLS OF REVOLUTION.   402
NITE ELEMENT METHOD FOR ANALYSING SYMMETRICALLY LOADED THIN SHELLS OF REVOLUTION.   1642
THE ANALYSIS OF THIN SHELLS WITH TRANSVERSE SHEAR STRAINS BY THE FINITE ELEMENT M   401
FINITE ELEMENT ANALYSIS OF THIN SHELLS.   1122
                                                                                          2034

311   PIECEWISE ANALYTIC INTERPOLATION AND APPROXIMATION IN TRIANGULATED POLYGONS

309   TRICUBIC POLYNOMIAL INTERPOLATION, PROC.

105   A TAPERED THICKNESS TRIM 6 ELEMENT FOR THE MATRIX DISPLACEMENT METHOD.

130   THE TUBA FAMILY OF PLATE ELEMENTS FOR THE MATRIX DISPLACEMENT METHOD.
131   THE TUBA FAMILY OF PLATE ELEMENTS FOR THE MATRIX DISPLACEMENT METHOD.
109   THE TUBA FAMILY OF PLATE ELEMENTS FOR THE MATRIX DISPLACEMENT METHOD.

28   OF THICK, CURVED, SHELLS WITH PARTICULAR REFERENCE TO TURBINE BLADES.

943   FECT OF QUADRATURE ERRORS IN THE NUMERICAL SOLUTION OF TWO-DIMENSIONAL BOUNDARY VALUE PROBLEMS BY VARIATIONAL TECHNIQUES
412   RITZ APPROXIMATIONS TO TWO-DIMENSIONAL BOUNDARY VALUE PROBLEMS.
585   APPLICATION OF THE FINITE ELEMENT METHOD TO TWO-DIMENSIONAL DIFFUSION PROBLEMS.
2020 IUM STRESS FIELD MODEL FOR FINITE ELEMENT SOLUTIONS OF TWO-DIMENSIONAL ELASTIC PROBLEMS.
1816 FINITE ELEMENT SOLUTION OF TWO-DIMENSIONAL EXTERIOR FIELD PROBLEMS.
394   TWO-DIMENSIONAL FINITE ELEMENT ANALYSIS OF SEMI-CONDUCTOR STEADY
386   TWO-DIMENSIONAL FINITE ELEMENT COMPUTER CODES FOR STRESS ANALYSIS
1203 APPLICATION OF THE FINITE ELEMENT METHOD IN TRANSIENT TWO-DIMENSIONAL HEAT CONDUCTION PROBLEMS SOLVED BY THE FINITE ELE
1796 APPLICATION OF THE FINITE ELEMENT METHOD TO TWO-DIMENSIONAL HYDRODYNAMICS USING LAGRANGIAN VARIABLES.
1382 APPLICATION OF PHASE SPACE FINITE ELEMENTS TO THE TWO-DIMENSIONAL MULTIGROUP NEUTRON DIFFUSION.
216   A NUMERICAL SOLUTION TECHNIQUE FOR A CLASS OF TWO-DIMENSIONAL NEUTRON TRANSPORT EQUATION IN X-Y GEOMETRY.
1571 A VARIATIONAL FINITE ELEMENT METHOD FOR TWO-DIMENSIONAL PROBLEMS IN FLUID DYNAMICS FORMULATED WITH THE US
692   EFINED FINITE ELEMENT ANALYSIS OF LINEAR AND NONLINEAR TWO-DIMENSIONAL STEADY VISCOUS FLOWS.
1903 AN ANALYSIS OF TWO-DIMENSIONAL STRUCTURES.
1851 TWO-DIMENSIONAL SURFACE RUN-OFF BY FINITE ELEMENTS.
1693 TWO-DIMENSIONAL TIME-DEPENDENT ANALYSIS OF PERFORATED END CAPS FO
      TRIPLET A TWO-DIMENSIONAL, MULTIGROUP, TRIANGULAR MESH, PLANAR GEOMETRY, EX

56   DISCONTINUITIES AND INTERACTIONS BY FINITE ELEMENTS IN TWO-DIMENSIONS.

1493 SIMULATION OF THE TENSIBLE BEHAVIOUR OF TWO-PHASE ALLOY CONTAINING A FIBRE UNDER TENSIBLE LOAD.

566   ON UNIFORM APPROXIMATION BY SPLINES.
1353 LOWER ESTIMATES FOR THE ERROR OF BEST UNIFORM APPROXIMATION.
471   MAXIMUM PRINCIPLE AND UNIFORM CONVERGENCE FOR THE FINITE ELEMENT METHOD.
1965 VISCOUS FLOW AROUND A BOX-GIRDER BRIDGE OSCILLATING IN UNIFORM FLOW.
43   IT ERROR BOUNDS FOR PERIODIC SPLINES OF ODD ORDER ON A UNIFORM MESH.
836   APPROXIMATION BY PERIODIC SPLINE INTERPOLATION ON UNIFORM MESHES.

383   FINITE ELEMENT SOLUTION OF UNSTEADY AND UNSATURATED FLOW IN POROUS MEDIA.
1506 A LAGRANGIAN FINITE ELEMENT SOLUTION TO UNSTEADY FLOW.
360   FINITE ELEMENT ANALYSIS OF THE UNSTEADY LINEARIZED BOLTZMANN EQUATION.
2079 UNSTEADY MHD DUCT FLOW BY THE FINITE ELEMENT METHOD.
1965 APPLICATIONS OF FINITE ELEMENT METHODS TO UNSTEADY VISCOUS FLOW AROUND A BOX-GIRDER BRIDGE OSCILLATING IN U

1556 INVESTIGATION OF THE CONVERGENCE RATE OF VARIATIONAL - DIFFERENCE SCHEMES FOR ELLIPTIC SECOND ORDER EQUATI
1417 VARIATIONAL ANALYSIS.
1634 VARIATIONAL AND ACCURATE SOLUTION OF THE ORR-SOMMERFELD EQUATION.
293   SINGULAR ELEMENTS IN VARIATIONAL AND FINITE ELEMENT METHODS.
8   INTRODUCTION TO VARIATIONAL AND FINITE ELEMENT TRANSPORT COMPUTATIONS.
1978 THE ROLE OF INTERPOLATION AND APPROXIMATION THEORY IN VARIATIONAL AND PROJECTIONAL METHODS FOR SOLVING PARTIAL DIFFEREN

374   FINITE ELEMENT METHOD IN LINEAR VISCOELASTICITY 2.

2046   FINITE ELEMENTS IN LINEAR VISCOELASTICITY.

1516   ND ITS APPLICATION TO A CLASS OF PROBLEMS IN NONLINEAR VISCOELASTICITY.

373   OLUTIONAL VARIATIONAL PRINCIPLES AND METHODS IN LINEAR VISCOELASTICITY.

461   INCREMENTAL THERMOMECHANICAL THEORY OF VISCOELASTOPLASTIC SOLIDS AND SOLUTION BY FINITE ELEMENTS.

460   ANALYSIS OF VISCOELASTOPLASTIC STRUCTURAL BEHAVIOUR OF ANISOTROPIC SHELLS BY

1965   APPLICATIONS OF FINITE ELEMENT METHODS TO UNSTEADY VISCOUS FLOW AROUND A BOX-GIRDER BRIDGE OSCILLATING IN UNIFORM FL

1567   CONVERGENCE OF FINITE ELEMENT SOLUTIONS IN VISCOUS FLOW PROBLEMS.

1213   FINITE ELEMENT METHOD FOR TRANSIENT LINEAR VISCOUS FLOW PROBLEMS.

805   COMPUTATIONALLY EFFICIENT FINITE ELEMENT ANALYSIS OF VISCOUS FLOW PROBLEMS.

1926   FINITE ELEMENT METHOD FOR INCOMPRESSIBLE SLOW VISCOUS FLOW WITH A FREE SURFACE.

770   NUMERICAL ANALYSIS OF THREE-DIMENSIONAL INCOMPRESSIBLE STEADY VISCOUS FLOWS.

1571   IONAL FINITE ELEMENT METHOD FOR TWO-DIMENSIONAL STEADY VISCOUS FLOWS.

1094   STEADY FLOW ANALYSIS OF INCOMPRESSIBLE VISCOUS FLUID BY THE FINITE ELEMENT METHOD.

136   FINITE ELEMENT ANALYSIS OF SLOW INCOMPRESSIBLE VISCOUS FLUID MOTION.

1553   FINITE ELEMENT ANALYSES THE ANALYSIS OF FLOW OF VISCOUS FLUIDS BY THE FINITE ELEMENT METHOD

218   FINITE ELEMENT SOLUTION ALGORITHM FOR VISCOUS FLUIDS INCOMPRESSIBLE FLUID DYNAMICS INT.

1342   NEW RECTANGULAR FINITE ELEMENT FOR SHEAR WALL ANALYSIS.

488   FINITE ELEMENT ANALYSES OF RETAINING WALL BEHAVIOUR.

1468   STRESS WAVE ANALYSIS IN LAYERED THERMOVISCOELASTIC MATERIALS BY EXTENDED

2030   VARIATIONAL METHODS FOR WAVE AND SHOCK PROPAGATION IN NONLINEAR HYPERELASTIC MATERIALS.

273   A DYNAMIC SHELL THEORY COUPLING THICKNESS STRESS WAVE EFFECTS WITH GROSS STRUCTURAL RESPONSE.

506   FINITE ELEMENT APPROACH TO STRESS WAVE PROBLEMS.

1905   TIDAL AND LONG WAVE PROPAGATION - A FINITE ELEMENT APPROACH.

305   A FINITE ELEMENT ANALYSIS FOR WAVE PROPAGATION IN ELASTIC-PLASTIC SOLIDS.

1575   FINITE ELEMENT STUDY OF HARMONIC WAVE PROPAGATION IN PERIODIC STRUCTURES.

1195   SCRETIZATION-DISPERSION IN FINITE ELEMENT MODELLING OF WAVE PROPAGATION IN SOLIDS.

1986   FINITE ELEMENT STRESS FORMULATION FOR WAVE PROPAGATION.

1814   A GENERAL HIGH-ORDER FINITE-ELEMENT WAVEGUIDE ANALYSIS PROGRAM.

1813   FINITE ELEMENT SOLUTION OF HOMOGENEOUS WAVEGUIDE PROBLEMS.

26   FINITE ELEMENT METHOD FOR WAVEGUIDE PROBLEMS.

32   WAVEGUIDE SOLUTIONS BY THE FINITE ELEMENT METHOD.

678   NTED VECTOR VARIATIONAL SOLUTION OF LOADED RECTANGULAR WAVEGUIDES.

544   NUMERICAL SOLUTION OF DIELECTRIC LOADED WAVEGUIDES.

289   IATIONAL PRINCIPLES FOR ELECTROMAGNETIC RESONATORS AND WAVEGUIDES.

31   FINITE ELEMENT METHODS FOR INHOMOGENEOUS WAVEGUIDES.

1416   DIFFERENTIABILITY THEOREMS FOR WEAK SOLUTIONS OF NONLINEAR ELLIPTIC DIFFERENTIAL EQUATIONS.

1284   FINITE ELEMENTS FORMULATED BY THE WEIGHTED DISCRETE LEAST SQUARES METHOD.

648   L2 OPTIMALITY AND WEIGHTED H1 PROJECTIONS INTO PIECEWISE POLYNOMIAL SPACES.

1462   RECENT DEVELOPMENTS IN DISCRETE WEIGHTED RESIDUAL METHODS.

384   A FINITE ELEMENT WEIGHTED RESIDUAL SOLUTION TO ONE-DIMENSIONAL FIELD PROBLEMS.

700   THE METHOD OF WEIGHTED RESIDUALS AND ITS RELATION TO CERTAIN VARIATIONAL PRINCI

701   THE METHOD OF WEIGHTED RESIDUALS- A REVIEW.

1235   SMALL SCALE YIELDING NEAR A CRACK IN PLAIN STRAIN: A FINITE ELEMENT ANALYSIS.